D1067692

Engineering Properties of Soils and Rocks

Engineering Properties of Soils and Rocks

Third edition

F. G. Bell

Butterworth-Heinemann Ltd
Linacre House, Jordan Hill, Oxford OX2 8DP

 PART OF REED INTERNATIONAL BOOKS

OXFORD LONDON BOSTON
MUNICH NEW DELHI SINGAPORE SYDNEY
TOKYO TORONTO WELLINGTON

First published 1981
Second edition 1983
Reprinted 1985
This edition 1992

British Library Cataloguing in Publication Data
Bell, F. G. (Frederic Gladstone)
Engineering properties of soils and rocks.
I. Title
624.15136028

ISBN 07506 0489 1

Library of Congress Cataloguing in Publication Data
Bell, F. G. (Frederic Gladstone)
Enginnering properties of soils and rocks/F. G. Bell.—3rd ed.
p. cm.
Includes bibliographical references and index.
ISBN 0 7506 0489 1
1. Soil mechanics. 2. Rock mechanics. I. Title.
TA710.B424 1992
624.1′513—dc20 91–39429
CIP

Composition by Genesis Typesetting, Laser Quay, Rochester, Kent
Printed and bound in Great Britain

Contents

Preface

As was stated in the Preface to the first edition, civil engineers, mining engineers and engineering geologists require a working knowledge of the engineering properties and behaviour of soils and rocks. Although this text provides such information and cites actual values of particular soils and rocks, the reader is reminded that these should be taken only as guides. Natural materials are variable, some highly variable; they are not made to specification like manufactured materials. The properties and behaviour of individual soil and rock types vary according to many factors, notably their composition, texture, degree of weathering, presence of fissures or discontinuities, etc. The text pays regard to these factors. In fact it is intended to complement texts on soil and rock mechanics.

The author has taken the opportunity to revise the text fully, with the result that this edition is twice as long as its predecessor. It is the author's hope that readers will find this volume, with its increased coverage, a valuable addition to their bookshelves, both now and throughout their professional lives.

F. G. Bell

Chapter 1

Properties and classification of soils

1.1 Origin of soil

Soil is an unconsolidated assemblage of solid particles between which are voids. These may contain water or air, or both. Soil is derived from the breakdown of rock material by weathering and/or erosion and it may have suffered some amount of transportation prior to deposition. It may also contain organic matter. The type of breakdown process(es) and the amount of transport undergone by sediments influence the nature of the macro- and microstructure of the soil, which in turn influence its engineering behaviour (Table 1.1).

Probably the most important methods of soil formation are mechanical and chemical weathering. The agents of weathering, however, are not capable of transporting material. Transport is brought about by gravity, water, wind or moving ice. If sedimentary particles are transported, then this affects their character, particularly their grain-size distribution, sorting and shape. For instance, stream channel deposits are commonly well graded, although the grain size

Table 1.1 *Effects of transportation on sediments*

	Gravity	*Ice*	*Water*	*Air*
Size	Various	Varies from clay to boulders	Various sizes from boulder gravel to muds	Sand size and less
Sorting	Unsorted	Generally unsorted	Sorting takes place both laterally and vertically. Marine deposits often uniformly sorted. River deposits may be well sorted	Uniformly sorted
Shape	Angular	Angular	From angular to well rounded	Well rounded
Surface texture	Striated surfaces	Striated surfaces	Gravel: rugose surfaces. Sand: smooth, polished surfaces. Silt: little effect	Impact produces frosted surfaces

characteristics may vary erratically with location. On the other hand, wind-blown deposits are usually uniformly sorted with well-rounded grains.

Changes occur in soils after they have accumulated. In particular, seasonal changes take place in the moisture content of sediments above the water table. Volume changes associated with alternate wetting and drying occur in cohesive soils with high plasticity indices (see below). Exposure of a soil to dry conditions means that its surface dries out and that water is drawn from deeper zones by capillary action. The capillary rise is associated with a decrease in pore pressure in the layer beneath the surface and a corresponding increase in effective pressure. This supplementary pressure is known as capillary pressure and it has the same mechanical effect as a heavy surcharge. Therefore surface evaporation from very compressible soils produces a conspicuous decrease in the void ratio of the layer undergoing desiccation. If the moisture content in this layer reaches the shrinkage limit then air begins to invade the voids and the soil structure begins to break down. Moreover if the plasticity index of the soil exceeds 20% then the seasonal variations in moisture content of the upper layers are accompanied by ground movement. The decrease in void ratio consequent upon desiccation of a cohesive sediment leads to an increase in its shearing strength. Thus if a dry crust is located at or near the surface above softer material it acts as a raft. The thickness of dry crusts often varies erratically.

Chemical changes which take place in the soil due, for example, to the action of weathering, may bring about an increase in its clay mineral content, the latter developing from the breakdown of less stable minerals. In such instances the plasticity of the soil increases whilst its permeability decreases. Leaching, whereby soluble constituents are removed from the upper, to be precipitated in a lower horizon, occurs where rainfall exceeds evaporation. The porosity may be increased in the zone undergoing leaching.

During sediment accumulation the stress at any given elevation continues to build up as the thickness of the overburden increases. As a result the properties of the sediment are continually changing, the void space, in particular, being reduced. If, subsequently, overburden is removed by erosion, or for that matter by extensive excavation, the void ratio tends to increase.

With continuing exposure soil develops a characteristic profile from the surface downwards. This development involves the accumulation and decay of organic matter, leaching, precipitation, oxidation or reduction and further mechanical and biological breakdown. The profile which forms is influenced by the character of the parental material but climatic conditions, vegetative cover, groundwater level and relief also play their part and the time factor allows the distinction between immature and mature soils.

1.2 Basic properties of soil

As remarked, a soil consists of an assemblage of particles between which are voids, and as such can contain three phases – solids, water and air. The interelationships

of the weights and volumes of these three phases are important since they help define the character of a soil.

One of the most fundamental properties of a soil is the void ratio, which is the ratio of the volume of the voids to that of the volume of the solids. The porosity is a similar property, it being the ratio of the volume of the voids to the total volume of the soil, expressed as a percentage. Both void ratio and porosity indicate the relative proportion of void volume in a soil sample.

Water plays a fundamental part in determining the engineering behaviour of any soil and the moisture content is expressed as a percentage of the weight of the solid material in the soil sample. The degree of saturation expresses the relative volume percentage of water in the voids.

The range of values of phase relationships for cohesive soils is much larger than for granular soils. For instance, saturated sodium montmorillonite at low confining pressure can exist at a void ratio of more than 25, its moisture content being some 900%. On the other hand saturated clays under high stress that exist at great depth may have void ratios of less than 0.2, with about 7% moisture content.

The unit weight of a soil is its weight per unit volume, whilst its specific gravity is the ratio of its weight to that of an equal volume of water. In soil mechanics the specific gravity is that of the actual soil particles.

The density of a soil is the ratio of its mass to that of its volume. A number of types of density are distinguished. The dry density is the mass of the solid particles divided by the total volume, whereas the bulk density is simply the mass of the soil (including its natural moisture content) divided by its volume. The saturated density is the density of the soil when saturated, whilst the submerged density is the ratio of effective mass to volume of soil when submerged. In fact the submerged density can be derived by subtracting the density of water from the saturated density.

The density of a soil is governed by the manner in which its solid particles are packed. For example, granular soils may be densely or loosely packed. Indeed a maximum and minimum density can be distinguished. The smaller the range of particle sizes present and the more angular the particles, the smaller the minimum density. Conversely, if a wide range of particle sizes is present the void space is reduced accordingly, hence the maximum density is higher. A useful way to characterize the density of a granular soil is by its relative density (D_r) which is defined as

$$D_r = \frac{e_{max} - e}{e_{max} - e_{min}} \tag{1.1}$$

where e is the naturally occurring void ratio, e_{max} is the maximum void ratio and e_{min} is the minimum void ratio. Five degrees of density have been distinguished (Table 1.2).

Table 1.2 *Grades of relative density and their description*

Class	*Relative density*	*Description*
1	Less than 0.2	Very loose
2	0.2–0.4	Loose
3	0.4–0.6	Medium dense
4	0.6–0.8	Dense
5	Over 0.8	Very dense

1.3 Particle-size distribution

The particle-size distribution expresses the size of particles in a soil in terms of percentages by weight of boulders, cobbles, gravel, sand, silt and clay. The United Soil Classification (Wagner, 1957) and Anon (1981) give the limits shown in Table 1.3 for these size grades.

In nature there is a deficiency of soil particles in the fine gravel and silt ranges, and boulders and cobbles are quantitatively speaking not significant. Sands and clays are therefore the most important soil types.

The results of particle-size analysis are given in the form of a series of fractions, by weight, of different size grades. These fractions are expressed as a percentage of the whole sample and are generally summed to obtain a cumulative percentage. Cumulative curves are then plotted on a semi-logarithmic paper to give a graphical representation of the particle-size distribution. The slope of the curve provides an indication of the degree of sorting. If, for example, the curve is steep as in curve A

Table 1.3 *Particle size distribution of soils*

Types of material		*Sizes* (mm)
Boulders		Over 200
Cobbles		60–200
Gravel	Coarse	20–60
	Medium	6–20
	Fine	2–6
Sand	Coarse	0.6–2
	Medium	0.2–0.6
	Fine	0.06–0.2
Silt	Coarse	0.02–0.06
	Medium	0.006–0.02
	Fine	0.002–0.006
Clay		Less than 0.002

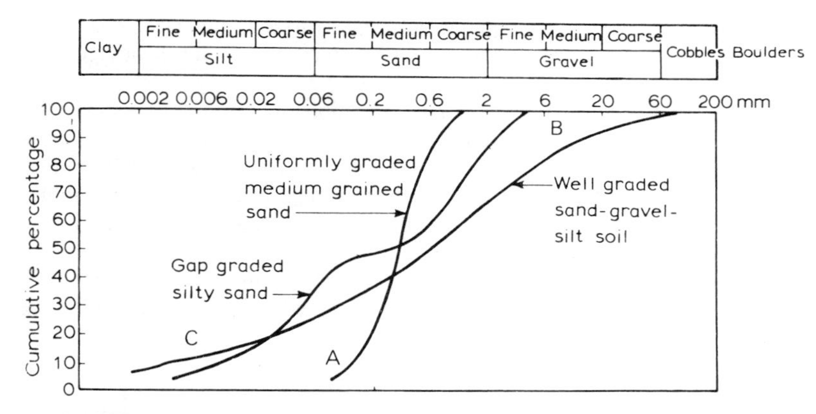

Figure 1.1 *Grading curves*

in Figure 1.1, then the soil is uniformly sorted, whilst curve B represents a well sorted soil. The sorting or uniformity of a particle size distribution has been expressed in a great many ways but one simple statistical measure which has been used by engineers is the coefficient of uniformity (U). This makes use of the effective size of the grains (D_{10}), that is, the size on the cumulative curve where 10% of the particles are passing and is defined as

$$U = \frac{D_{60}}{D_{10}} \tag{1.2}$$

Similarly D_{60} is the size on the curve at which 60% of the particles are passing. A soil having a coefficient of uniformity of less than 2 is considered uniform whilst one with a value of 10 is described as well graded. In other words the higher the coefficient of uniformity, the larger is the range of particle sizes. The coefficient of curvature (C_c) is obtained from the expression

$$C_c = \frac{D_{30}^{\,2}}{D_{60}D_{10}} \tag{1.3}$$

A well-graded soil has a coefficient of curvature between 1 and 3.

1.4 Consistency limits

The Atterberg or consistency limits of cohesive soils are founded on the concept that such soils can exist in any of four states depending on their water content. These limits are also influenced by the amount and character of the clay mineral content. In other words a cohesive soil is solid when dry but as water is added, it first turns to a semi-solid, then to a plastic, and finally to a liquid state. The water content at the boundaries between these states is referred to as the shrinkage limit (SL), the plastic limit (PL) and the liquid limit (LL) respectively.

The shrinkage limit is defined as the percentage moisture content of a soil at the point where it suffers no further decrease in volume on drying. The plastic limit is the percentage moisture content at which a soil can be rolled, without breaking, into a thread 3 mm in diameter, any further rolling causing it to crumble. Unfortunately the inadequacy of control involved in the test means that the results obtained are not consistent for any particular clay.

Turning to the liquid limit, this is defined as the minimum moisture content at which a soil will flow under its own weight. Clays may be classified according to their liquid limit as shown in Table 1.4.

Table 1.4 *Plasticity according to liquid limit*

Description	*Plasticity*	*Range of liquid limit*
Lean or silty	Low plasticity	Less than 35
Intermediate	Intermediate plasticity	35–50
Fat	High plasticity	50–70
Very fat	Very high plasticity	70–90
Extra fat	Extra high plasticity	Over 90

The consistency of cohesive soils depends on the interaction between the clay particles. Any decrease in water content results in a decrease in cation layer thickness and an increase in the net attractive forces between particles. For a soil to exist in the plastic state the magnitudes of the net interparticle forces must be such that the particles are free to slide relative to each other with cohesion between them being maintained. The plasticity of fine grained soils refers to their ability to undergo irrecoverable deformation at constant volume without cracking or crumbling.

The numerical difference between the liquid and plastic limits is referred to as the plasticity index (PI). This indicates the range of moisture content over which the material exists in a plastic condition. The plasticity index has been divided into five classes which are as shown in Table 1.5.

Table 1.5 *Plasticity of soils (after Anon, 1979)*

Class	*Plasticity index (%)*	*Description*
1	Less than 1	Non-plastic
2	1–7	Slightly plastic
3	7–17	Moderately plastic
4	17–35	Highly plastic
5	Over 35	Extremely plastic

The liquidity index of a soil is defined as its moisture content in excess of the plastic limit, expressed as a percentage of the plasticity index. It describes the moisture content of a soil with respect to its index limits and indicates in which part of its plastic range a soil lies, that is, its nearness to the liquid limit.

The consistency index is the ratio of the difference between the liquid limit and natural moisture content to the plasticity index. It can be used to classify the different types of consistency of cohesive soils, as shown in Table 1.6.

Table 1.6 *Consistency of cohesive soils*

Description	*Consistency index*	*Approximate undrained shear strength* (kPa)	*Field identification*
Hard		Over 300	Indented with difficulty by thumbnail, brittle
Very stiff	Above 1	150–300	Readily indented by thumbnail, still very tough
Stiff	0.75–1	75–150	Readily indented by thumb but penetrated only with difficulty. Cannot be moulded in the fingers
Firm	0.5–0.75	40–75	Can be penetrated several centimetres by thumb with moderate effort, and moulded in the fingers by strong pressure
Soft	Less than 0.5	20–40	Easily penetrated several centimetres by thumb, easily moulded
Very soft		Less than 20	Easily penetrated several centimetres by fist, exudes between fingers when squeezed in fist

The plasticity of a soil is influenced by the amount of its clay fraction, since clay minerals greatly influence the amount of attracted water held in the soil. With this in mind Skempton (1953) defined the activity of a clay as

$$\text{Activity} = \frac{\text{Plasticity index}}{\text{\% by mass finer than 0.002 mm}} \tag{1.4}$$

He suggested three classes of activity, namely, active, normal and inactive which he further subdivided into five groups as follows:

(1) inactive with activity less than 0.5,
(2) inactive with activity range 0.5–0.75,
(3) normal with activity range 0.75–1.25,
(4) active with activity range 1.25–2,
(5) active with activity greater than 2.

It would appear that there is only a general correlation between the clay mineral content of a deposit and its activity, that is, kaolinitic and illitic clays are usually inactive whilst montmorillonitic clays range from inactive to active. Usually active clays have a relatively high water-holding capacity and a high cation exchange capacity. They are also highly thixotropic, have a low permeability and a low resistance to shear.

1.5 Soil classification

Any system of soil classification involves grouping the different soil types into categories which possess similar properties and in so doing providing the engineer with a systematic method of soil description. Such a classification should provide some guide to the engineering performance of the soil type and should provide a means by which soils can be identified quickly.

Although soils include materials of various origins, for purposes of engineering classification it is sufficient to consider their simple index properties, which can be assessed easily, such as their particle size distribution, consistency limits or density.

Casagrande (1948) advanced one of the first comprehensive engineering classifications of soil. In the Casagrande system the coarse-grained soils are distinguished from the fine on a basis of particle size. Gravels and sands are the two principal types of coarse-grained soils and in this classification both are subdivided into five subgroups on a basis of grading (Table 1.7). Well graded soils are those in which the particle size distribution extends over a wide range without excess or

Table 1.7 *Symbols used in the Casagrande soil classification*

Main soil type		*Prefix*
Coarse-grained soils	Gravel	G
	Sand	S
Fine-grained soils	Silt	M
	Clay	C
	Organic silts and clays	O
Fibrous soils	Peat	Pt
Subdivisions		*Suffix*
For coarse-grained soils	Well graded, with little or no fines	W
	Well graded with suitable clay binder	C
	Uniformly graded with little or no fines	U
	Poorly graded with little or no fines	P
	Poorly graded with appreciable fines or well graded with excess fines	F
For fine-grained soils	Low compressibility (plasticity)	L
	Medium compressibility (plasticity)	I
	High compressibility (plasticity)	H

Table 1.8 *Unified soil classification (after Wagner, 1957)*

Field identification procedures (excluding particles larger than 76 mm and basing fractions on estimated weights)				*Group symbols*[1]	*Typical names*	*Information required for describing soils*		*Laboratory classification criteria*		
Coarse-grained soils More than half of material is *larger* than No. 200 sieve size[2]	Gravels More than half of coarse fraction is larger than No. 7 sieve size[3]	Clean gravels (little or no fines)	Wide range in grain size and substantial amounts of all intermediate particle sizes	*GW*	Well-graded gravels, gravel–sand mixtures, little or no fine	Give typical name; indicate approximate percentages of sand and gravel; maximum size; angularity, surface condition, and hardness of the coarse grains; local or geologic name and other pertinent descriptive information: and symbols in parentheses	Use grain size curve in identifying the fractions as given under field identification	Determine percentages of gravel and sand from grain size curve. Depending on fines (fraction smaller than No. 200 sieve size) coarse-grained soils are classified as follows: Less than 5%: *GW, GP, SW, SP*. More than 12%: *GM, GC, SM, SC*. 5% to 12%: *Borderline* cases require use of dual symbols	$C_u = \dfrac{D_{60}}{D_{10}}$ greater than 4 $C_c = \dfrac{(D_{30})^2}{D_{10} \times D_{60}}$ between 1 and 3	
			Predominantly one size or a range of sizes with some intermediate sizes missing	*GP*	Poorly graded gravels, gravel–sand mixtures, little or no fine				Not meeting all gradation requirements for *GW*	
		Gravels with fines (appreciable amount of fines)	Nonplastic fines (for identification procedures see *ML* below)	*GM*	Silty gravels, poorly graded gravel–sand–silt mixtures	For undisturbed soils add information on stratification, degree of compactness, cementation, moisture conditions and drainage characteristics			Atterberg limits below 'A' line, or *PI* less than 4	Above 'A' line with *PI* between 4 and 7 are *borderline* cases requiring use of dual symbols
			Plastic fines (for identification procedures, see *CL* below)	*GC*	Clayey gravels, poorly graded gravel–sand–clay mixtures				Atterberg limits above 'A' line with *PI* greater than 7	
	Sands More than half of coarse fraction is smaller than No. 7 sieve size[3]	Clean sands (little or no fines)	Wide range in grain sizes and substantial amounts of all intermediate particle sizes	*SW*	Well graded sands, gravelly sands, little or no fines	Example: *Silty sand*, gravelly; about 20% hard, angular gravel particles 12.5 mm maximum size; rounded and subangular sand grains coarse to fine, about 15% nonplastic fines with low dry strength; well compacted and moist in place; alluvial sand; (*SM*)			$C_u = \dfrac{D_{60}}{D_{10}}$ greater than 6 $C_c = \dfrac{(D_{30})^2}{D_{10} \times D_{60}}$ between 1 and 3	
			Predominantly one size or a range of sizes with some intermediate sizes missing	*SP*	Poorly graded sands, gravelly sands, little or no fines				Not meeting all gradation requirements for *SW*	
		Sands with fines (appreciable amount of fines)	Nonplastic fines (for identification procedures, see *ML* below)	*SM*	Silty sands, poorly graded sand–silt mixtures				Atterberg limits below 'A' line, or *PI* less than 5	Above 'A' line with *PI* between 4 and 7 are *borderline* cases requiring use of dual symbols
			Plastic fines (for identification procedures, see *CL* below)	*SC*	Clayey sands poorly graded sand–clay mixtures				Atterberg limits above 'A' line with *PI* greater than 7	

Table 1.8 *continued*

Fine-grained soils More than half of material is *smaller* than No. 200 sieve size[2]	Identification procedures on fraction smaller than No. 40 sieve size								Plasticity index 0 10 20 30 40 50 60; Liquid limit 0 10 20 30 40 50 60 70 80 90 100; Comparing soils at equal liquid limit; Toughness and dry strength increase with increasing plasticity index; A line; CH; CL; OL or ML; OH or MH; CL; CL-ML; ML
		DRY STRENGTH (crushing characteristics)	*DILATANCY (reaction to shaking)*	*TOUGHNESS (consistency near plastic limit)*					
	Silts and clays liquid limit less than 50	None to slight	Quick to slow	None	*ML*	Inorganic silts and very fine sands, rock flour, silty or clayey fine sands with slight plasticity	Give typical name: indicate degree and character of plasticity, amount and maximum size of coarse grains; colour in wet condition, odour if any, local or geologic name, and other pertinent descriptive information, and symbol in parentheses		
		Medium to high	None to very slow	Medium	*CL*	Inorganic clays of low to medium plasticity, gravelly clays, sandy clays, lean clays	For undisturbed soils add information on structure, stratification, consistency in undisturbed and remoulded states, moisture and drainage conditions		
		Slight to medium	Slow	Slight	*OL*	Organic silts & organic silt–clays of low plasticity			
	Silts and clays liquid limit greater than 50	Slight to medium	Slow to none	Slight to medium	*MH*	Inorganic silts micaceous or diatomaceous fine sandy or silty soils, elastic silts	Example: *Clayey silt*, brown: slightly plastic: small percentage of fine sand, numerous vertical root holes: firm and dry in place; loess; (*ML*)		
		High to very high	None	High	*CH*	Inorganic clays of high plasticity, fat clays			
		Medium to high	None to very slow	Slight to medium	*OH*	Organic clays of medium to high plasticity			
Highly organic soils		Readily identified by colour, odour, spongy feel and frequently by fibrous texture			*Pt*	Peat and other highly organic soils			

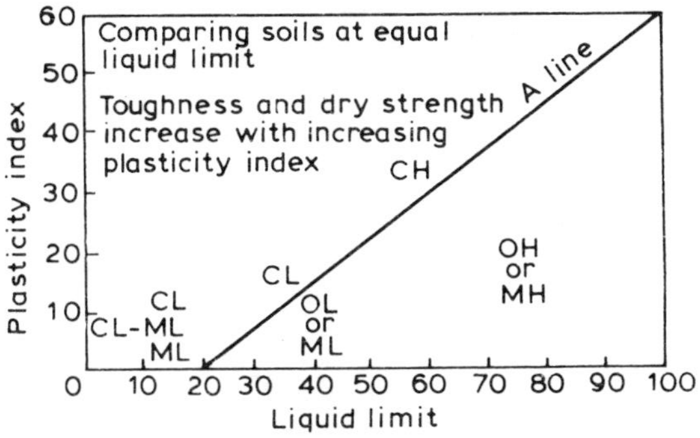

Footnotes to Table 1.8

[1] *Boundary classifications*. Soils possessing characteristics of two groups are designed by combinations of group symbols. For example *GW–GC*, well graded gravel-sand mixture with clay binder.

[2] All sieve sizes on this chart are US standard. The No. 200 sieve size is about the smallest particle visible to the naked eye.

[3] For visual classification, the 6.3 mm size may be used as equivalent to the No. 7 sieve size.

Field identification procedure for fine-grained soils or fractions

These procedures are to be performed on the minus No. 40 sieve size particles, approximately 0.4 mm. For field classification purposes, screening is not intended, simply remove by hand the coarse particles that interfere with the tests.

Dilatancy (reacting to shaking):

After removing particles larger than No. 40 sieve size, prepare a pat of moist soil with a volume of about one cubic centimetre. Add enough water if necessary to make the soil soft but not sticky. Place the pat in the open palm of one hand and shake horizontally, striking vigorously against the other hand several times. A positive reaction consists of the appearance of water on the surface of the pat which changes to a livery consistency and becomes glossy. When the sample is squeezed between the fingers, the water and gloss disappear from the surface, the pat stiffens and finally it cracks and crumbles. The rapidity of appearance of water during shaking and of its disappearance during squeezing assist in identifying the character of the fines in a soil.

Dry strength (crushing characteristics):

After removing particles larger than No. 40 sieve size, mould a pat of soil to the consistency of putty, adding water if necessary. Allow the pat to dry completely by oven, sun or air drying, and then test its strength by breaking and crumbling between the fingers. This strength is a measure of the character and quantity of the colloidal fraction contained in the soil. The dry strength increases with increasing plasticity. High dry strength is characteristic for clays of the CH group. A typical inorganic silt possesses only very slight dry strength. Silty fine sands and silts have about the same slight dry strength, but can be distinguished by the feel when powdering the dried specimen. Fine sand feels gritty whereas a typical silt has the smooth feel of flour.

Toughness (consistency near plastic limit):

After removing particles larger than the No. 40 sieve size, a specimen of soil about one cubic centimetre in size, is moulded to the consistency of putty. If too dry, water must be added and if sticky, the specimen should be spread out in a thin layer and allowed to lose some moisture by evaporation. Then the specimen is rolled out by hand on a smooth surface or between the palms into a thread about 3 mm in diameter. The thread is then folded and re-rolled repeatedly. During this manipulation the moisture content is gradually reduced and the specimen stiffens, finally loses its plasticity, and crumbles when the plastic limit is reached.

After the thread crumbles, the pieces should be lumped together and a slight kneading action continued until the lump crumbles.

The tougher the thread near the plastic limit and the stiffer the lump when it finally crumbles, the more potent is the colloidal clay fraction in the soil. Weakness of the thread at the plastic limit and quick loss of coherence of the lump below the plastic limit indicate either inorganic clay of low plasticity, or materials such as kaolin-type clays and organic clays which occur below the A-line. Highly organic clays have a very weak and spongy feel at the plastic limit.

Table 1.9 *Engineering use chart (after Wagner 1957)*

Typical names of soil groups	*Group symbols*	*Important properties*				*Relative desirability for various uses (Graded from 1 (highest) to 14 (lowest))*									
						Rolled earth dams			*Canal sections*		*Foundations*		*Roadways*		
													Fills		
		Permeability when compacted	*Shearing strength when compacted and saturated*	*Compressibility when compacted and saturated*	*Workability as a construction material*	*Homogeneous embankment*	*Core*	*Shell*	*Erosion resistance*	*Compacted earth lining*	*Seepage important*	*Seepage not important*	*Frost heave not possible*	*Frost heave possible*	*Surfacing*
Well graded gravels, gravel-sand mixtures, little or no fines	*GW*	Pervious	Excellent	Negligible	Excellent	–	–	1	1	–	–	1	1	1	1
Poorly graded gravels, gravel-sand mixtures, little or no fines	*GP*	Very pervious	Good	Negligible	Good	–	–	2	2	–	–	3	3	3	–
Silty gravels, poorly graded gravel-sand-silt mixtures	*GM*	Semi-pervious to impervious	Good	Negligible	Good	2	4	–	4	4	1	4	4	9	5
Clayey gravels, poorly graded gravel-sand-clay mixtures	*GC*	Impervious	Good to fair	Very low	Good	1	1	–	3	1	2	6	5	5	1
Well-graded sands, gravelly sands, little or no fines	*SW*	Pervious	Excellent	Negligible	Excellent	–	–	3 if gravelly	6	–	–	2	2	2	4
Poorly graded sands, gravelly sands, little or no fines	*SP*	Pervious	Good	Very low	Fair	–	–	4 if gravelly	7 if gravelly	–	–	5	6	4	–

Silty sands, poorly graded sand-silt mixtures	*SM*	Semi-pervious to impervious	Good	Low	Fair	4	5	–	8 if gravelly	5 erosion critical	3	7	8	10	6
Clayey sands, poorly graded sand-clay mixtures	*SC*	Impervious	Good to fair	Low	Good	3	2	–	5	2	4	8	7	6	2
Inorganic silts and very fine sands, rock flour, silty or clayey fine sands with slight plasticity	*ML*	Semi-pervious to impervious	Fair	Medium	Fair	6	6	–	–	6 erosion critical	6	9	10	11	–
Inorganic clays of low to medium plasticity, gravelly clays, sandy clays, silty clays, lean clays	*CL*	Impervious	Fair	Medium	Good to fair	5	3	–	9	3	5	10	9	7	7
Organic silts and organic silt-clays of low plasticity	*OL*	Semi-pervious to impervious	Poor	Medium	Fair	8	8	–	–	7 erosion critical	7	11	11	12	–
Inorganic silts, micaceous or diatomaceous fine sandy or silty soils, elastic silts	*MH*	Semi-pervious to impervious	Fair to poor	High	Poor	9	9	–	–	–	8	12	12	13	–
Inorganic clays of high plasticity, fat clays	*CH*	Impervious	Poor	High	Poor	7	7	–	10	8 volume change critical	9	13	13	8	–
Organic clays of medium to high plasticity	*OH*	Impervious	Poor	High	Poor	10	10	–	–	–	10	14	14	14	–
Peat and other highly organic soils	*Pt*	–	–	–	–	–	–	–	–	–	–	–	–	–	–

deficiency in any particular sizes, whereas in uniformly graded soils the distribution extends over a very limited range of particle sizes. In poorly graded soils the distribution contains an excess of some particle sizes and a deficiency of others.

Each of the main soil types and subgroups is given a letter, a pair of which is combined in the group symbol, the former being the prefix, the latter the suffix. A plasticity chart (Table 1.8) is also used when classifying fine-grained soils. On this chart the plasticity index is plotted against liquid limit. The A line is taken as the boundary between organic and inorganic soils, the latter lying above the line. Subsequently the Unified Soil Classification (Table 1.8) was developed from the Casagrande system. The engineering uses of these various soils are given in Table 1.9.

The British Soil Classification for engineering purposes (Anon, 1981) also uses particle size as a fundamental parameter and is very much influenced by the Casagrande system. Boulders, cobbles, gravels, sands, silts and clays are distinguished as individual groups, each group being given the following symbol and size range:

(1) boulders (B), over 200 mm,
(2) cobbles (Cb), 60–200 mm,
(3) gravel (G), 2–60 mm,
(4) sand (S), 0.06–2 mm,
(5) silt (M), 0.002–0.06 mm,
(6) clay (C), less than 0.002 mm.

The gravel, sand and silt ranges may be further divided into coarse, medium and fine categories (see Section 1.3). Sands and gravels are granular materials, ideally possessing no cohesion, whereas silts and clays are cohesive materials. Mixed soil types can be indicated as shown in Table 1.10.

These major soil groups are again divided into subgroups on a basis of grading in the case of cohesionless soils, and on a basis of plasticity in the case of fine material. Granular soils are described as well graded (W) or poorly graded (P). Two further types of poorly graded granular soils are recognized, namely, uniformly graded

Table 1.10 *Mixed soil types*

Term	*Composition of the coarse fraction*
Slightly sandy GRAVEL	Up to 5% sand
Sandy GRAVEL	5–20% sand
Very sandy GRAVEL	Over 20% sand
GRAVEL/SAND	About equal proportions of gravel and sand
Very gravelly SAND	Over 20% gravel
Gravelly SAND	5–20% gravel
Slightly gravelly SAND	Up to 5% gravel

(Pu) and gap-graded (Pg). Silts and clays are generally subdivided according to their liquid limits (LL) into low (under 35%), intermediate (35–50%) and high (50–70%) subgroups. Very high (70–90%) and extremely high (over 90%) categories have also been recognized. As in the Casagrande classification each subgroup is given a combined symbol in which the letter describing the predominant sized fraction is written first (e.g. GW = well graded gravels; CH = clay with high liquid limit).

Any group may be referred to as organic if it contains a significant proportion of organic matter, in which the case the letter O is suffixed to the group symbol (e.g. CVSO = organic clay of very high liquid limit with sand). The symbol Pt is given to peat.

In many soil classifications boulders and cobbles are removed before an attempt is made at classification, for example, their proportions are recorded separately in the British Soil Classification. Their presence should be recorded in the soil description, a plus sign being used in symbols for soil mixtures, for example, G + Cb for gravel with cobbles. The British Soil Classification has proposed that very coarse deposits should be classified as follows:

(1) *Boulders*. Over half of the very coarse material is of boulder size (over 200 mm). May be described as cobbly boulders if cobbles are an important second constituent in the very coarse fraction.
(2) *Cobbles*. Over half of the very coarse material is of cobble size (200 mm–60 mm). May be described as bouldery cobbles if boulders are an important second constituent in the very coarse fraction.

Mixtures of the very coarse material and soil can be described by combining the terms for the very coarse constituent and the soil constituent as in Table 1.11.

Table 1.11 *Mixtures of very coarse materials and soil*

Term	*Composition*
BOULDERS (or COBBLES) with a little finer material[1]	Up to 5% finer material
BOULDERS (or COBBLES) with some finer material[1]	5–20% finer material
BOULDERS (or COBBLES) with much finer material[1]	20–50% finer material
FINER MATERIAL[1] with many BOULDERS (or COBBLES)	50–20% boulders (or cobbles)
FINER MATERIAL[1] with some BOULDERS (or COBBLES)	20–5% boulders (or cobbles)
FINER MATERIAL[1] with occasional BOULDERS (or COBBLES)	Up to 5% boulders (or cobbles)

[1] Give the name of the finer material (in parentheses when it is the minor constituent), e.g. sandy GRAVEL with occasional boulders; cobbly BOULDERS with some finer material (sand with some fines).

Table 1.12 *Field identification and description of soils (after Anon, 1981)*

	Basic soil type	*Particle size* (mm)	*Visual identification*	*Particle nature and plasticity*
Very coarse soils	BOULDERS		Only seen complete in pits or exposures.	Particle shape: Angular Subangular Subrounded Rounded Flat Elongate
		200		
	COBBLES		Often difficult to recover from boreholes	
		60		
Coarse soils (over 65% sand and gravel sizes)	GRAVELS	coarse 20 medium 6 fine	Easily visible to naked eye; particle shape can be described; grading can be described Well graded: wide range of grain sizes, well distributed. Poorly graded: not well graded. (May be uniform: size of most particles lies between narrow limits; or gap graded: an intermediate size of particle is markedly under-represented.)	
		2		Texture: Rough Smooth Polished
	SANDS	coarse 0.6 medium 0.2 fine	Visible to naked eye; very little or no cohesion when dry; grading can be described. Well graded: wide range of grain sizes, well distributed. Poorly graded: not well graded. (May be uniform: size of most particles lies between narrow limits; or gap graded: an intermediate size of particle is markedly under-represented.)	
		0.06		
Fine soils (over 35% silt and clay sizes)	SILTS	coarse 0.02 medium 0.006 fine	Only coarse silt barely visible to naked eye; exhibits little plasticity and marked dilatancy; slightly granular or silky to the touch. Disintegrates in water; lumps dry quickly; possess cohesion but can be powdered easily between fingers.	Non-plastic or low plasticity
		0.002		
	CLAYS		Dry lumps can be broken but not powdered between the fingers; they also disintegrate under water but more slowly than silt; smooth to the touch; exhibits plasticity but no dilatancy; sticks to the fingers and dries slowly; shrinks appreciably on drying usually showing cracks. Intermediate and high plasticity clays show these properties to a moderate and high degree, respectively.	Intermediate plasticity (Lean clay) High plasticity (Fat clay)
Organic soils	ORGANIC CLAY, SILT or SAND	Varies	Contains substantial amounts of organic vegetable matter.	
	PEATS	Varies	Predominantly plant remains usually dark brown or black in colour, often with distinctive smell; low bulk density.	

Composite soil types (mixtures of basic soil types)

Scale of secondary constituents with coarse soils

Term		*% of clay or silt*
slightly clayey / slightly silty	GRAVEL or SAND	under 5
– clayey / – silty	GRAVEL or SAND	5 to 15
very clayey / very silty	GRAVEL or SAND	15 to 35
Sandy GRAVEL / Gravelly SAND	Sand or gravel as important second constituent of the coarse fraction	

For composite types described as:
clayey: fines are plastic, cohesive;
silty: fines non-plastic or of low plasticity

Scale of secondary constituents with fine soils

Term		*% of sand gravel*
sandy / gravelly	CLAY or SILT	35 to 65
–	CLAY:SILT	under 35

Examples of composite types

(Indicating preferred order for description)

Loose, brown, subangular very sandy, fine to coarse GRAVEL with small pockets of soft grey clay

Medium dense, light brown, clayey, fine and medium SAND

Stiff, orange brown, fissured sandy CLAY

Firm, brown, thinly laminated SILT and CLAY

Plastic, brown, amorphous PEAT

mpactness/strength		Structure				Colour
rm	Field test	Term	Field identification	Interval scales		
ose	By inspection of voids and particle packing.	Homo-geneous	Deposit consists essentially of one type.	*Scale of bedding spacing*		Red Pink Yellow Brown Olive Green Blue White Grey Black, etc.
nse		Inter-stratified	Alternating layers of varying types or with bands or lenses of other materials. Interval scale for bedding spacing may be used.	*Term*	*Mean spacing (mm)*	
				Very thickly bedded	Over 2000	
				Thickly bedded	2000–600	
ose	Can be excavated with a spade; 50 mm wooden peg can be easily driven.	Hetero-geneous	A mixture of types.	Medium bedded	600–200	
				Thinly bedded	200–60	
nse	Requires pick for excavation; 50 mm wooden peg hard to drive.	Weathered	Particles may be weakened and may show concentric layering.	Very thinly bedded	60–20	Supple-mented as necessary with:
				Thickly laminated	20–6	
htly ented	Visual examinati n; pick removes soil in lumps which can be abraded.			Thinly laminated	Under 6	Light Dark Mottled, etc.
						and
t or se	Easily moulded or crushed in the fingers.	Fissured	Break into polyhedral fragments along fissures. Interval scale for spacing of discontinuities may be used.			
n or se	Can be moulded or crushed by strong pressure in the fingers.					Pinkish Reddish Yellowish Brownish, etc.
y soft	Exudes between fingers when squeezed in hand.	Intact	No fissures.			
t	Moulded by light finger pressure.	Homo-geneous	Deposit consists essentially of one type.	*Scale of spacing of other discontinuities*		
1	Can be moulded by strong finger pressure.	Inter-stratified	Alternating layers of varying types. Interval scale for thickness of layers may be used.	*Term*	*Mean spacing (mm)*	
:	Cannot be moulded by fingers. Can be indented by thumb.	Weathered	Usually has crumb or columnar structure.	Very widely spaced	Over 2000	
y	Can be indented by thumb nail.			Widely spaced	2000–600	
1	Fibres already compressed together.			Medium spaced	600–200	
				Closely spaced	200–60	
ngy	Very compressible and open structure.	Fibrous	Plant remains recognizable and retain some strength.	Very closely spaced	60–20	
tic	Can be moulded in hand, and smears fingers.	Amor-phous	Recognizable plant remains absent.	Extremely closely spaced	Under 30	

Table 1.13 *British Soil Classification System for Engineering Purposes (after Anon, 1981)*

Soil groups[1]			*Subgroups and laboratory identification*					
GRAVEL and SAND may be qualified Sandy GRAVEL and Gravelly SAND, etc. where appropriate			*Group symbol*[2,3]		*Subgroup symbol*[2]	*Fines (% less than 0.06 mm)*	*Liquid limit (%)*	*Name*
COARSE SOILS less than 35% of the material is finer than 0.06 mm	GRAVELS More than 50% of coarse material is of gravel size (coarser than 2 mm)	Slightly silty or clayey GRAVEL	G	GW	GW	0–5		Well graded GRAVEL
				GP	GPu GPg			Poorly graded/Uniform/Gap graded GRAVEL
		Silty GRAVEL	G-F	G-M	GWM GPM	5–15		Well graded/Poorly graded silty GRAVEL
		Clayey GRAVEL		G-C	GWC GPC			Well graded/Poorly graded clayey GRAVEL
		Very silty GRAVEL	GF	GM	GML, etc.	15–35		Very silty GRAVEL; subdivide as for GC
		Very clayey GRAVEL		GC	GCL GCI GCH GCV GCE			Very clayey GRAVEL (clay of low, intermediate, high, very high, extremely high plasticity)
	SANDS More than 50% of coarse material is of sand size (finer than 2 mm)	Slightly silty or clayey SAND	S	SW	SW	0–5		Well graded SAND
				SP	SPu SPg			Poorly graded/Uniform/Gap graded SAND
		Silty SAND	S-F	S-M	SWM SPM	5–15		Well graded/Poorly graded silty SAND
		Clayey SAND		S-C	SWC SPC			Well graded/Poorly graded clayey SAND
		Very silty SAND	SF	SM	SML, etc.	15–35		Very silty SAND' subdivided as for SC
		Very clayey SAND		SC	SCL SCI SCH SCV SCE			Very clayey SAND (clay of low, intermediate, high, very high, extremely high plasticity)

Table 1.13 *continued*

FINE SOILS more than 35% of the material is finer than 0.06 mm	Gravelly or sandy SILTS and CLAYS 35–65% fines	Gravelly SILT	FG	MG	MLG, etc.			Gravelly SILT; subdivide as for CG
		Gravelly CLAY[4]		CG	CLG CIG CHG CVG CEG		<35 35–50 50–70 70–90 >90	Gravelly CLAY of low plasticity of intermediate plasticity of high plasticity of very high plasticity of extremely high plasticity
		Sandy SILT[4]	FS	MS	MLS, etc.			Sandy SILT; subdivide as for CG
		Sandy CLAY		CS	CLS, etc.			Sandy CLAY; subdivide as for CG
	SILTS and CLAYS 65–100% fines	SILT (M-SOIL)	F	M	ML, etc.			SILT; subdivide as for C
		CLAY[5,6]		C	CL CI CH CV CE		<35 35–50 50–70 70–90 >90	CLAY of low plasticity of intermediate plasticity of high plasticity of very high plasticity of extremely high plasticity
ORGANIC SOILS		Descriptive letter 'O' suffixed to any group or subgroup symbol.			Organic matter suspected to be a significant constituent. Example MHO: Organic SILT of high plasticity.			
PEAT		Pt Peat soils consist predominantly of plant remains which may be fibrous or amorphous.						

[1] The name of the soil group should always be given when describing soils, supplemented, if required, by the group symbol, although for some additional applications (e.g. longitudinal sections) it may be convenient to use the group symbol alone.
[2] The group symbol or subgroup should be placed in brackets if laboratory methods have not been used for identification, e.g. (GC).
[3] The designation FINE SOIL, or FINES, F, may be used in place of SILT, M, or CLAY, C, when it is not possible or not required to distinguish between them.
[4] GRAVELLY if more than 50% of coarse material is of gravel size. SANDY if more than 50% of coarse material is of sand size.
[5] SILT (M-SOIL), M, is material plotting below the 'A' line, and has a restricted plastic range in relation to its liquid limit, and relatively low cohesion. Fine soils of this type include clean silt-sized materials and rock flour, micaceous and diatomaceous soils, pumice, and volcanic soils, and soils containing halloysite. The alternative term 'M-soil' avoids confusion with materials of predominantly silt size, which form only a part of the group.
Organic soils also usually plot below the 'A' line on the plasticity chart, when they are designated ORGANIC SILT, MO.
[6] CLAY, C, is material plotting above the 'A' line, and is fully plastic in relation to its liquid limit.

The British Soil Classification can be made either by rapid assessment in the field or by full laboratory procedure (Tables 1.12 and 1.13 respectively). The classification is made, like that of the Unified System, on a basis of grading the soil according to particle-size distribution and in the case of silts and clays by determining plasticity.

As far as soil description is concerned Anon (1981) recommends that the principal features to be included should be

(1) *Mass characteristics*
  (a) Field strength or compactness (Table 1.6) and indication of moisture condition.
  (b) Bedding.
  (c) Discontinuities.
  (d) Degree of weathering.
(2) *Material characteristics*
  (a) Colour (Table 1.12).
  (b) Particle shape (see below) and composition.
  (c) Soil name (in capitals, e.g. SAND), grading and plasticity.
(3) *Geological formation, age and type of deposit.*
(4) *Classification* (optional).
  Soil group symbol.

In particular instances it may be necessary to describe the shape of soil particles. The terms shown in Table 1.14 have been suggested by Anon (1981).

Table 1.14 *Description of particle shape*

Angularity	angular subangular subrounded rounded
Form	equidimensional flat elongated flat and elongated irregular
Surface texture	rough smooth

Since it is rarely possible to carry out significant soil tests on made ground, good descriptions are highly important. According to Anon (1981) such descriptions should include information on the following, as well as on the soil constituents:

(1) mode of origin of the material,
(2) presence of large objects such as concrete, masonry or old motor cars,
(3) presence of voids or collapsible hollow objects,

(4) chemical waste, and dangerous or poisonous substances,
(5) organic matter, with a note on the degree of decomposition,
(6) smell,
(7) striking colour tints,
(8) any dates readable on buried newspapers,
(9) signs of heat or internal combustion under ground, e.g. steam emerging from boreholes.

1.6 Shear strength of soil

The shear strength of a soil is the maximum resistance which it can offer to shear stress. When this maximum has been reached the soil is regarded as having failed, its strength having been fully mobilized. However, the shear strength value determine experimentally is not a unique constant which is characteristic of the material but varies with the method of testing. Shear displacement also continues to take place after the shear strength is exceeded. Shear displacements occur across a well-defined single plane of rupture or across a shear zone.

The stress on any plane surface can be resolved into the normal stress, σ_n, which acts perpendicular to the surface, and the shearing stress, τ, which acts along the surface, the magnitude of the resistance being given by Coulomb's equation

$$\tau = c + \sigma_n \tan\phi \qquad (1.5)$$

where ϕ is the angle of shearing resistance and c is the cohesion.

The stress that controls changes in the volume and the strength of a soil is known as the effective stress (σ'). When a load is applied to a saturated soil it is either carried by the pore water, which gives rise to an increase in the pore pressure (u), or the soil skeleton, or both. The effect that a load has on a soil is therefore affected by the drainage conditions but it has been shown that for most practical cases the effective stress is equal to the intergranular stress and can be determined from the equation

$$\sigma' = \sigma - u \qquad (1.6)$$

where σ is the total stress. Hence shear strength depends upon effective stress and not total stress. Accordingly Coulomb's equation must be modified in terms of effective stress and becomes

$$\tau = c' + \sigma' \tan\phi' \qquad (1.7)$$

The internal frictional resistance of a soil, for example, as is developed in a sand, is generated by friction when the grains in the zone of shearing are caused to slide, roll and rotate against each other. Local crushing may occur at the points of contacts which suffer the highest stress. The total resistance to rolling is the sum of the behaviour of all the particles and is influenced by the confining stress, the coefficient of friction and angles of contact between the minerals as well as their surface roughness. However, the angle of internal friction does not depend solely on the internal friction between grains because a proportion of the shearing stress on the plane of failure is utilized in overcoming interlocking, that is, it is also dependent upon the initial void ratio or density of a given soil. It is also influenced by the size and shape of the grains. The larger the grains, the wider is the zone which is affected. The more angular the grains, then the greater is the frictional resistance to their relative movement. Electrical forces of attraction and repulsion may also be involved in shearing resistance.

It is commonly believed that interparticle friction is influenced by the composition of the particles involved and the state of their surface chemistry. However, after an investigation of quartz (siliceous) and calcareous grains, Frossard (1979) proposed that interparticle friction appears to increase with angularity, independently of grain composition. He also noted that sphericity has a notable influence on volumetric strain, particularly on the maximum rate of dilatancy. A review of the effects of stress history on the deformation of sand has been given by Lambrechts and Leonards (1978). They also referred to how minor differences in packing can influence stress–strain behaviour.

In clay soils the cohesion which is developed by the molecular attractive forces between the minute soil particles is mainly responsible for the resistance offered to shearing. Because molecular attractive forces depend to a large extent on the mineralogical composition of the particles and on the type of concentration of electrolytes present in the pore water, the magnitude of the true cohesion also depends on these factors.

Lambe's concept of the shear strength in clay postulated the existence of forces of attraction and repulsion between the particles with a net repulsive force in accordance with physico-chemical principles (Lambe, 1960). Hence the equilibrium of internal stresses in a clay soil can be expressed as

$$\sigma' = \sigma - u = R - A$$

where σ' is the effective stress, σ is the total stress, u is the pore water pressure, R is the total force of repulsion and A is the total force of attraction per unit area between the particles. Unfortunately the R and A forces cannot be measured. The relationship is demonstrated by the fact that clay behaviour differs when immersed in solutions with different electrolyte concentrations. In other words, the electrical attractive forces decrease and the electrical repulsive forces increase with an increase in dielectric constant of the pore medium. Further consideration has been given to this modified concept of effective stress by Sridharan and Rao (1979). They showed that the drained strengths of kaolinitic and montmorillonitic clays decreased when the dielectric constant of the pore fluid was increased.

The shear strength of an undisturbed clay is generally found to be greater than that obtained when it is remoulded and tested under the same conditions and at the same water content. The ratio of the undisturbed to the remoulded strength at the same moisture content is defined as the sensitivity of a clay. Skempton and Northey

(1952) proposed the following grades of sensitivity:

(1) insensitive clays, under 1,
(2) low sensitive clays, 1–2,
(3) medium sensitive clays, 2–4,
(4) sensitive clays, 4–8,
(5) extra-sensitive clays, 8–16,
(6) quick clay, over 16.

Clays with high sensitivity values have little strength after being disturbed. Indeed if they suffer slight disturbance this may cause an initially fairly strong material to behave as a viscous fluid. High sensitivity seems to result from the metastable arrangement of equidimensional particles. The strength of the undisturbed clay is chiefly due to the strength of the framework developed by these particles and the bonds between their points of contact. If the framework is destroyed by remoulding the clay loses most of its strength and any subsequent regain in strength due to thixotropic hardening does not exceed a small fraction of its original value. Sensitive clays generally possess high moisture contents, frequently with liquidity indices well in excess of unity. A sharp increase in moisture content may cause a great increase in sensitivity, sometimes with disastrous results. Heavily overconsolidated clays are insensitive. The effect of remoulding on clays of various sensitivities is illustrated in Figure 1.2.

Some clays with moderate to high sensitivity show a regain in strength when, after remoulding, they are allowed to rest under unaltered external conditions. Such soils are thixotropic. Thixotropy is the property of a material which allows it to undergo an isothermal gel-to-sol-to-gel transformation upon agitation and subsequent rest. This transformation can be repeated indefinitely without fatigue and the gelation time under similar conditions remains the same. The softening and subsequent recovery of thixotropic soils appears to be due, first, to the destruction, and then, secondly, to the rehabilitation of the molecular structure of the adsorbed layers of the clay particles. For example, the loss of consistency in soils containing Na montmorillonite occurs since large volumes of water are adsorbed upon and held between the colloidal clay particles. Furthermore the ionic forces attracting the colloidal clay particles together have a definite arrangement which is an easily destroyable microstructure when subjected to agitation. When the material is at rest the ions and water molecules tend to reorientate themselves and strength is thereby recovered.

1.7 Consolidation

The theory of consolidation, as advanced by Terzaghi (1925) has enabled engineers to determine the amount and rate of settlement which is likely to occur when structures are erected on cohesive soils. When a layer of soil is loaded, some of the pore water is expelled from its voids, moving slowly away from the region of high stress as a result of the hydrostatic gradient created by the load. The void ratio

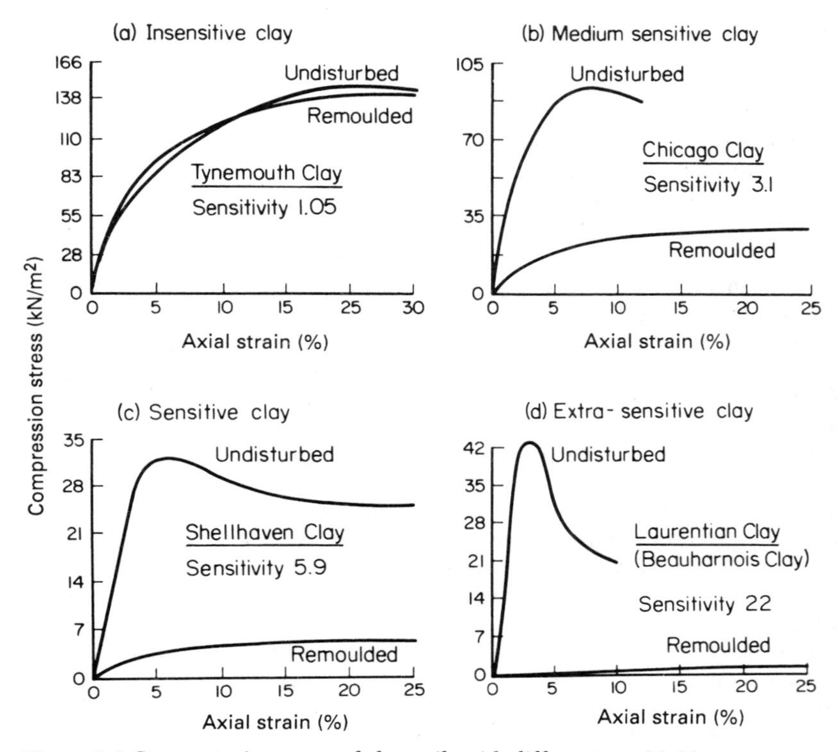

Figure 1.2 *Stress–strain curves of clay soils with different sensitivities*

accordingly decreases and settlement occurs. This is termed primary consolidation. Further settlement, usually of minor degree, may occur due to the rearrangement of the soil particles under stress, this being referred to as secondary consolidation. However, in reality primary and secondary consolidation are not distinguishable.

Terzaghi (1943) showed that the relationship between unit load and the void ratio for a sediment can be represented by plotting the void ratio, e, against the logarithm of the unit load, p (Figure 1.3). The shape of the e/log p curve is related to the stress history of a clay. In other words the e/log p curve for a normally consolidated clay is linear and is referred to as the virgin compression curve. On the other hand if the clay is overconsolidated the e/log p curve is not straight and the preconsolidation pressure can be derived from the curve. The preconsolidation pressure refers to the maximum overburden pressure to which a deposit has been subjected. Overconsolidated clay is appreciably less compressible than normally consolidated clay.

The compressibility of a clay can be expressed in terms of the compression index (C_c) or the coefficient of volume compressibility (m_v). The compression index is the slope of the linear section of the e/log p curve and is dimensionless.

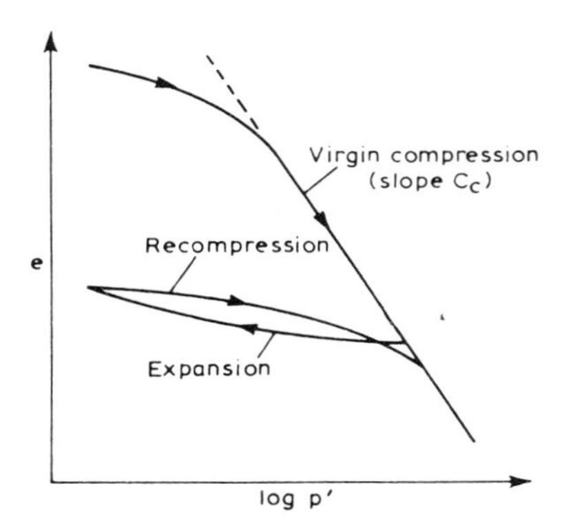

Figure 1.3 *Results of consolidation test,* e *log* p *curve*

The value of C_c for cohesive soils ranges between about 0.075 for silty clays to more than 1.0 for highly colloidal bentonic clays. An approximation of the degree of compressibility is as follows

Range of C_c	*Degree of compressibility*	*Soil type*
Over 0.3	Very high	Soft clay
0.3–0.15	High	Clay
0.15–0.075	Medium	Silty clay
Less than 0.075	Low	Sandy clay

Hence the compressibility index increases with increasing clay content and so with increasing liquid limit. Indeed Skempton (1944) found that C_c for normally consolidated clays is closely related to their liquid limit, the relationship between the two being expressed as

$$C_c = 0.009(\text{LL} - 10) \tag{1.9}$$

The coefficient of volume compressibility is defined as the volume change per unit volume per unit increase in load, its units are the inverse of pressure (m^2/MN) (i.e. the strain per unit stress increased in one-dimensional compression). The value of m_v for a given soil depends upon the stress range over which it is determined; Anon (1975) recommends that it should be calculated for a pressure increment of 100 kN/m^2 in excess of the effective overburden pressure on the soil at the depth in question.

The compressibility of a soil is dependent on the average rate of compression and the soil structure has a substantial time-dependent resistance to compression. At the instant when a load, p, on a layer of clay is suddenly increased by Δp, the thickness of the layer remains unchanged. Hence the application of the load, Δp, produces an equal increase, Δu, in the hydrostatic pressure of the pore water. As time proceeds, the excess pore pressure is gradually dissipated and finally disappears, whilst the grain-to-grain pressure simultaneously increases from an initial value p to $p + \Delta p$. The ratio between the decrease of the void ratio, Δe, at time, t, and the ultimate decrease, Δe_1 represents the degree of consolidation, U, at time, t

$$U = 100 \frac{\Delta e}{\Delta e_1} \tag{1.10}$$

With a given thickness, H, of a layer of clay the degree of consolidation at time, t, depends exclusively on the coefficient of consolidation, c_v,

$$c_v = \frac{k}{\gamma_w m_v} \tag{1.11}$$

where k is the coefficient of permeability and γ_w is the unit weight of water. The coefficient of consolidation which determines the rate at which settlement takes place is calculated for each load increment and either a mean value or that value appropriate to the pressure range in question is used (c_v is measured in m^2/year). With increasing values of p both k and m_v decrease. The coefficient of consolidation decreases for normally consolidated clays from about 3.15 m^2/year for very lean clays to about 0.03 m^2/year for highly colloidal clays. At any value of c_v, the time (t) at which a given degree of consolidation, U, is reached, increases in simple proportion to the square of the thickness, H, of the layer.

References

Anon (1975). *Methods of Test for Soils for Civil Engineering Purposes, BS 1377*, British Standards Institution, London.

Anon (1979). 'Classification of rocks and soils for engineering geological mapping. Part I: Rock and soil materials', *Bull. Int. Ass. Engg. Geol.*, No. 19, 364–71.

Anon (1981). *Code of Practice on Site Investigation, BS5930*, British Standards Institution, London.

Casagrande, A. (1948). 'Classification and identification of soils', *Trans ASCE,* **113**, 901–2.

Frossard, A. (1979). 'Effect of sand grain shape on interparticle friction: indirect measurements by Rowe's stress dilatancy theory', *Geotechnique,* **29**, 341–50.

Lambe, T. W. (1960). 'A mechanistic picture of the shear strength in clay', *ASCE, Proc. Res. Conf., Shear Strength of Cohesive Soils*, Boulder, Colorado, 555–80.

Lambrechts, J. R. and Leonards, G. A. (1978). 'Effects of stress history on the deformation of sand', *Proc. ASCE, J. Geot. Engg. Div.,* **GT11**, 1371–87.

Skempton, A. W. (1944). 'Notes on the compressibility of clays', *Q. J. Geol. Soc. London,* **100**, 119–35.

Skempton, A. W. (1953). 'The colloidal activity of clays', *Proc. 3rd Int. Conf. Soil Mech. Foundation Engg, Zurich,* **1**, 57–60.

Skempton, A. W. and Northey, R. D. (1952). 'The sensitivity of clays', *Geotechnique,* **2**, 30–53.

Sridharan, A. and Rao, V. G. (1979). 'Shear strength behaviour of saturated clays and the role of the effective stress concept', *Geotechnique,* **29**, 177–93.

Terzaghi, K. (1925) *Erdbaumechanik auf Bodenphysikalischer Grundlage*, Deuticke, Vienna.

Terzaghi, K. (1943). *Theoretical Soil Mechanics*, Wiley, New York.

Wagner, A. A. (1957). 'The use of the Unified Soil Classification System for the Bureau of Reclamation', *Proc. 4th Int. Conf. Soil Mech. Foundation Engg*, London, **1**, 125–34.

Chapter 2

Coarse-grained soils

The composition of a gravel deposit reflects not only the source rocks of the area from which it was derived but is also influenced by the agents responsible for its formation and the climatic regime in which it was or is being deposited. The latter two factors have a varying tendency to reduce the proportion of unstable material. Relief also influences the nature of a gravel deposit, for example, gravel production under low relief is small and the pebbles tend to be chemically inert residues such as vein quartz, quartzite, chert and flint. By contrast high relief and rapid erosion yield coarse, immature gravels.

Sands consist of a loose mixture of mineral grains and rock fragments. Generally they tend to be dominated by a few minerals, the chief of which is quartz. There is a presumed dearth of material in those grades transitional to gravel on the one hand and silt on the other ((Glossop and Skempton, 1945).

Alluvial sands include those found in alluvial fans, in riverchannels, on flood plains and in deltaic deposits. Marine sands occur as beach deposits, as offshore bars and barriers, in tidal deltas and in tidal flats. They are mainly deposited on the continental shelf but some sand is carried over the continental edge by turbidity currents. Wind-blown sands accumulate as coastal dunes and as extensive dune fields in deserts. Sands are also produced as a result of glacial and flucio-glacial action.

Sands tend to be close packed and usually the grains show some degree of orientation, presumably related to the direction of flow of the transporting medium. They vary greatly in maturity, the ultimate end product being a uniformly sorted quartz sand with rounded grains.

2.1 Fabric of granular soils

The engineering behaviour of a soil is a function of its structure or fabric, which in turn is a result of the geological conditions governing deposition and the subsequent stress history. The macrostructure of a soil includes its bedding, laminations, fissures, joints and tension cracks, all of which can exert a dominant influence on the shear strength and drainage characteristics of a soil mass. The microstructure of a sand or gravel refers to its particle arrangement which in turn involves the concept of packing – in other words the spatial density of particles in the aggregate (Kahn, 1956).

The conceptual treatment of packing begins with a consideration of the arrangement of spherical particles of equal size. These can be packed either in a disorderly or systematic fashion. The closest type of systematic packing is rhombohedral packing whereas the most open type is cubic packing, the porosities approximating to 26% and 48% respectively (Table 2.1). Put another way, the void

Table 2.1 *Some values of the common properties of gravels, sands and silts*

	Gravels	*Sands*	*Silts*
Relative density	2.5–2.8	2.6–2.7	2.64–2.66
Bulk density (Mg/m^3)	1.45–2.3	1.4–2.15	1.82–2.15
Dry density (Mg/m^3)	1.4–2.1	1.35–1.9	1.45–1.95
Porosity (%)	20–50	23–35	–
Void ratio	–	–	0.35–0.85
Liquid limit (%)	–	–	24–35
Plastic limit (%)	–	–	14–25
Coefficient of consolidation (m^2/yr)	–	–	12.2
Effective cohesion (kN/m^2)	–	–	75
Shear strength (kN/m^2)	200–600	100–300	–
Angle of friction (deg)	34–35	32–42	32–36

ratio of a well sorted and perfectly cohesionless aggregate of equidimensional grains can range between extreme values of about 0.35 and 1.00. If the void ratio is more than unity the microstructure will be collapsible or metastable. If a large number of spheres of equal size is arranged in any systematic packing pattern then there is a certain diameter ratio for smaller spheres which can just pass through the throats between the larger spheres into the interstices, for example, in rhombohedral packing this critical diameter is 0.154 D (D being the diameter of the larger spheres). However, a considerable amount of disorder occurs in most coarse grained deposits and, according to Graton and Frazer (1935), there are colonies of tighter and looser packing within any deposit.

In a single grain structure individual particles are bulky and pore passages have average diameters of the same order of magnitude as smaller particle diameters. There is virtually no effective combination of particles to form aggregates. Each particle functions individually in the soil framework, and particles are in contact with one another, so that the movement of any individual grain is influenced by the position of adjacent grains. For most equilibrium conditions in coarse-grained soil the soil framework serves exclusively as the stressed member.

Size and sorting have a significant influence on the engineering behaviour of granular soils; generally speaking the larger the particles, the higher the strength. Deposits consisting of a mixture of different sized particles are usually stronger than those which are uniformly graded. For example, Holtz and Gibbs (1956) showed that the amount of gravel in a sand-gravel mixture has a significant effect on shear strength, which increases considerably as the gravel content is increased up to 50 or 60%. Beyond this point the material becomes less well graded and the density does not increase. However, the mechanical properties of such sediments depend mainly on their density index (formerly relative density) which in turn depends on packing. Indeed Holtz and Gibbs (1956) maintained that densely packed sands are almost incompressible whereas loosely packed deposits, located above the water table, are relatively compressible but otherwise stable. If the relative density of a sand varies erratically this can give rise to differential settlement. Generally settlement is relatively rapid. However, when the stresses are large enough to produce appreciable grain fracturing, there is a significant time lag.

Greater settlement is likely to be experienced in granular soils where foundation level is below the water table rather than above. Additional settlement may occur if the water table fluctuates or the ground is subject to vibrations. Although the relative density may decrease in a general manner with decreasing grain size there is ample evidence to show, for example, that water-deposited sands with similar grain size can vary between wide limits. Hence factors other than grain size, such as rate of deposition and particle shape, influence the density index.

2.2 Deformation of granular soil

Two basic mechanisms contribute towards the deformation of granular soil, namely, distortion of the particles, and the relative motion between them. These mechanisms are usually interdependent. At any instant during the deformation process different mechanisms may be acting in different parts of the soil and these may change as deformation continues. Interparticle sliding can occur at all stress levels, the stress required for its initiation increasing with initial stress and decreasing void ratio. Crushing and fracturing of particles begins in a minor way at small stresses, becoming increasingly important when some critical stress is reached. This critical stress is smallest when the soil is loosely packed and uniformly graded, and consists of large, angular particles with a low strength. Usually fracturing becomes important only when the stress level exceeds 3.5 MN/m^2.

The internal shearing resistance of a granular soil is generated by friction developed when grains in the zone of shearing are caused to slide, roll and rotate against each other. Frossard (1979) pointed out that interparticle friction in sand is governed by the mineralogy of the grains and the physico-chemical condition of their surfaces. None the less he admitted that interparticle friction increased with increasing angularity of the grains and that this is independent of their mineralogy (see below). At the commencement of shearing in a sand some grains are moved into new positions with little difficulty. The normal stress acting in the direction of movement is small but eventually these grains occupy positions in which further sliding is more difficult. By contrast other grains are so arranged in relation to the grains around them that sliding is difficult. They are moved without sliding by the movements of other grains. The frictional resistance of the former is developed as the grains become impeded whereas in the latter case it is developed immediately. The resistance to rolling represents the sum of the behaviour of all the particles, and the resistance to sliding is essentially attributable to friction which, in turn, is proportional to the confining stress.

Frictional resistance is built up gradually and consists of establishing normal stresses in the intergranular structure as the grains push or slide along (Cornforth, 1964). At the same time sliding allows the structure to loosen in dilatant soils which reduces normal stress. The maximum shearing resistance is a function of these two factors. The packing and external stress conditions govern the amount of sliding by individual grains in mobilizing shearing resistance. According to Cornforth the latter factor is the more important and in fact is really a strain condition. He therefore concluded that the strain condition during shear is a major factor contributing to the strength of sand.

As noted, the nature of the pre-stressing and the magnitude of the associated strains affect the behaviour of granular soil. Ladd *et al.* (1977) maintained that prestressing with small strains gives rise to stiffening and that in an anisotropic sand, prestressing leads to stiffening in a preferred direction. However, if stressing produces large strains, then the fabric of the sand is altered radically. This yields a stiff structure for subsequent loadings in the same direction in conditions of drained shear. On the other hand, in undrained conditions liquefaction may be brought about which destroys the original fabric of the soil. Such sands are readily liquefied subsequently.

Although density may have a significant effect on the initial compressibility of sand in that compressibility decreases with increasing density, Lambrechts and Leonards (1978) maintained that its importance is overshadowed by the past loading history. Indeed, they went on to claim that no other factor has such a significant influence on the compressibility of granular soils. The deformation response of a sand is also directly related to the composition of its grains, and the nature of the bedding also strongly influences the behaviour of sands. Lambrechts and Leonards further pointed out that a well graded sand, with the same relative density, is more compressible than one which is uniformly graded: that increasing grain size means that compresibility declines somewhat: that as the proportion of angular grains increases, so the compressibility increases: and that the compressibility decreases as the surface roughness of grains increases.

2.3 Strength and distortion

Interlocking grains contribute a large proportion of the strength in densely packed granular soils and shear failure occurs by overcoming the frictional resistance at the grain contacts. Conversely, interlocking has little or no effect on the strength of very loosely packed coarse-grained soils in which the mobility of the grains is greater (Borowicka, 1973). Interlocking decreases as the confining stress increases because the particles become rounded at the points of contact, sharp corners being crushed and particles may even fracture. Even though this results in a denser material it is still easier for shear deformations to occur.

Figure 2.1 shows that dense sand has a high peak strength and that, when it is subjected to shear stress, it expands up to the point of failure, after which a slight

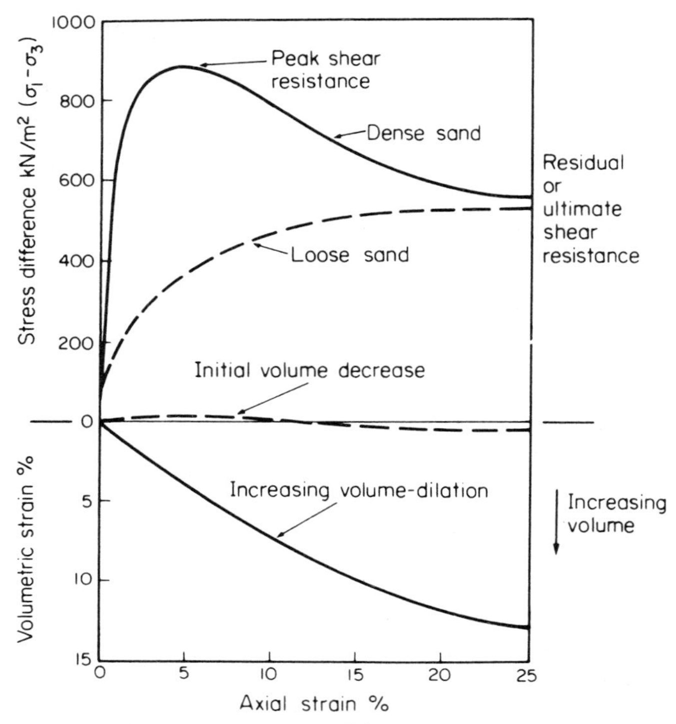

Figure 2.1 *Stress–strain curves of dense and loose sand*

decrease in volume may occur. Loose sand, on the other hand, compacts under shearing stress and its residual strength may be similar to that of dense sand, and tends to remain so. Hence a constant void ratio is obtained, that is, the critical volume condition which has a critical angle of friction and a critical void ratio. These are independent of initial density being a function of the normal effective stress at which shearing occurs. For a discussion of the shear strength of sands see Rowe (1969) and Frossard (1979).

Both curves in Figure 2.1 indicate strains which are approximately proportional to stress at low stress levels, suggesting a large component of elastic distortion. If the stress is reduced the unloading stress–strain curve indicates that not all the strain is recovered on unloading. The hysteresis loss represents the energy lost in crushing and repositioning of grains. At higher shear stresses the strains are proportionally greater indicating greater crushing and reorientation. Indeed Arnold and Mitchell (1973) showed that as a sample of sand is subjected to cyclic loading, the unloading response involves an increasing degree of hysteresis; in other words they found that on unloading recoverable deformation in sand under triaxial conditions was small. Because irrecoverable strains were larger than the elastic strains this led them to suggest that total strains could, in fact, be regarded as irrecoverable. As would be expected loose sand with larger voids and fewer points of contact exhibits greater strains and less recovery when unloaded than dense sand (Lambrechts and Leonards, 1978; Lade, 1978).

The angle of shearing resistance is also influenced by the grain-size distribution and grain shape (Holtz and Gibbs, 1956); the larger the grains the wider the zone affected, the more angular the grains the greater the frictional resistance to their relative movement, since they interlock more thoroughly than do rounded ones and they therefore produce a larger angle of shearing resistance (Table 2.2). A well-graded granular soil experiences less breakdown on loading than a uniformly sorted soil of the same mean particle size since in the former type there are more interparticle contacts and hence the load per contact is less than in the latter. In gravels the effect of angularity is less because of particle crushing. Lambrechts and Leonards (1978) found that stress history, which may influence the stress–strain behaviour of a sand significantly, has little effect on its shear strength.

Table 2.2 *Effect of grain shape and grading on the peak friction angle of cohesionless soil (after Terzaghi, 1955)*

Shape and grading	*Loose*	*Dense*
1. Rounded, uniform	30°	37°
2. Rounded, well graded	34°	40°
3. Angular, uniform	35°	43°
4. Angular, well graded	39°	45°

Kirkpatrick (1957); Sutherland and Mesdary (1969) and Arnold and Mitchell (1973) all studied the failure state of sand in a three-dimensional stress system. They found that generally the Mohr–Coulomb law for sand based on triaxial tests under-predicts the failure strength of the material in other stress conditions.

The presence of water in the voids of a granular soil does not usually produce significant changes in the value of the angle of internal friction. However, if stresses develop in the pore water they may bring about changes in the effective stresses between the particles whereupon the shear strength and the stress–strain relationships may be radically altered. Whether or not pore pressures develop depends upon the drainage characteristics of the soil mass and its tendency to undergo volume changes when subjected to stress. If the pore water can readily drain from the soil mass during the application of stress then the granular material behaves as it does when dry. On the other hand if loading takes place rapidly, particularly in fine-grained sands which do not drain as easily, then pore pressures are not dissipated. Since the water cannot readily escape from the voids of loosely packed, fine-grained sands no volume decrease can occur and so the pressure increases in the pore water. If the sample is loose enough nearly all the stress difference may be carried by the pore water so that very little increase occurs in the effective stress. In dense sands, if the stress and drainage conditions prevent the water flowing into the sand as it is stressed then the usual volume increase characteristic of dense dry sand does not occur and a negative pore pressure develops.

The relationship between unit load on a potential surface of sliding (p) and the shearing resistance per unit area (s) can be approximately expressed by the equation:

$$s = (p - u) \tan \phi \tag{2.1}$$

where u is the hydrostatic pressure of the pore liquid prior to the application of the shearing force. In the light of what has been said above concerning changing pore pressures this expression can be modified to:

$$s = (p - u - \Delta u) \tan \phi \tag{2.2}$$

where Δu represents the change in pressure.

Dusseault and Morgenstern (1979) introduced the term 'locked sands' to describe certain peculiar sands which were first recognized in the Athabasca Oil Sands in Canada, and are older than Quaternary age. They are characterized by their high quartzose mineralogy, lack of interstitial cement, low porosity, brittle behaviour and high strength, with residual shear strengths (ϕ_r) varying between 30° and 35°.

Locked sands possess very high densities. Indeed the relative densities of locked sands exceed 100%, that is, their porosities are less than those which can be obtained by laboratory tests for achieving minimum porosity. Dusseault and Morgenstern attributed this to the peculiar fabric of these sands. This has been developed by diagenetic processes which reduced the porosity of the sands by solution and recrystallization of quartz as crystal overgrowths. Hence locked sands have an interlocked texture with a relatively high incidence of long and interpenetrative grain contacts. Sutured contacts are present and all grains have rugose surfaces.

At low stress levels locked sands undergo high rates of dilation. They have peak frictional strengths considerably in excess of those of dense sand. Dilatancy becomes suppressed as the level of stress increases, since the asperities on the surfaces of individual grains are sheared through rather than causing dilation. The failure envelopes of locked sands are steeply curved as a result of the changing energy level required for shearing asperities as the level of stress increases.

If undisturbed, locked sands are capable of supporting large loads with only small deformations.

References

Arnold, M. and Mitchell, P. W. (1973). 'Sand deformation in three-dimensional stress state', *Proc. 8th Int. Conf. Soil Mech. Found. Engg., Moscow,* **1**, 11–18.

Borowicka, H. (1973). 'Rearrangement of grains by shear tests with sand', *Proc. 8th Int. Conf. Soil Mech. Found. Engng., Moscow,* **1**, 71–7.

Cornforth, D. H. (1964). 'Some experiments on the influence of strain conditions on the strength of sand', *Geotechnique,* **14**, 143–67.

Dusseault, M. B. and Morgenstern, N. R. (1979). 'Locked sands', *Q. J. Engng. Geol.,* **12**, 117–32.

Frossard, E. (1979). 'Effect of sand grain shape on the interparticle friction; indirect measurements by Rowe's stress dilatancy theory', *Geotechnique,* **29**, 341–50.

Glossop, R. and Skempton, A. W. (1945). 'Particle size in silts and sands', *J. Inst. Civ. Engrs.,* **25**, Paper No. 5492, 81–105.

Graton, L. C. and Frazer, H. J. (1935). 'Systematic packing of spheres with particular relation to porosity and permeability', *J. Geol.,* **43**, 785–909.

Holtz, W. G. and Gibbs, H. J. (1956). 'Shear strength of pervious gravelly soils', *Proc. ASCE, J. Soil Mech. Found. Div.,* **82**, Paper No. 867.

Kahn, J. S. (1956). 'The analysis and distribution of the properties of packing in sand size particles', *J. Geol.,* **64**, 385–95, 578–606.

Kirkpatrick, W. M. (1957). 'The condition of failure of sands', *Proc. 4th Int. Conf. Soil Mech. Found. Engng., London,* **1**, 172–85.

Ladd, C. C., Foott, R., Ishihara, K., Schlosser, F. and Poulos, H. G. (1977). 'Stress deformation and stress characteristics', State-of-the-Art Report. *Proc. 9th Int. Conf. Soil Mech. & Found. Engg.*, Tokyo, **3**, 421–94.

Lade, V. P. (1978). 'Prediction of undrained behaviour of sand', *Proc. ASCE, J. Geotech. Engg. Div.*, **104**, GT6, 721–36.

Lambrechts, J. R. and Leonards, G. A. (1978). 'Effects of stress history on deformation of sand', *Proc. ASCE, J. Geotech. Engng. Div.,* **104**, GT11, 1371–89.

Rowe, P. W. (1969) 'The relation between the shear strength of sands in triaxial compression, plane strain and direct shear', *Geotechnique,* **19**, 75–86.

Sutherland, H. and Mesdary, M. (1969). 'The influence of the intermediate principal stress on the strength of sand', *Proc. 7th Int. Conf. Soil Mech. Found. Engng.,* Mexico City, **1**, 391–9.

Terzaghi, K. (1955). 'The influence of geological factors in the engineering properties of sediments', *Econ. Geol.*, 50th Ann. Vol., 557–618.

Chapter 3

Cohesive soils

3.1 Silts

Silts are clastic sediments, that is, they are derived from pre-existing rock types chiefly by mechanical breakdown processes. They are mainly composed of fine quartz material. Silts may occur in residual soil horizons but in such instances they are usually not important. However, silts are often found in alluvial, lacustrine and marine deposits. As far as alluvial sediments are concerned silts are typically present in flood plain deposits, they may also occur on terraces which border such plains. These silts tend to interdigitate with deposits of sand and clay. Silts are also present with sands and clays in estuarine and deltaic sediments. Lacustrine silts are often banded and may be associated with varved clays, which themselves contain a significant proportion of particles of silt size. Marine silts are also frequently banded and have high moisture contents. Wind blown silts are usually uniformly sorted.

Grains of silt are often rounded with smooth outlines which influence their degree of packing. The latter, however, is more dependent on the grain size distribution within a silt deposit, uniformly sorted deposits not being able to achieve such close packing as those in which there is a range of grain size. This, in turn, influences the porosity and void ratio values as well as the bulk and dry densities (Table 2.1).

Dilatancy is characteristic of fine sands and silts. The environment is all important for the development of dilatancy since conditions must be such that expansion can take place. What is more it has been suggested that the soil particles must be well wetted and it appears that certain electrolytes exercise a dispersing effect, thereby aiding dilatancy. The moisture content at which a number of sands and silts from British formations become dilatant usually varies between 16 and 35%. According to Boswell (1961), dilatant systems are those in which the anomalous viscosity increases with increase of shear.

Schultze and Kotzias (1961) showed that consolidation of silt was influenced by grain size, particularly the size of the clay fraction, porosity and natural moisture content. Primary consolidation accounted for 76% of the total consolidation exhibited by the Rhine silts tested by Schultze and Kotzias, secondary consolidation contributing the remainder. It was noted that unlike many American silts, which are often unstable when saturated, and undergo significant settlements when loaded, the Rhine silts in such a condition were usually stable. The difference no doubt lies in the respective soil structures. Most American silts are, in fact, loess soils which have a more open structure than the reworked river silts of the Rhine with a void ratio of less than 0.85. None the less, in many silts, settlement continues to take place after construction has been completed and may exceed 100 mm. Settlement may continue for several months after completion because the rate at which water can drain from the voids under the influence of applied stress is slow.

Schultze and Horn (1961) found that the direct shear test proved unsuitable for the determination of the shear strength of silt, this had to be obtained by triaxial testing. They demonstrated that the cohesion of silt was a logarithmic function of the water content and that the latter and the effective normal stress determined the shear strength. The angle of friction was dependent upon the plasticity index.

In a series of triaxial tests carried out on silt, Penman (1953) showed that in drained tests with increasing strain, the volume of the sample first decreased, then increased at a uniform rate and ultimately reached a stage where there was no further change. The magnitude of the dilatancy which occurred when the silt was sheared, and was responsible for these volume changes, increased with increasing density as it does in sands. The expansion was caused by the grains riding over each other during shearing. The strength of the silt was attributed mainly to the friction between the grains and the force required to cause dilatancy against the applied pressures.

These drained tests indicated that the angle of shearing resistance (ϕ) decreased with increasing void ratio and with increasing lateral pressure. Grain interlocking was responsible for the principal increase in the angle of shearing resistance and increased with increasing density. A fall in pore water pressure occurred in the undrained tests during shearing, and there was an approximately linear relation between the maximum fall in pore pressure and the void ratio. Provided the applied pressures were sufficiently high the drop in pore pressure governed the ultimate strength. The fall in the pore pressure was dependent on the density of the silt: the greater the density, the greater the fall in pore pressure. At a given density the ultimate strength was independent of applied pressure (above a critical pressure) and so the silt behaved as a cohesive material ($\phi = 0$). Below this critical pressure silt behaved as a cohesive and frictional material. An exceptional condition occurred when a highly dilatable sample was placed under low cell pressure. When the pore pressure fell below atmospheric pressure gas was liberated by the pore water and the sample expanded.

Frost heave is commonly associated with silty soils and loosely packed silts can exhibit quick conditions (see Chapter 12).

3.2 Loess

Most loess is of aeolian origin and has accumulated over North America, Europe, Russia and China during the last two to three million years. It is mainly of silt size. Wind-blown deposits of loess are characterized by a lack of stratification and uniform sorting and occur as blanket deposits. Nonetheless it is claimed that some deposits of loess are of alluvial origin (e.g. in Poland, see Grabowska-Olszewska, 1988). Such deposits are stratified. The most notable occurrences of loess are found in China, where in Gansu Province thicknesses in excess of 300 m have been recorded. Formerly, because deposits of loess show a close resemblance to fine-grained glacial debris, their origin was assigned a glacial association. In other words winds blowing from the arid interiors of the northern continents during glacial times picked up fine glacial outwash material and carried it many hundreds of kilometres before deposition took place. Deposition is presumed to have occurred over steppe lands, the grasses having left behind fossil root-holes which typify loess. The lengthy transport accounts for the uniform sorting of loess. Although this may represent the origin of some loess deposits, it cannot be assigned to that of all deposits. The loess of China is silty material which was removed by wind from desert basins, wadis and playas which lie to the north and west. They probably still contribute wind-blown material to the loess belt of China. The collapsible silty material found in the arid south-west of the United States also continues to accumulate at present and has no glacial association.

In spite of having a variety of origins and appreciable differences in provenance, thickness and age, loess is a remarkably uniform soil in terms of its dominant minerals and geotechnical behaviour. In other words loess deposits have similar grain size distribution, mineral composition, open texture, low degree of saturation and bonding of grains which is not resistant to water. For example, all the samples of loess from China tested by Derbyshire and Mellors (1988) contained quartz, feldspars, micas and calcite. Chlorite and vermiculite were present in many samples and kaolinite, frequently poorly ordered, was found in some. Mixed-layer minerals have been widely reported and montmorillonite also is present. Illite is often the most abundant clay mineral. In the deposits of loess in Poland reported by Grabowska-Olszewska (1988), quartz is the principal mineral accounting for 40–70%, with feldspars around 15%, and micas, goethite and clay minerals (up to 30%). The clay fraction is composed of mixed-layer clays (50–80%), illite (5–40%), kaolinite (2–5%) and fine quartz (2–10%). The clay minerals could have been transported and deposited by wind or they could have been formed in place as a result of weathering of feldspars. Both origins, no doubt, are true. The content of mixed-layer clays in the younger loess is usually less than in the older material (50–56% compared with 65–75%). This increase occurs at the expense of illite, which in the older loess tends not to exceed 15%.

Calcium carbonate occurs in loess in China in the form of root linings and at grain contacts. It also may occur as concretions, which may be fairly soft and about the size of a pea, or as loess dolls. The carbonate content tends to vary between 4 and 20%. Up to 20% calcium carbonate may be present in unweathered loess in Poland (generally 6–12%) but in weathered loess the amount is significantly less (usually less than 6%), indeed it may be absent.

The texture of loess takes the form of a loose skeleton built of grains and microaggregates (assemblages of clay or clay and silty clay particles). The sand and silt-sized particles are sub-angular and sub-rounded and separate from each other, being connected by bonds and 'bridges', with uniformly distributed pores. The bridges are formed of clay-sized materials, be they clay minerals, fine quartz, feldspar or calcite. These clay-sized materials also occur as coatings to grains. Silica and iron oxide may be concentrated as cement at grain contacts and amorphous overgrowths of silica occur on grains of quartz and feldspar. As sand and silt particles are not in contact, the mechanical behaviour of loess is governed by the structure and quality of the bonds.

In a detailed examination of the micro-structure of loess soils in Russia, Larionov (1965) found that the coarser grains were never in contact with each other, being carried in a fine granular dispersed mass. Hence the strength of the soil is largely determined by the character of this fine mass. The ratio of coarser grains to fine dispersed fraction varies not only quantitatively but morphologically. Consequently three micro-structures were recognized, namely, granular, where a filmy distribution of the fine dispersed fraction predominates; aggregate, consisting mainly of aggregates; and granular–aggregate, having an intermediate character. Larionov suggested that generally loess soils with granular microstructure have less water resistance than aggregate types, they also have lower cohesion and higher permeability. They are therefore more likely to collapse on wetting than the aggregate types.

Grabowska-Olszewska (1988) maintained that carbonates do not form coatings surrounding quartz grains in Polish loess but they occur as irregularly distributed growths. Since carbonates do not bind coarser grains of quartz their effect on the strength of cementing bonds is insignificant. In fact Grabowska-Olszewska claimed that the bonds between grains in Polish loess are mainly due to electrostatic forces and so depend on the moisture content and distribution of the pores. Most of the pores are less than 1 μm in size (usually from 80 to over 90%), whilst pores between 1 and 10 μm in size normally account for only a few per cent.

Large scale piping occurs in the Upper Pleistocene loess of China, piping systems greater than one metre in diameter and many tens of metres in length being referred to as loess karst (Derbyshire and Mellors, 1988). Piping and sinkhole development also was reported in loess at Kamloops, British Columbia, by Hardy (1950). It also occurs in Poland (Grabowska-Olszewska, 1988) and in central Nebraska and western Iowa (Lutenegger and Hallberg, 1988). In addition, loess may exhibit sub-vertical columnar jointing.

3.2.1 Geotechnical properties of loess

Loess deposits generally consist of 50–90% particles of silt size. For example, in the loess deposits of the Missouri Basin, investigated by Clevenger (1958), most of the grains were between 0.019 and 0.074 mm (Figure 3.1a). In fact he distinguished sandy, silty and clayey loess. Over three-quarters of the loess examined fell within the silty variety, a fifth being clayey loess, and the rest being sandy loess. The loess deposits of China have an average grain size in the coarse and medium silt range (Derbyshire and Mellors, 1988). They are generally poorly to very poorly sorted (i.e. uniformly sorted) and consistently fine skewed (Figure 3.1b). The loess tends to become finer with increasing age. In the case of Polish loess the particle size distribution is generally sand grains, 3–38% (usually 11–19%); silt grains, 63–88% (usually 73–76%); and clay particles, 6–10% (usually 8–10%).

Loess soils have a low range of values of specific gravity, for instance, those of Chinese loess vary from 2.65 to 2.7. Similarly the range of dry density is very low to

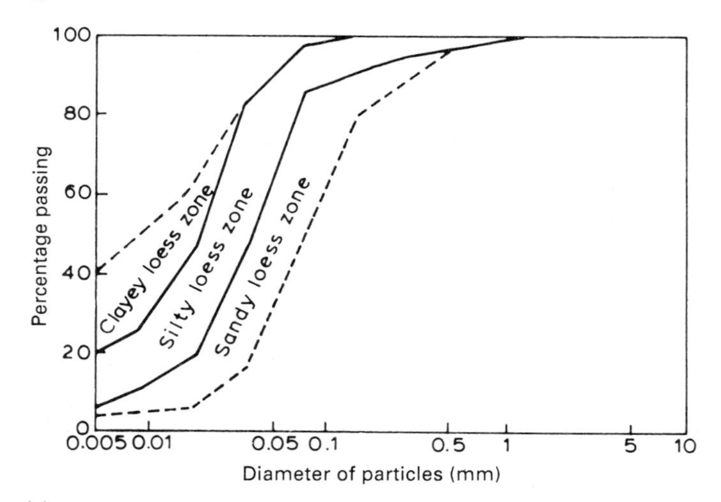

(a)

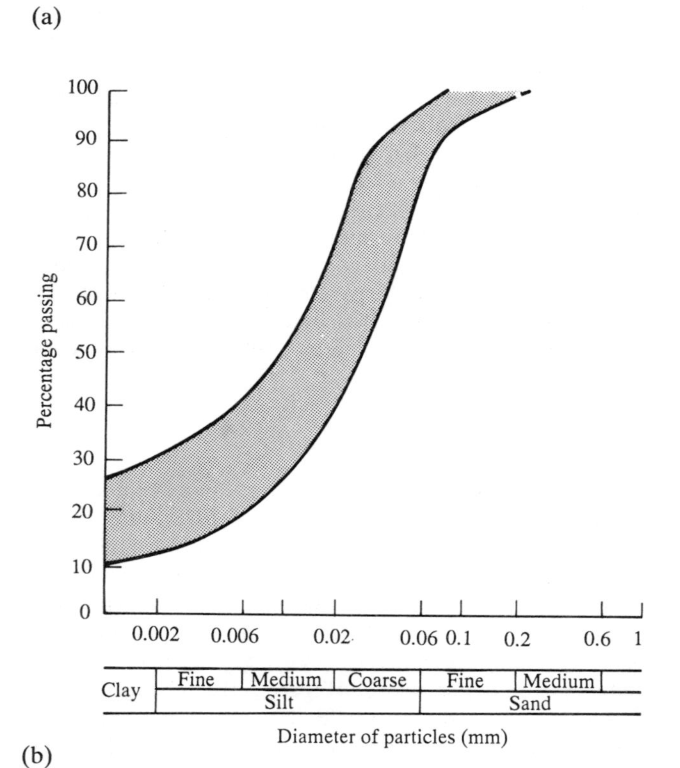

(b)

Figure 3.1 *Particle size distribution:* (a) *Loess of Missouri Basin (after Clevenger, 1958);* (b) *Loess of Lanzhou Province, China (from Derbyshire and Mellors, 1988)*

low (e.g. in Chinese loess it varies from 1.4 to 1.46 Mg/m^3; and in loess at Kamloops, British Columbia, the range extends from 1.2 to 1.4 Mg/m^3), as is that of void ratio (0.81–0.89 for Chinese loess) and porosity (44.8–47.1% for Chinese loess). Luteneggar and Hallberg (1988) observed that the bulk densities of unstable loess (such as Peorian Loess), tend to range between 1.34 and 1.55 Mg/m^3 (Table 3.1). Clevenger (1958) found that the undisturbed densities of loess in the Missouri Basin were a little lower, ranging from around 1.2 to 1.36 Mg/m^3. If this material is wetted or consolidated (or reworked), the density increases, sometimes to as high as 1.6 Mg/m^3.

Loess soils are slightly to moderately plastic, the plasticity increasing as clay content increases. Gibbs and Holland (1960) indicated that liquid limits varied from 25 to 35%, with exceptional values for clayey loess ranging up to 45% (Figure 3.2).

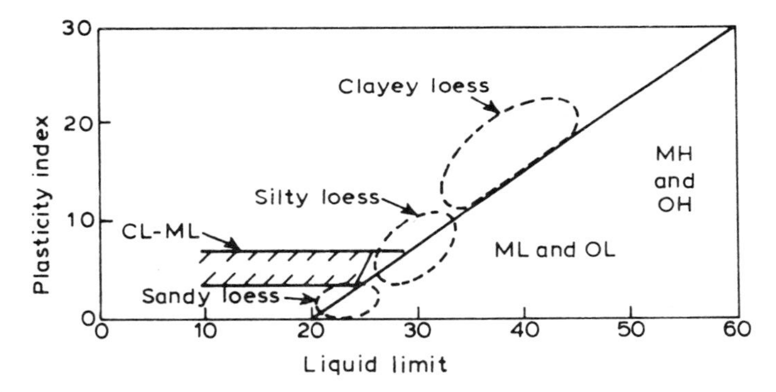

Figure 3.2 *Plasticity of Missouri Basin loess (after Clavenger, 1958)*

The values of plasticity index reported were between 5 and 22 per cent (Table 3.1). Values of Atterberg limits, quoted by Derbyshire and Mellors (1988), for loess from China were plastic limit 12–22%, liquid limit 39–32%, and plasticity index 10–17%. All values plotted above the ‘A’ line on the plasticity chart which indicates that they are inorganic soils of low to medium plasticity. Low plasticity indices suggest that the soil must be very sensitive to changes in moisture content.

Gibbs and Holland (1960) mentioned activity values for loess in the range 0.4–0.7. Obviously activity is influenced by clay content and in clayey loess values are higher. For example, the average activity of loess soils from Iowa with clay contents between 16 and 32% is about 1.3 (Lutenegger and Hallberg, 1988). Below 16% clay content the activity falls to about 0.5. The latter type of loess exhibits a greater tendency for instability.

The strength of loess is dependent on the initial porosity and moisture content, the degree of deterioration of the bonds and the increase in granular contacts under consolidation, as well as the changes in moisture content. When loess with many macropores and high water content is loaded, the cementing bonds are first broken which results in a lowering of the cohesion and softening of the soil. With further

Table 3.1 *Some geotechnical properties of loess soils*

Property	*Shaansi Province, China** *Sandy loess*	*Clayey loess*	*Lanzhow Province, China†*	*Czechoslovakia, near Prague‡*	*South Polish Uplands§*
Natural moisture content (%)	9–13	13–20	11–10	21	3–26
Specific gravity					2.66–2.7
Bulk density (Mg/m^3)	1.59–1.68	1.4–1.85			1.54–2.12
Dry density (Mg/m^3)			1.4–1.5		1.46–1.73
Void ratio	0.8–0.92	0.76–1.11	1.05		
Porosity (%)				44–50	35–46
Grain-size distribution (%)					
Sand	20.5–35.2	12–15	20–24		
Silt	54.8–69.0	64–70	57–65		
Clay	8.0–15.5	17–24	16–21		
Plastic limit (%)			10–14	20	
Liquid limit (%)	26–28	30–31	27–30	36	
Plasticity index (%)	8–10	11–12	10–14	16	
Activity				1.32	
Coefficient of collapsibility	0.007–0.016	0.003–0.023		0.006–0.011	0.0002–0.06
Angle of friction					7–36

* From Lin and Wang (1988)
† From Tan (1988)
‡ From Feda (1988)
§ From Grabowska-Olszewska (1988).

loading the grains are brought more and more into contact, thereby increasing friction, so giving rise to a hardening effect. As far as the angle of shearing resistance of loess is concerned, this usually varies between 30° and 34°. As the liquidity index of loess increases the shear strength decreases, becoming essentially zero at around a liquidity index of one (Figure 3.3). In triaxial tests carried out by

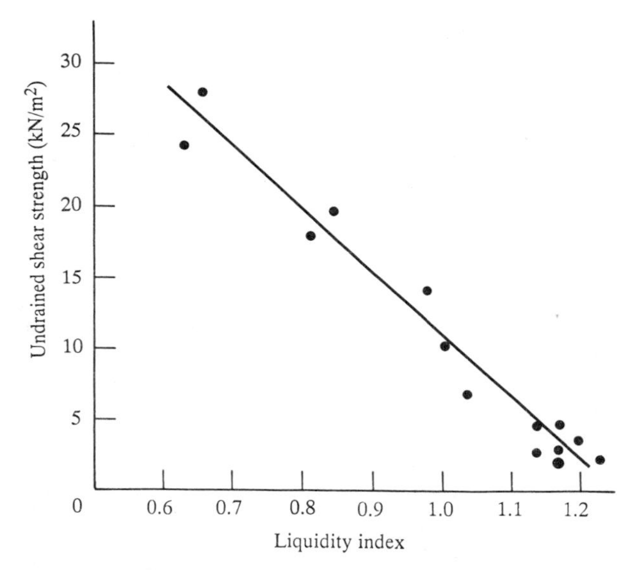

Figure 3.3 *Strength vs. liquidity index for a loessial silt (from Lutenegger and Hallberg, 1988)*

Tan (1988), when the confining pressure was less than 100 kN/m^2, the stress–strain curves showed a peak followed by shear softening. The loess failed by brittle fracture. When the confining pressure exceeded 100 kN/m^2, no peak was observed and the strain increased with increasing stress. The strain hardening was attributed to the increase of frictional forces due to the increased number of grain contacts due to compression and shear. Under the latter confining conditions the vertical strain could exceed 10%, after which failure occurred by plastic bulging. Samples underwent a volume increase when the confining pressure was less than 100 kN/m^2 and were consolidated when it exceeded this figure.

Loess can support heavy structures with small settlements if loads do not exceed the apparent preconsolidation stress and natural moisture content is low. On the other hand loess can compress substantially if the apparent preconsolidation stress is exceeded. Primary and secondary compression occur similar to that of saturated clay. Primary settlements generally occur rapidly, with much of the settlement occurring during the actual application of load. Loess also may exhibit creep deformation under loading.

Loess has a much higher vertical than horizontal permeability, which is enhanced by long vertical voids in the loess structure, which are regarded as fossil rootholes. Deposits of loess are better drained (their permeability ranges from 10^{-5} to 10^{-8} m/s) than are true silts.

Unlike silt, loess does not appear to be frost susceptible, this being due to its more permeable character, but like silt it can exhibit quick conditions and it is difficult, if not impossible, to compact. Lutenegger and Hallberg (1988) suggested that the widespread occurrence of landslides in loess during earthquakes may be in part due to the liquefaction of loess. Because of its porous structure a 'shrinkage' factor must be taken into account when estimating earthwork in loess.

The presence of vertical rootholes explains why vertical slopes are characteristic of loess landscapes. These may remain stable for long periods and when failure occurs it generally does so in the form of a vertical slide. By contrast, an inclined slope can undergo rapid erosion.

3.2.2 Collapse of soils

Soils which are liable to collapse possess porous textures with high void ratios and have relatively low densities. They have sufficient void space in their natural state to hold their liquid limit moisture at saturation. At their natural low moisture content these soils possess high apparent strength but they are susceptible to large reductions in void ratio upon wetting. In other words the metastable texture collapses as the bonds between the grains break down when the soil is wetted. Collapse on saturation normally takes only a short period of time, although the more clay a loess contains the longer this period tends to be.

Gao (1988) commented that the structural stability of loess soils in China is related not only to the origin of the material, to its mode of transportation and to depositional environment, but also to weathering. For instance, the weakly weathered loess of the north-west of the loess plateau has a high potential for collapse, whilst the weathered material of the south-east of the plateau is relatively stable and the features associated with collapsible loess are gradually disappearing. Furthermore, in more finely textured loess deposits, high capillarity potential plus high perched groundwater conditions have caused loess to collapse naturally through time, thereby reducing its porosity. This reduction in porosity, combined with high liquid limit, makes the possibility of collapse less likely. In general, highly collapsible loess occurs in regions (primarily major river valleys) near the source of the loess where its thickness is at a maximum and where the landscape and/or the climatic conditions are not conducive to development of long-term saturated conditions within the soil.

Similarly Grabowska-Olszewska (1988) found that in Poland collapse is most frequent in the youngest loess and that it is almost exclusively restricted to loess which contains slightly more than 10% particles of less than 0.002 mm in size. Such loess is of aeolian origin and is characterized by a random texture and a carbonate content of less than 5%. These soils are more or less unweathered and possess a pronounced vertical pattern of jointing. The size of the pores is also important. Grabowska-Olszewska maintained that collapse occurs as a result of reduction taking place in pores greater than 1 μm, and more especially in those exceeding 10 μm in size.

From the evidence of loess found in China and Poland, it appears that collapse does not occur in all these soils. Those older deposits in which the fabric and mineral composition have been altered by weathering are not nearly as susceptible as young unweathered soils.

Several collapse criteria have been proposed for predicting whether a soil is liable to collapse upon saturation. For instance, Gibbs and Bara (1962) suggested the use of dry density and liquid limit as criteria. Their method is based on the premise that a soil which has enough void space to hold its liquid limit moisture content at saturation is susceptible to collapse on wetting. Such soils plot above the lines in Figure 3.4. This criterion applies only if the soil is uncemented and the liquid limit

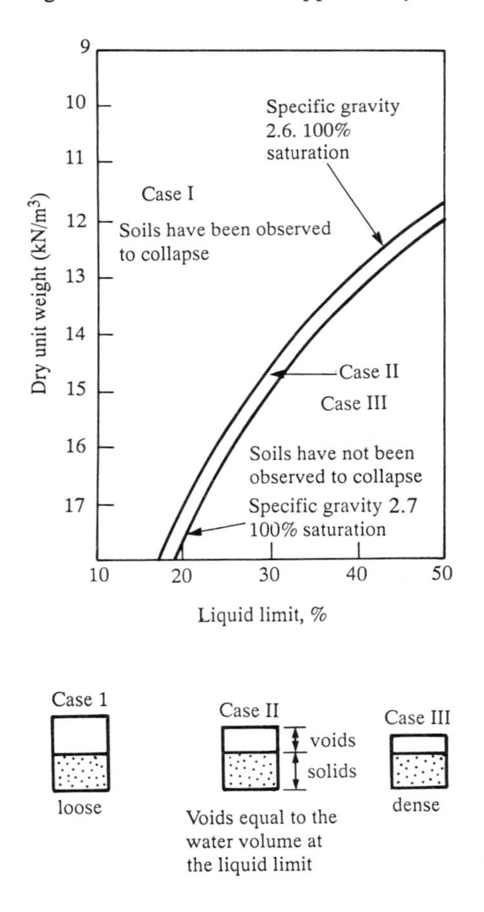

Figure 3.4 *Criterion to distinguish between collapsible and non-collapsible soils (after Gibbs and Bara, 1962)*

is above 20%. When the liquidity index in such soils approaches or exceeds one, then collapse may be imminent. As the clay content of loess increases, the liquid limit also increases, as does the bulk density, hence the saturation moisture content becomes less than the liquid limit. Such deposits are relatively stable.

Collapse criteria have been proposed which depend upon the void ratio at the liquid limit (e_l) and the plastic limit (e_p). According to Audric and Bouquier (1976) collapse is probable when the natural void ratio is higher than a critical void ratio (e_c) which depends on e_l and e_p. They quoted the Denisov and Feda criteria as providing fairly good estimates of the likelihood of collapse:

$$e_c = e_l \quad \text{(Denisov, 1963)} \tag{3.1}$$

$$e_c = 0.85e_l + 15e_p \quad \text{(Feda, 1966).} \tag{3.2}$$

Audric and Bouquier described a series of consolidated undrained triaxial tests they had carried out, at natural moisture content and after wetting, on loess soil from Roumare in Normandy. They distinguished collapsible and non-collapsible types of loess. The main feature of the collapsible soils was the soil structure, which was formed by soil domains with a low number of grain contacts. The authors noted that when collapsible loess was tested, the deviator stress reached a peak at rather small values of axial strain and then decreased with further strain. The pore water pressures continued to increase after a peak deviator stress had been reached. By contrast in non-collapsible soils the deviator stress continued to increase and there was only a small increase of pore water pressure. As expected the shear strength of collapsible loess was always less than that of the non-collapsible type.

The coefficient of collapsibility, i_c, is derived from an oedometer test and can be defined in terms of void ratio, height of specimen tested or strain as follows:

$$i_c = \frac{\Delta e}{1 + e_l} \tag{3.3}$$

where Δe is the decrease in void ratio brought about by wetting and e_l is the initial void ratio prior to wetting, or

$$i_c = \frac{h_p - h_p'}{h_1} \tag{3.4}$$

where h_p is the height of the specimen under pressure p, h_p' is the height of the sample under the same pressure but in the saturated condition, and h_1 is the initial height of the specimen, or

$$i_c = \frac{\Delta\epsilon}{1 - \epsilon_1} \tag{3.5}$$

where $\Delta\epsilon$ represents the additional strain upon wetting and ϵ_1 is the strain at the same stress before wetting. The Chinese Code on Building Construction in Regions of Collapsible Loess regards a collapsible loess as one which on wetting undergoes a unit collapse of 0.015 or greater (Lin and Wang, 1988). In the United States values greater than 0.02 are taken as indicative of soils susceptible to collapse (Lutenegger and Hallberg, 1988). However, investigation of Polish loess suggests that the coefficient of collapsibility can be related to natural moisture content. For instance, Figure 3.5 shows that loess with natural moisture content less than 6% is potentially unstable, that in which the natural moisture content exceeds 19% can be regarded as stable, whilst that with values between these two figures exhibits intermediate behaviour. As in many areas of northern and central Shaansi province loess tends to collapse under its own weight of overburden when the terrain is flooded, Lin and Wang (1988) defined the limit self-weight collapse as

$$i_{cz} = \frac{h_z - h_z'}{h_1} \tag{3.6}$$

where h_z is the sample height under the overburden pressure at the sample depth z, h_z' is the same height under the same overburden pressure in saturated conditions.

Jennings and Knight (1975) developed the double oedometer test for assessing the response of a soil to wetting and loading at different stress levels (i.e. two

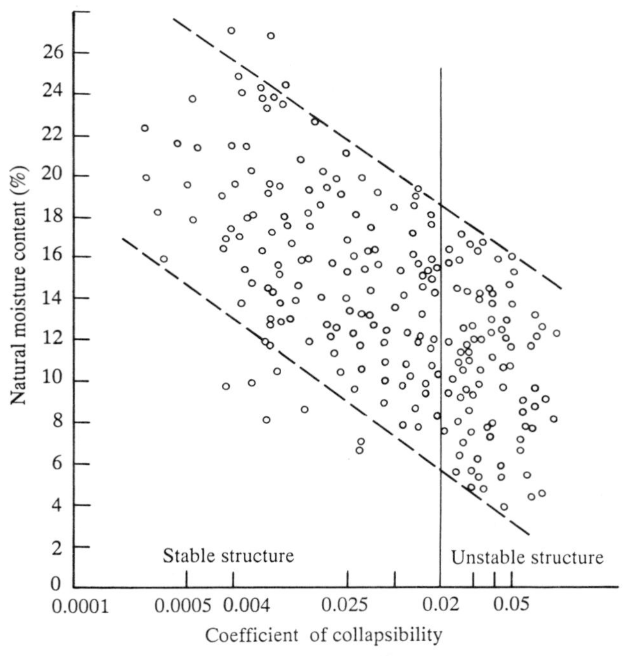

Figure 3.5 *Interrelation between the coefficient of collapsibility of loess deposits and their natural moisture content (from Grabowska-Olszewska, 1988)*

oedometer tests are carried out on identical samples, one being tested at its natural moisture content, whilst the other is tested under saturated conditions the same loading sequence being used). Houston *et al.* (1988) modified this test procedure. They tested an undisturbed sample at natural moisture content in an oedometer. A load of 5 kN/m^2 is placed on the sample and the dial gauge is zeroed. Then the vertical stress is incrementally increased every half hour until less than 0.1% compression occurs in an hour. This process is continued until the stress on the sample is equal to or somewhat greater than the overburden pressure. At this point the sample is inundated and the resulting collapse strain recorded. The collapsed sample then is subjected to further loading to develop the inundated compression curve (Figure 3.6). The amount of collapse of a layer of soil is simply obtained by multiplying the thickness of the layer by the amount of collapse strain. Table 3.2 provides an indication of the potential severity of collapse.

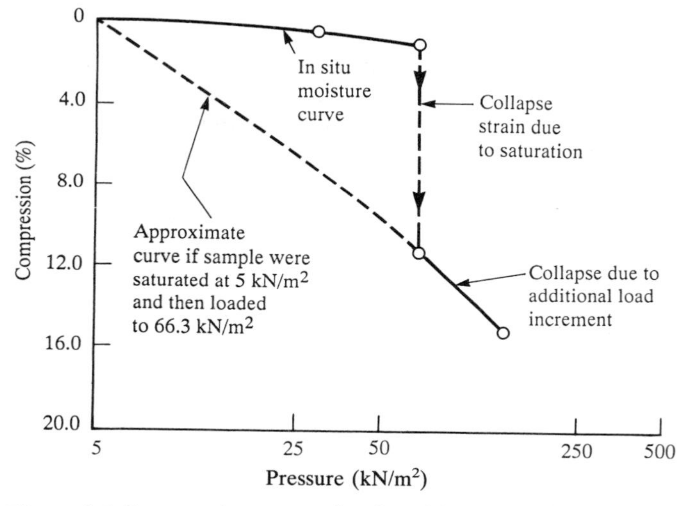

Figure 3.6 *Compression curves developed from modified Jennings and Knight laboratory tests (after Houston* et al., *1988)*

Table 3.2 *Collapse percentage as an indication of potential severity (After Jennings and Knight, 1975)*

Collapse (%)	*Severity of problem*
0–1	No problem
1–5	Moderate trouble
5–10	Trouble
10–20	Severe trouble
Above 20	Very severe trouble

From Figure 3.7 it can be seen that the compression curve for a saturated loess resembles that for a normally consolidated clay. There is a limiting value of pressure, defined as the collapse pressure (p_{cs}), beyond which deformation increases appreciably. The collapse pressure varies with the degree of saturation so that a soil at natural moisture content also has a collapse pressure (p_{cn}). For collapsible soils $p_{cn} > p_{cs}$. Popescu (1986) defined truly collapsible soils as those in which $p_{cs} < p_o$, p_o being the overburden pressure. In other words such soils collapse when saturated since the soil structure cannot then support the weight of overburden. When the saturation collapse pressure exceeds the overburden pressure soils are capable of supporting a certain level of stress on saturation and Popescu defined them as conditionally collapsible soils. The maximum load which such saturated soils can support is the difference between the saturation collapse and overburden pressures.

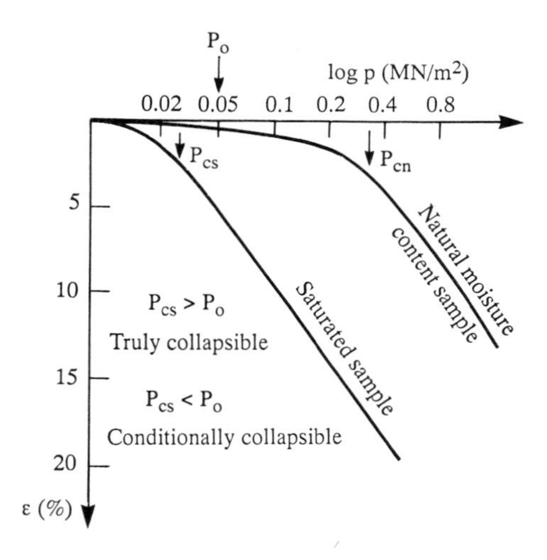

Figure 3.7 *Collapse pressure definition (after Popescu, 1986)*

3.3 Clay deposits

Clay deposits are principally composed of fine quartz and clay minerals. The latter represent the commonest breakdown products of most of the chief rock forming silicate minerals. Clay minerals are phyllosilicates and their atomic structure, which significantly affects their engineering behaviour, can be regarded as consisting of two fundamental units. One of these units is composed of two sheets of closely packed oxygens or hydroxyls in which atoms of aluminium, magnesium or iron are arranged in octohedral coordination (Figure 3.8a). The other unit is formed of linked SiO_4 tetrahedrons which are arranged in layers (Figure 3.8b). In the common clay minerals these fundamental units are arranged in the respective

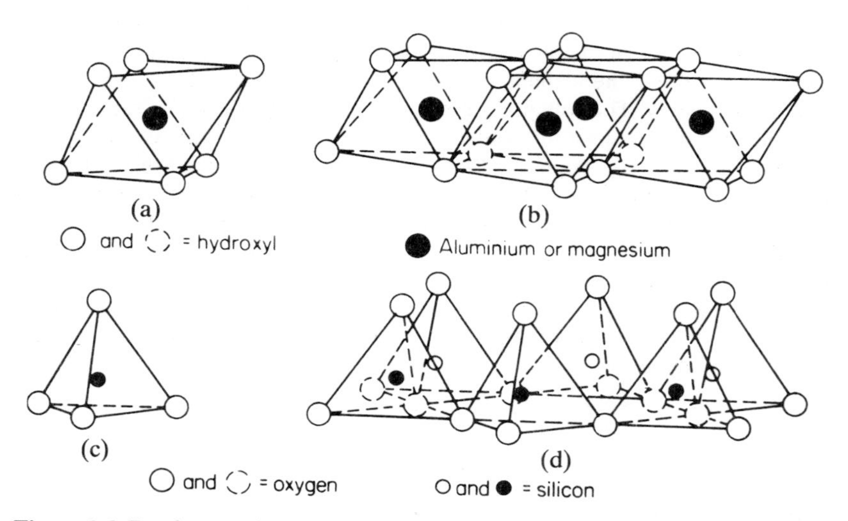

Figure 3.8 *Fundamental units comprising the structure of clay minerals:* (a) *a single octahedral unit in Gibbsite A1 is surrounded by six oxygens, whereas in brucite they surround Mg in sixfold coordination;* (b) *the sheet structure of the octahedral units;* (c) *the silica tetrahedron;* (d) *the sheet structure of silica tetrahedons arranged in a hexagonal network (after Grim, 1962)*

atomic lattices shown in Figures 3.9a–c. The chemical composition of the clay minerals varies according to the amount of aluminium which is substituted for silicon in the atomic structure and also with the replacement of magnesium by other ions. Nevertheless they are basically hydrated aluminium silicates.

The three major clay minerals are kaolinite, illite and montmorillonite. Kaolinite is principally formed as an alteration product of feldspars, feldspathoids and muscovite as a result of weathering under acidic conditions. It is the most important clay mineral in china clays, ball clays and fireclays as well as most residual and transported clay deposits. Illite is a common mineral in most clays and shales, and is present in various amounts in tills and loess, but less common in soils. It develops due to the weathering of feldspars, micas and ferromagnesium silicates, or may form from other clay minerals upon diagenesis. Irrespective of the process responsible for its formation, this appears to be favoured by an alkaline environment.

Both kaolinite and illite have non-expansive lattices whilst that of montmorillonite is expansive. In other words montmorillonite is characterized by its ability to swell and by its notable cation exchange properties. The basic reason why montmorillonite can readily absorb water into the interlayer spaces in its sheet structure is simply that the bonding between them is very weak. Montmorillonite forms when basic igneous rocks, in badly drained areas, are weathered. An alkaline environment also favours its formation.

The shape, size and specific surface all influence the engineering behaviour of clay minerals. Clay minerals have a plate-like shape and are very small in size. For example, an individual particle of montmorillonite is typically 1000 Å by 10 Å thick, whilst kaolinite is 10 000 Å by 1000 Å thick. The specific surface refers to the surface area in relation to the mass and the smaller the particle, the larger the specific surface (Table 3.3). The specific surface provides a good indication of the relative influence of electrical forces on the behaviour of a particle.

Table 3.3 *Size and specific surface of soil particles*

Soil particle	*Size* (mm)	*Specific surface* (m^2/g)	*Ion exchange capacity* (milli-equivalents/100 g)
Sand grain	1	0.002	
Kaolinite	$d = 0.3$ to 3^{-3} thickness = 0.3 to 0.1 d	10–20	3–15
Illite	$d = 0.1$ to 2^{-3} thickness = 0.1 d	80–100	20–40
Montmorillonite	$d = 0.1$ to 1^{-3} thickness = 0.1 d	800	60–100

The surface of a clay particle has a net charge. This means that a clay particle is surrounded by a strongly attracted layer of water. As the diapolar water molecules do not satisfy the electrostatic balance at the surface of the clay particle, some metal cations are also adsorbed. The ions are usually weakly held and therefore can be replaced readily by others. Consequently they are referred to as exchangeable ions. The ion exchange capacity of soils normally ranges up to 40 milli-equivalents per 100 g. However, for some clay soils it may be greater as can be inferred from the ion exchange capacity of kaolinite, illite and montmorillonite (Table 3.3). The type of adsorbed cations influences the behaviour of the soil in that the greater their valency, the better the mechanical properties. For instance, clay soils containing montmorillonite with sodium cations are characterized by high water absorption and considerable swelling. If these are replaced by calcium, a cation with a high valency, both these properties are appreciably reduced. The thickness of the adsorbed layer influences the soil permeability, that is, the thicker the layer, the lower the permeability since a greater proportion of the pore space is occupied by strongly held adsorbed water. As the ion exchange capacity of a cohesive soil increases so does its plasticity index, the relationship between the two being almost linear.

3.3.1 Microstructure of clay soils

The microstructure of cohesive soils is largely governed by the clay minerals present and the forces acting between them. Because of the complex electrochemistry of clay minerals the spatial arrangement of newly sedimented

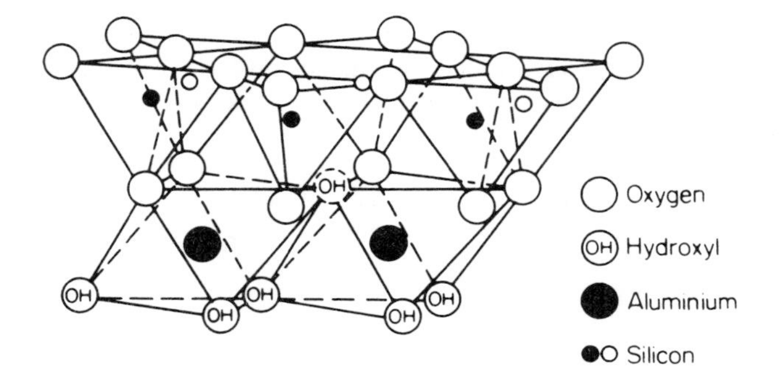

(a)

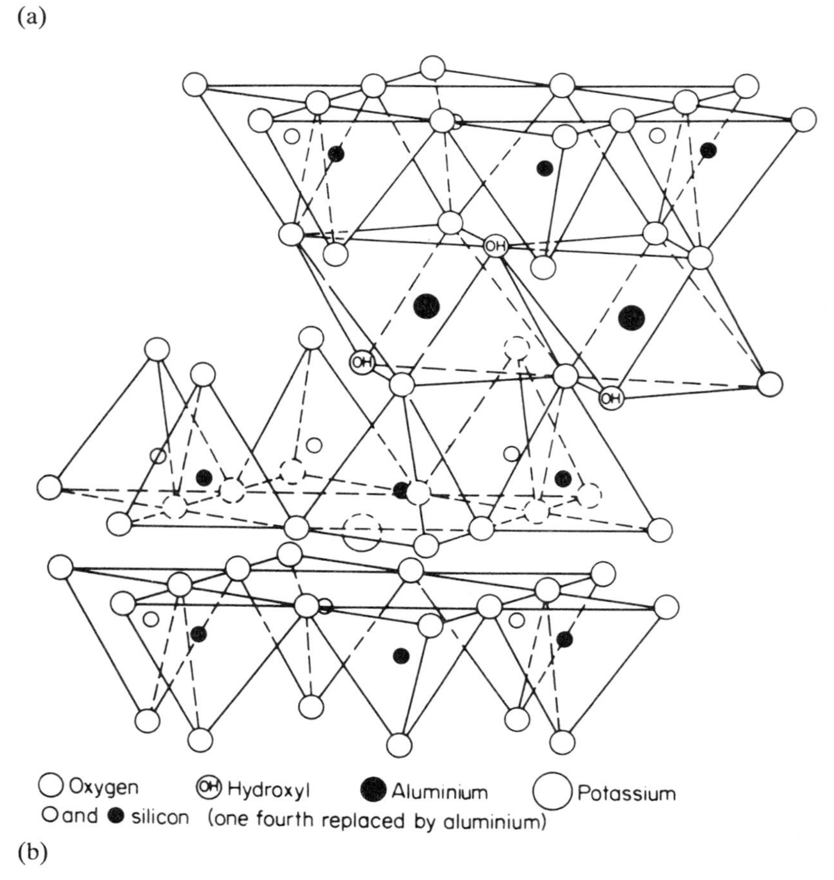

(b)

Figure 3.9 (a) *Diagrammatic sketch of the kaolinite structure;* (b) *the above structure is that of muscovite, which is regarded as essentially the same as that of illite;* (c) *diagrammatic sketch of the montmorillonite structure*

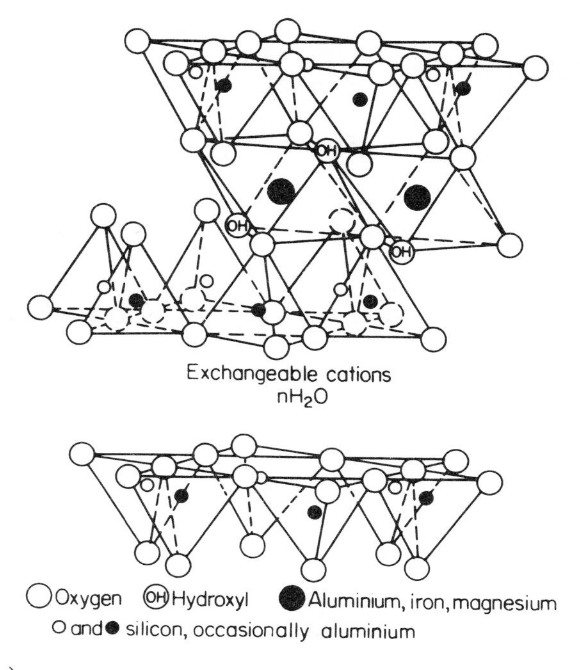

(c)

Figure 3.9 *continued*

particles is very much influenced by the composition of the water in which deposition takes place. Single clay mineral platelets may associate in an edge-to-edge (EE), edge-to-face (EF) face-to-face (FF) or random type of arrangement depending on the interparticle balance between the forces of attraction and repulsion, and the amount or absence of turbulence in the water in which deposition occurs. Since sea water represents electrolyte-rich conditions it causes clay particles to flocculate. Flocculation refers to the attraction of particles to one another in a loose, haphazard arrangement (Figure 3.10a). Considerable free water is trapped in the large voids. Flocculent soils are light in weight and very compressible but are relatively strong and insensitive to vibration because the particles are lightly bound by their edge-to-face attraction. They are sensitive to remounding which destroys the bond between the particles so that the free water is released to add to the adsorbed layers at the former points of contact.

Flocculation does not occur amongst clay particles deposited in fresh water. In this case they assume a more or less parallel, close-packed type of orientation. This has been referred to as a dispersed microstructure (Figure 3.10b). Any bulky grains are distributed throughout the mass and cause localized departures from the pattern. Soils having a dispersed structure are usually dense and watertight. Typical void ratios are often as low as 0.5.

Estuarine clays, because they have been deposited in marine through brackish to freshwater conditions can contain a mixture of flocculated and dispersed microstructures. The aggregate microstructure (Figure 3.10c) was proposed by Van Olphen (1963).

The original microstructure of a clay deposit is subsequently modified by overburden pressures due to burial, which bring about consolidation. Consolidation tends to produce a preferred orientation with the degree of reorientation of clay particles being related to both the intensity of stress and the electrochemical environment, dispersion encouraging and flocculation discouraging clay particle parallelism. For instance, Barden (1972) maintained that lightly consolidated marine clay retains a random open structure, that medium consolidated brackish water clay has a very high degree of orientation, and that extremely heavily consolidated marine clay develops a fair degree of orientation (Figure 3.11).

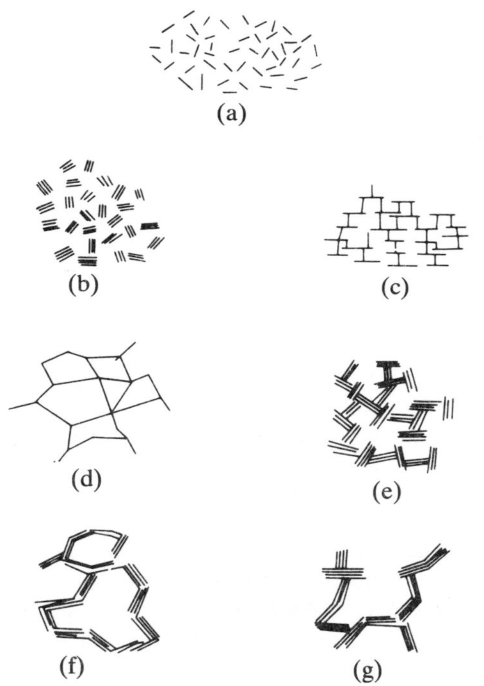

Figure 3.10 *Modes of particle association in clay suspensions (after Van Olphen, 1963):* (a) *dispersed and flocculated;* (b) *aggregated but deflocculated;* (c) *edge-to-face flocculated but dispersed;* (d) *edge-to-edge flocculated but dispersed;* (e) *edge-to-face flocculated and aggregated;* (f) *edge-to-edge flocculated and aggregated;* (g) *edge-to-face and edge-to-edge flocculated and aggregated*

(a) Magnification × 25 500

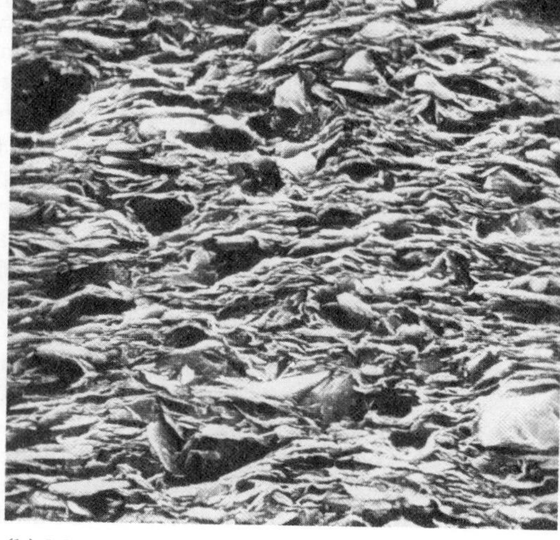

(b) Magnification × 2800

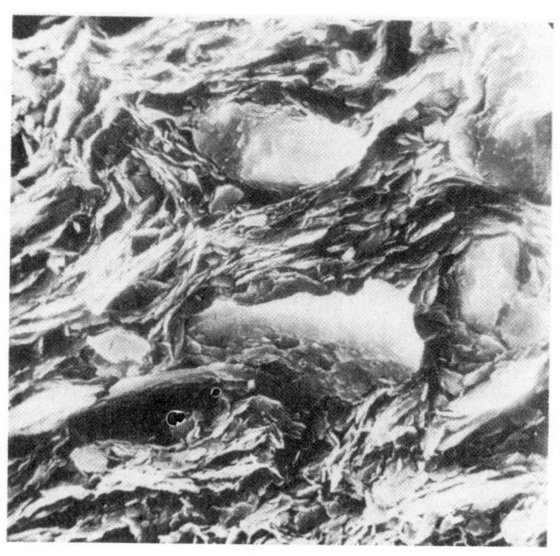

(c) Magnification × 2600

Figure 3.11 (a)*Photomicrograph of lightly overconsolidated, post-glacial, marine-brakish water clay from the Clyde estuary;* (b) *photomicrograph of lightly overconsolidated brackish water clay from the Clyde estuary;* (c) *photomicrograph of heavily overconsolidated marine clay from Luanda, Angola (all photographs courtesy Dr K. Collins, Strathclyde University)*

It is generally accepted that turbostratic groups or domains occur in consolidated clays, although it has been suggested that ill-defined domains are present in unconsolidated clays. These domains consist of aggregations of clay particles which have a preferred orientation but between the aggregates the orientation is random. With increasing overburden pressure it appears that the number of clay particles in each domain increases and that there is an increase in domain orientation. For example, it has been shown that most of the Oxford Clay is made up of small domains, up to 50^{-3} mm across, these showing some alignment with the bedding.

The engineering properties of clay may be influenced by chemical interactions between particles. Olson (1974) sedimented particles in solutions with different pH concentrations and showed that at pH 9 a dispersed structure was developed in which there was a maximum opportunity for diffuse layer forces to influence the failure envelope. At pH 5 edge-to-edge flocculation occurred. A predominant edge-to-edge or edge-to-face particle arangement minimizes double layer repulsions.

Olson (1974) found that the failure envelopes for clays in the calcium form were independent of the electrolyte concentration, indicating that diffuse double layer forces do not develop with sufficient force to influence the shearing strength measurably. For clays in the sodium form, no diffuse double-layer effects were found for kaolinite, possibly because of the low level of isomorphous substitution in the kaolinite lattice. Sodium illite had lower strength than calcium illite, that depended on pore water electrolyte concentration. Hence a distinct diffuse double-layer effect was noted for sodium illite. The strength of sodium montmorillonite is so low that even the effects of substantial changes in diffuse double layer forces would be masked. It is believed that sodium montmorillonite breaks down to flakes only one unit cell thick (10 Å) and these flakes have little physical strength. The particles are thought to be so flexible that essentially all particle interaction must be of the face-to-face type, which leads to low intergranular stresses and a strong probability that the particles are always separated by at least one layer of adsorbed water, if not a full diffuse layer. Calcium montmorillonite is believed to aggregate into particles, a number of unit cells thick, which have significant strengths. Olson suggested that location of the effective stress failure envelope was controlled mainly by the size and shape of the individual particles. Particles that are large and more or less equidimensional lead to high strengths regardless of chemical effects, whereas particles which are very thin with high diameter-to-thickness ratios, such as sodium montmorillonite, have very low strengths.

3.3.2 Engineering properties of clay soils

The principal minerals in a deposit of clay tend to influence its engineering behaviour. For instance, Burnett and Fookes (1974) demonstrated that the clay fraction of the London Clay in the London Basin increased eastwards which leads to an increase in its plasticity but a reduction in its undrained shear strength in that direction. Similarly Russell and Parker (1979) found that in the Oxford Clay the undrained shear strength was related to the amount and type of clay minerals present together with the presence of cementing agents. In particular, they noted that the strength was reduced with increasing content of mixed layer-clay and montmorillonite in the clay fraction. The increasing presence of cementing agents, especially calcite and pyrite, enhanced the strength of the clay. Samuels (1975) found significant variations in the plastic properties of the Gault Clay which were largely dependent on the amount of calcite present.

Geological age also has an influence on the engineering behaviour of a clay deposit. In particular the water content and plasticity normally decrease in value with increasing depth, whereas the strength and elastic modulus increase. This has been demonstrated in the case of the Oxford Clay by Jackson and Fookes (1974) and by Burland *et al.* (1977) (Figure 3.12). The London Clay undergoes similar changes with depth (Marsland, 1973). Cripps and Taylor (1981) referred to a general reduction in the porosity and water content of clays with increasing age and previous overburden.

The engineering performance of clay deposits is also very much affected by the total moisture content and by the energy with which this moisture is held. For instance, the moisture content influences their density, consistency and strength. The energy with which moisture is held influences their volume change characteristics since swelling, shrinkage and consolidation are affected by permeability and moisture migration. Furthermore, moisture migration may give rise to differential movement in clay soils. The gradients which generate moisture migration in clays may arise from variations in temperature, extent of saturation, and chemical composition or concentration of pore solutions. Clays in Britain show a range of water content, it varying from 3 to 30% (Table 3.4).

There is no particular value of plastic limit that is characteristic of an individual clay mineral, indeed the range of values for montmorillonite is large. This is due to the inherent variations of structure and composition within the crystal lattice and the variations in exchangeable-cation composition. Generally the plastic limits for the three clay minerals noted decrease in the order montmorillonite, illite and kaolinite. As far as montmorillonite is concerned, if the exchangeable ions are Na and Li then these give rise to high plastic limits. In the case of the other two clay minerals the exchangeable cations produce relatively insignificant variation in the plastic limit. On the other hand, poorly crystalline kaolinite of small particle size has a substantially higher plasticity than that of relatively coarse, well-organized particles.

Similarly there is no single liquid limit that is characteristic of a particular clay mineral; indeed the range of limits is much greater than that of the plastic limits. Again the highest liquid limits are obtained with Li and Na montmorillonite; then follow in decreasing order Ca, Mg, K, Al montmorillonite; illite; poorly crystalline kaolinite; and well crystallized kaolinite. Indeed the liquid limit of Li and Na montmorillonite cannot be determined accurately because of their high degree of thixotropy. But the character of the cation is not the sole factor influencing the liquid limit, the structure and composition of the silicate lattice are also important. The liquid limits for illites fall in the range 60–90% whilst those for kaolinites vary from about 30 to 75%. Again the crystallinity of the lattice and particle size are the controlling factors, for instance, poorly crystallized fine grained samples may be

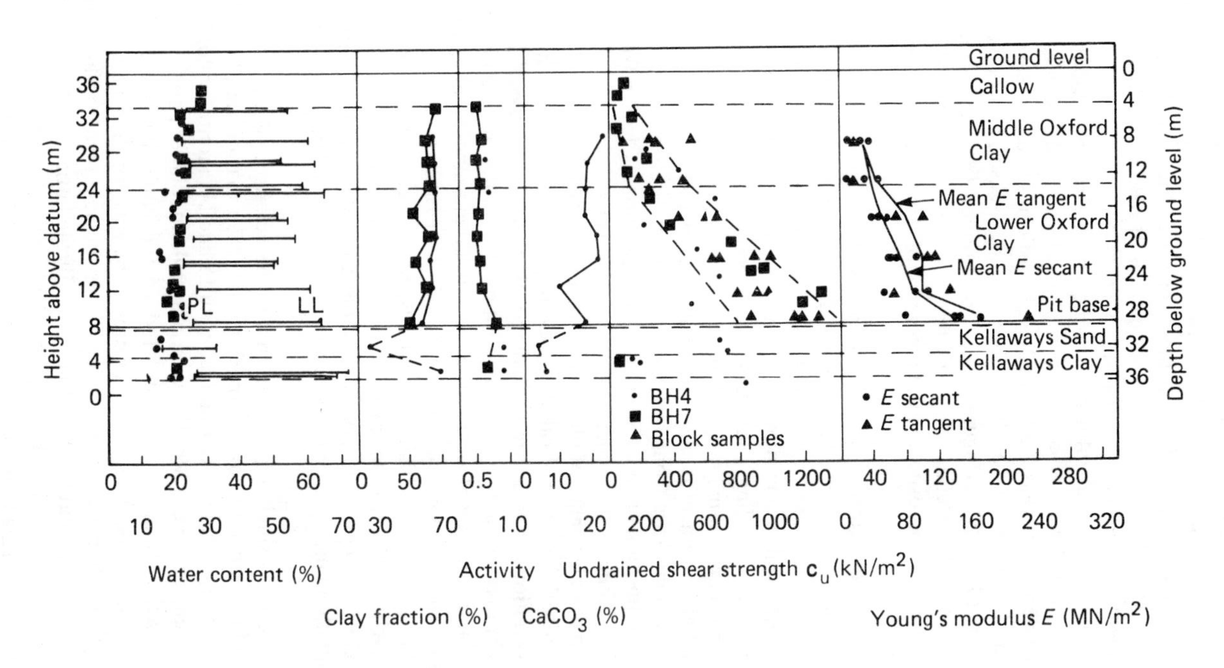

Figure 3.12 *Variation with depth of the geotechnical properties of the Oxford Clay and Kellaways Beds (after Burland et al.*, 1977)

over 100%. The presence of 10% montmorillonite in an illitic or kaolinitic clay can cause a substantial increase in their liquid limits.

The plasticity indices of Na and Li montmorillonite clays have exceedingly high values, ranging between 300 and 600%. Montmorillonites with other cations have values varying from about 50 to 300% with most of them in the range of 75–125%. As far as the latter are concerned there is no systematic variation with cation composition. In the case of illitic clays the plasticity indices range from 23 to 50%. The values for well-crystallized illite are extremely low, indeed they are almost non-plastic. The presence of montmorillonite in these clays substantially increases their indices. The range of plasticity indices for kaolinitic clays varies from about 1 to 40%, generally being around 25%. As can be inferred from above, the limit values increase with a decrease in particle size and the liquid limit tends to increase somewhat more than the plastic limit.

Variations in the plasticity of the London Clay appear to be mainly due to changes in the total clay mineral content (Burnett and Fookes, 1974). The plasticity of British clays ranges from low to extra high (Table 3.4). For example, the Weald Clay has a low plasticity because it has a small clay fraction. By contrast, the plasticity of the Gault Clay is extra high (moisture content 21–29%; plastic limit 26–31%; liquid limit 64–86%; plasticity index 40–54%; activity 0.81–1.05; see Samuels, 1975).

It would appear that there is only a general correlation between the clay mineral composition of a deposit and its activity. In other words, kaolinitic and illitic clays are usually inactive, whilst montmorillonitic clays range from inactive to active. Burnett and Fookes (1974), for instance, found that the activity of the London Clay was related to the proportion of clay minerals present and, more particularly, varied according to the montmorillonite content. Usually active clays have a relatively high water-holding capacity and a high cation exchange capacity. They are also highly thixotropic, have a low permeability and a low resistance to shear.

3.3.3 Volume changes in clay soils

One of the most notable characteristics of clays from the engineering point of view is their susceptibility to slow volume changes which can occur independently of loading due to swelling or shrinkage. Differences in the period and magnitude of precipitation and evaporation are the major factors influencing the swell–shrink response of an active clay beneath a structure. Poor surface drainage and leakage from underground pipes can produce concentrations of moisture in clay. Trees with high water demand and uninsulated hot-process foundations may dry out a clay causing shrinkage. Cold stores may cause desiccation of soil (see below). The density of a clay soil also influences the amount of swelling it is likely to undergo. Expansive clay minerals absorb moisture into their lattice structure, tending to expand into adjacent zones of looser soil before volume increase occurs. In a densely packed soil having small void space, the soil mass has to swell to accommodate the volume change of the expansive clay particles.

Grim (1962) distinguished two modes of swelling in clay soils, namely, intercrystalline and intracrystalline swelling. Interparticle swelling takes place in

Table 3.4 *Engineering properties of some British clays of Tertiary and Mesozoic age (From Cripps and Taylor, 1986, 1987)*

1. Tertiary clays

Engineering property	*Barton Clay*		*Bracklesham Beds, unweathered*	*London Clay*		*Woolwich and Reading Beds, unweathered*	*Ball Clays*
	weathered	*unweathered*		*weathered*	*unweathered*		
Liquid limit (%)	(59) 45–82		[a] 52–68	(80) 66–100	(70) 50–105	(52) 42–67	61–92
Plastic limit (%)	(24) 21–29		[a] 15–22	(28) 22–34	(28) 24–35	(24) 15–30	25–35
Plasticity index (%)	(35) 21–55		[a] 41	(44) 36–55	(47) 41–65	20–37	31–57
Void ratio				[h] 0.69–1.41	[h] 0.60–0.83		
Clay fraction <2 μm (%)	(42) 35–70			42–72	48–61		71–94
Natural water content (%)	(23) 17–32		[a] 19–26	[b] 23–49	19–28	(22) 15–27	18–31
Bulk density (Mg/m^3)			[a] 2.07	1.70–2.00	1.92–2.04		1.92–2.18
Undrained shear strength (kN/m^2)	(40) 20–210	(112–150) 50–350	[a] 143	(100–175) 40–190	(100–400) [d] 80–800	(400) [f] 34–814	[s] 79–1245
Effective cohesion (kN/m^2)	7–11	[c] 8–24	(24) [a] 0–55	12–18	17–252		34–62
Effective angle of friction (degrees)	18–24	[c] 27–39	(25) [a] 18–32	17–23	20–29		20–26
Residual shear strength (degrees)	15			(14) 10.5–22	[t] 9.4–17		10–15
Secant modulus of elasticity (MN/m^2)				10–35	[g] 25–141		
Coefficient of volume change (m^2/MN) [q]			[a] $m_v = 0.065–0.5$ [a] $m_s = 0.023–0.028$	$m_v = 0.05–0.18$	$m_v = 0.01–0.002$ $m_s = 0.094–0.003$	$m_v = (0.04)$ $m_v = 0.104–0.002$	
Coefficient of consolidation (m^2/yr)				0.2–2.0	0.3–60	[r] 0.75–95	
Permeability (m/s)					[m] $2.2 \cdot 10^{-10}$ [p] $3 \cdot 10^{-10}–3 \cdot 10^{-8}$		
Effective stress ratio (K_o)				0.5–4.4	1.1–2.8		

[a] State of weathering not known.
[b] Upper limit value for mudflow.
[c] $c' = 24\ kN/m^2$, $\phi' = 39°$ – intact strength of 'medium hard lumps'.
[d] Depth up to 30 m.
[f] Depth 50–70 m.
[g] Depth up to 46 m.
[h] Calculated from SG, ω and γ_b values.
[j] Value quoted.
[m] Laboratory test.
[p] In situ test.
[q] m_v = coefficient of volume compression. m_s = coefficient of volume expansion.
[r] Values may include determination on sands.
[s] Includes intact clay.
[t] Ring shear test.

Table 3.4 *continued*

2. Cretaceous clays

	Gault Clay		*Speeton Clay, weathered*	*Fuller's Earth, unweathered*	*Atherfield Clay, weathered*	*Weald and Atherfield Clay, weathered*	*Weald Clay*		*Wadhurst Clay*
	weathered	*unweathered*					*weathered*	*unweathered*	
Liquid limit (%)	70–92	[f] 60–120 / 90–110	50		[r, s] (35–85) 30–95	(72) 41–90	42–82	36–55	
Plastic limit (%)	25–34	18–36			[s] 18–32	(30) 23–41	26–31	25–30	
Plasticity index (%)	[b] 44–80	27–55	28		[r, s] 17–55	(42) 17–58	26–40	11–28	
Clay fraction <2 μm (%)	[f] 50–62	38–60		69	52–59	(51) 17–71	45–74	20–32	
Void ratio		[q] 0.68		[q] 1.31				[q] 0.18	
Natural water content (%)	(35) 30–42	(22) 18–30		46	[a] 19–36	27–46	23–34	5	[c] 11–32
Specific gravity of particles	(2.69) 2.61–2.75	2.64–2.71		2.8				2.69	
Bulk density (Mg/m^3)	(1.99) 1.93–2.01	1.96–2.13		1.77		γ_d = [n] 1.21–1.41		2.41	
Undrained shear strength (kNm^2)	[h] (60) 17–76	(300–550) 56–1280				[j] (42) 20–85			[c] (100–220) 40–180
Effective cohesion (kN/m^2)	(12) 10–14	25–124			0–21	21–34	13		
Effective angle of friction (degrees)	23–25	(25) 19–53			24	18–24	13–18		
Residual shear strength (degrees)	13–14	12–19	14		12–16	6–14	(133) 10–20		[f] 12–14
Coefficient of volume compressibility (m^2/MN)	(0.075–0.12)	0.005–0.08		[k] 0.007				0.002	
Coefficient of consolidation (m^2/yr)	1.0–3.6	(0.19) 0.09–1.4		[k] 0.02		(0.39) 0.2–0.78		25.1	[d] 1–35000
Permeability (m/s)	[m] $1.8 \cdot 10^{-11}$ / $6.8 \cdot 10^{-11}$	$0.02 \cdot 10^{-10}$		$4 \cdot 10^{-15}$				[m] $13 \cdot 10^{-10}$–$35 \cdot 10^{-10}$ / $1 \cdot 10^{-10}$–$5 \cdot 10^{-10}$	[b] $2 \cdot 10^{-10}$

[a] Depth 7–14 m.
[b] Remoulded specimen.
[c] Depth 0–14 m.
[d] Depth 5–9 m.
[f] Upper Gault, ω_1 = 60–120%. Lower Gault, ω_1 = 90–110%.
[h] 100 mm diameter specimens.
[j] Shear vane tests.
[k] Borehole sample.
[m] In situ test.
[n] Dry density.
[p] Weathered material.
[q] Value quoted.
[r] Upper value for clay.
[s] Lower value for silt.

Table 3.4 *continued*

3. Jurassic clays

Property	*Kimmeridge Clay, unweathered*	*Ampthill Clay, weathered*	*Middle Oxford Clay, unweathered*	*Lower Oxford Clay*	
				weathered	*unweathered*
Liquid limit (%)	(79) 70–81	79	(58) [h] 53–63		(60) 45–75
Plastic limit (%)	(27) 24–46	30	[h] 21–23		(33) 18–47
Plasticity index (%)	(53) 24–59	49	[h] 31–39		(32) 12–50
Clay fraction, <2 μm (%)	57	[c] 64	(64) [h] 60–68		(52) 30–70
Void ratio	54			[j] 0.41–0.18	[j] 0.42–0.96
Natural water content (%)	(21) 18–22	23–88	(25) [h] 22–28	(29) [d] 17–36	(21) [d] 15–32
Specific gravity of particles	2.49				2.53–2.73
Bulk density (Mg/m^3)	2.08	[c] 2.10	(2.03)	1.71–2.1	2.03 1.84–2.05
Undrained shear strength (kN/m^2)	(130–470) 70–500		(110–360) [h] 45–510	52–93	(360–1100) 96–1300
Effective cohesion (kN/m^2)	[a] 14–67	23–48	10	0–20	10–216
Effective angle of friction (degrees)	[a] 14–23	17–32	31	21.5–28	23–40
Residual shear strength (degrees)	10–18	10–14	15		(18) 12.5–18.5
Coefficient of volume compressibility (m^2/MN)	[a] 0.22 [b] 0.002	[c] 0.09	(0.077) [h] 0.046–0.12		[b] 0.003–0.12
Coefficient of consolidation (m^2/yr)	0.25–2.7		(0.16) [h] 0.40–1.27		(0.93) 0.49–>40
Modulus of elasticity (Mn/m^2)			(40) [h] 10–70		(50–135) 45–230
Permeability (m/s)	[a] $1\cdot10^{-7}$ $1\cdot10^{-10}$–$1\cdot10^{-12}$				$5\cdot10^{-10}$

[a] Weathered value.
[b] Deep borehole.
[c] Ampthill and Oxford Clay.
[d] Division not specified.
[h] May include some weathered material.

Fuller's Earth Clay		*Upper Lias Clay*		*Lower Lias Clay*	
weathered	*unweathered*	*weathered*	*unweathered*	*weathered*	*unweathered*
(51–95) 40–95	100	(60) 56–72	(54) 39–72	(59–62) [k] 56–75	(59) 53–63
(21) 20–51	61	(28) 28–36	(26) 18–45	(23–25) [k] 22–40	(27) 25–29
(35) 20–62	39	(32) 24–42	(33) 19–40	(36) [k] 33–39	(32)
38–65	68	(48) 28–68	(38) 18–65	(53) 50–56	
	0.93	[j] 0.65–1.10	[j] 0.43–0.80	[j] 0.70–1.01	[j] 0.47–0.73
(28–35.5) 17–4	33	(20–38) 20–38	(13–18) 12–27	(20–32) [k] 25–47	(18) [k] 10–22
	2.70	2.70–2.73	2.65–2.67		(2.72) 2.67–2.75
	1.86	1.79–1.96	1.87–2.09	1.96–2.05	[k] 1.94–2.00
(50) 10–120		(30–150) 20–180	20–1200	[k] 28–57	12.5 MN/m^2 *8* [k] 95–179
0–2		1–17		0–5	
17.4		18–25		24–27	
12–13.5		5–13.5		12.5–16	
	[b] 0.005		[b] 0.002–0.003	0.2	
	[b] 0.02		[b] 0.42–0.67	(100) [m] 1000–10000	
		10 GN/m^2	35 GN/m^2 *67* 22–52	10 GN/m^2	35 GN/m^2 *67* 22–52
	[b] $2 \cdot 10^{-10}$	($6.2 \cdot 10^{-10}$) $3.2 \cdot 10^{-10}$–$1.2 \cdot 10^{-9}$	$3.6 \cdot 10^{-7}$–$1.0 \cdot 10^{-12}$		

[j] From SG, γ_b and ω values.
[k] May include some Middle Lias Clay.
[m] In situ test.
NB Figures given in brackets represent average values.

any type of clay deposit irrespective of its mineralogical composition, and the process is reversible. In relatively dry clays the particles are held together by relict water under tension from capillary forces. On wetting the capillary force is relaxed and the clay expands. In other words intercrystalline swelling takes place when the uptake of moisture is restricted to the external crystal surfaces and the void spaces between the crystals. Intracrystalline swelling, on the other hand, is characteristic of the smectite family of clay minerals, of montmorillonite in particular. The individual molecular layers which make up a crystal of montmorillonite are weakly bonded so that on wetting water enters not only between the crystals but also between the unit layers which comprise the crystals. Swelling in Na montmorillonite is the most notable and can amount to 2000% of the original volume, the clay then having formed a gel. Hydration volume changes are frequently assessed in terms of the free-swell capacity (see Chapter 9).

Mielenz and King (1955) showed that generally kaolinite has the smallest swelling capacity of the clay minerals and that nearly all of its swelling is of the interparticle type. Illite may swell by up to 15% but intermixed illite and montmorillonite may swell some 60–100%. Swelling in Ca montmorillonite is very much less than in the Na variety, it ranges from about 50 to 100%. The large swelling capacity of montmorillonites means that they give the most trouble in foundation work.

The swelling potential of clay soils depends not only on the type and amount of clay minerals present but also on the soil structure. Cemented and undisturbed expansive clay soils often have a high resistance to deformation and may be able to absorb significant amounts of the swelling pressure. On the other hand remoulded expansive clays tend to swell more than their undisturbed equivalents. Soils with a flocculated structure swell more than those which possess a preferred orientation. In the latter the maximum swelling occurs normal to the direction of clay particle. orientation. As expansive clays tend to possess extremely low permeabilities, moisture transition is slow and an appreciable period of time may be involved in the swelling process. Consequently moderately expansive clays with a smaller potential to swell but with higher permeabilities than clays having a greater swell potential may swell more during a single wet season than the more expansive clay.

Cycles of wetting and drying are responsible for slaking in argillaceous sediments which can bring about an increase in their plasticity index and augment their ability to swell. The air pressure in the pore spaces helps the development of the swell potential under cyclic wetting and drying conditions. On wetting the pore air pressure in a dry clay increases and it can become large enough to cause breakdown, which at times can be virtually explosive. The rate of wetting is important, slow wetting allowing the air to diffuse through the soil water so that the pressure does not become large enough to disrupt the soil. In weakly bonded clay soils cyclic wetting and drying brings about a change in the swell potential as a result of the breakdown of the bonds between clay minerals and the alteration of the soil structure.

Bjerrum (1967) suggested that clay soils with high salt contents and a network of cracks undergo an increase in swell potential. This is due to slaking consequent upon osmotic pressures being developed as rain water infiltrates into the cracks.

Freeze–thaw action, like osmotic swelling, also affects the swell potential. Freezing can give rise to large internal pressures at the freezing front in fine-grained soils and if this front advances slowly enough the soil immediately beneath it can become quite desiccated. On melting the desiccation zone is saturated. Hence the swell potential is increased by the freeze pressures.

According to Schmertmann (1969) some clays increase their swell behaviour when they undergo repeated large shear strains due to mechanical remoulding. He introduced the term swell sensitivity for the ratio of the remoulded swelling index to the undisturbed swelling index and suggested that such a phenomenon may occur in unweathered highly overconsolidated clay when the bonds, of various origins, which hold clay particles in bent positions, have not been broken. When these bonds are broken by remoulding the clays exhibit significant swell sensitivity.

An internal swelling pressure will reduce the effective stress in a clay soil and therefore will reduce its shearing strength. As a consequence Hardy (1966) suggested that the Coulomb equation for swelling soils should take the swelling pressure, as well as pore pressure, into account, the expression becoming

$$\tau = c' + (\sigma - u - p_s)\tan\phi' \quad (3.7)$$

where τ is the shearing strength, c' is the cohesion, σ is the total stress, u is the pore water pressure, p_s is the swelling pressure and ϕ' is the angle of shearing resistance

Holtz and Gibbs (1956) showed that expansive clays can be recognized by their plasticity characteristics, and Vijayvergiya and Ghazzaly (1973) developed a method of estimating the amount of potential swell by using an oedometer. A specimen of clay is placed in an oedometer to derive the quantity of swell as a function of natural moisture content and liquid limit (Figure 3.13). Subsequently, Brackley (1980) suggested that the maximum movement due to swelling beneath a building founded on expansive clay could be obtained from the expression

$$\text{Swell}\ (\%) = \frac{(\text{PI} - 10)}{10}\log_{10} S/p \quad (3.8)$$

where PI is the plasticity index, S is the soil suction at the time of construction (in kN/m^2) and p is the overburden plus foundation pressure acting on each layer of soil (in kN/m^2). However, one of the most widely used soil properties to predict swell potential is the activity of a clay (Figure 3.14).

A review of the various methods which have been used to determine the amount of swelling that an expansive clay is likely to undergo when wetted has been provided by O'Neill and Poormoayed (1980). They quoted the United States Army Engineers Waterways Experimental Station (USAEWES) classification of potential swell (Table 3.5) which is based on the liquid limit (LL), plasticity index (PI) and initial (*in situ*) suction (S_i). The latter is measured in the field by a psychrometer.

Construction damage, according to Popescu (1979, 1986), which occurs in areas of Romania where expansive clay forms the surface cover is mainly caused by

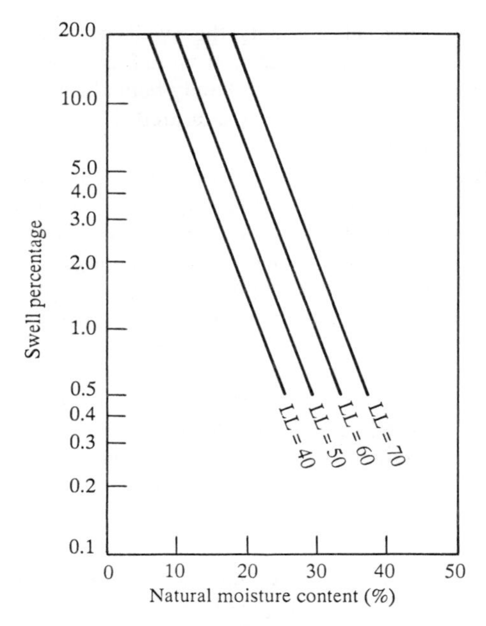

Figure 3.13 *Prediction of percentage swell for clays, LL = liquid limit (after Vijayvergiya and Ghazzaly, 1973)*

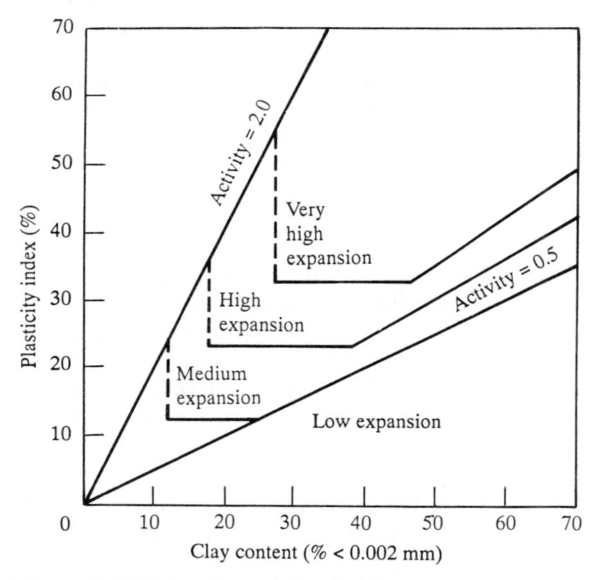

Figure 3.14 *Estimation of the degree of expansiveness of a clay soil (after Williams and Donaldson, 1980)*

Table 3.5 *USAEWES classification of swell potential (from O'Neill and Poormoayed, 1980)*

Liquid limit (%)	*Plastic limit* (%)	*Initial (in situ) suction* (kN/m^2)	*Potential swell* (%)	*Classification*
Less than 50	Less than 25	Less than 145	Less than 0.5	Low
50–60	25–35	145–385	0.5–1.5	Marginal
Over 60	Over 35	Over 385	Over 1.5	High

seasonal swelling and shrinkage of the foundation subsoil. These effects are notable in regions which experience alternating wet and dry seasons. The expansive clays of Romania are normally residual soils formed from basic igneous rocks. This accounts for the fact that the clay fraction of these soils usually contains between 40 and 80% montmorillonite. Popescu noted that maximum seasonal changes in moisture content in these expansive clays were around 20% at 0.4 m depth, 10% at 1.2 m, and less than 5% at 1.8 m. The corresponding cyclic movements of the ground surface are about 100–120 mm. Desiccation of clays during the dry season leads to the soil cracking. The cracks can gape up to 150 mm and frequently extend to 2 m in depth. Popescu stated that the depth of the active zone in expansive clays in Romania varies from about 2.0 to 2.5 m. In Australia, India, South Africa and the United States it may extend to 3 m depth, whereas in Israel depths of up to 6 m have been recorded.

Popescu (1979) recognized three distinct stages in the shrinkage process, the extent of each depending on the soil structure, particularly the number of interparticle bonds causing resistance. In both the initial and residual stages of shrinkage the reduction in soil volume is less than the volume of water lost. Only during the normal shrinkage stage are the two equal. This can give rise to differential movement. The anisotropic character of a clay can also lead to differential movements.

Williams and Jennings (1977) found that soil structure has a major influence on the pattern of the shrinkage process, as well as on the total amount of shrinkage. They indicated that in clay soils with the same initial moisture content and density, the more random the particle arrangement, the less the total shrinkage.

Volume changes in clays also occur as a result of loading and unloading which bring about consolidation and heave respectively. When clay is first deposited in water, its water content is very high and it may have void ratios exceeding 2. As sedimentation continues overburden pressure increases and the particles are rearranged to produce a new packing mode of greater stability as most of the water is expelled from the deposit. Expulsion of free and adsorbed water occurs until a porosity of about 30% is reached. Any further reduction in pore volume is attributable to mechanical deformation of particles and diagenetic processes. As material is removed from a deposit by erosion the effective overburden stress is reduced and elastic rebound begins. Part of the rebound or heave results from an increase in the water content of the clay. Cyclic deposition and erosion has resulted in multiple loading and unloading of many clay deposits.

The heave potential arising from stress release depends upon the nature of the diagenetic bonds within the soil, that is, the post-depositional changes such as precipitation of cement and recrystallization that have occurred (Bjerrum, 1967). It would appear that significant time-dependent vertical swelling may arise, at least in part, from either of two fundamentally different sources, namely, localized shear stress failures or localized tensile stress failures. Localized shear stress failures are associated with long-term deformations of soils having well-developed diagenetic bonds such as clay shales. When an excavation is made in a clay with weak diagenetic bonds elastic rebound will cause immediate dissipation of some stored strain energy in the soil. However, part of the strain energy will be retained due to the restriction on lateral straining in the plane parallel to the ground surface. The lateral effective stresses will either remain constant or decrease as a result of plastic deformation of the clay as time passes (Bjerrum, 1972). These plastic deformations can result in significant time-dependent vertical heaving. However, creep of weakly bonded soils is not a common cause of heaving in excavations.

The relationships between the stresses, failure mechanisms and time-dependent heaving are complex in clay soils with well-developed diagenetic bonds. According to Obermeier (1974) heaving is in part related to crack development, cracking giving rise to an increase in volume, the rate of crack growth being of particular significance. The initial rate of heaving is probably controlled by cracking due to tensile failures and to plastic deformations arising from shear failures. Furthermore, Obermeier maintained that because of the breakdown of diagenetic bonds there is an increase in the lateral stresses parallel to the ground surface.

When a load is applied to a clay soil its volume is reduced, this being due to a reduction in the void ratio (Padfield and Sharrock, 1983). If such a soil is saturated, then the load is initially carried by the pore water which causes a pressure, the hydrostatic excess pressure, to develop. The excess pressure of the pore water is dissipated at a rate which depends upon the permeability of the soil mass and the load is eventually transferred to the soil structure. The change in volume during consolidation is equal to the volume of the pore water expelled and corresponds to the change in void ratio of the soil. In clay soils, because of their low permeability, the rate of consolidation is slow. Primary consolidation is brought about by a reduction in the void ratio. Further consolidation may occur due to a rearrangement of the soil particles. This secondary compression is usually much less significant. However, it should not be assumed that secondary compression always follows primary consolidation. In other words, the two processes can occur at the same time (Ladd *et al.*, 1977).

The compressibility of a clay is related to its geological history, that is, to whether it is normally consolidated or overconsolidated. Furthermore clays which have undergone volume increase due to swelling or heave are liable to suffer significantly increased gross settlement when they are loaded. The various factors which influence the compressibility of a clay soil have been reviewed by Wahls (1962). Values of the coefficient of volume compressibility of some British clays are listed in Table 3.4. Cripps and Taylor (1981) indicated that there was a slight trend towards a decrease in this parameter with increasing age or depth of burial.

3.3.4 Normally consolidated and overconsolidated clays

A normally consolidated clay is one that at no time in its geological history has been subject to pressures greater than its existing overburden pressure, however, an overconsolidated clay is one that has. The major factor in overconsolidation is removal of material that once existed above a clay deposit by erosion. Berre and Bjerrum (1973) carried out a series of triaxial and shear tests on normally consolidated clay, the confining conditions simulating the overburden pressures in the field. They demonstrated that the clay could sustain a shear stress in addition to the *in situ* value, undergoing relatively small deformation as long as the shear did not exceed a given critical value. This critical shear value represents the maximum shear stress which can be mobilized under undrained conditions, and governs the bearing pressure such a clay can carry with limited amount of settlement. Generally this critical shear value varies with plasticity and the rate at which load is applied. An overconsolidated clay is considerably stronger at a given pressure than a normally consolidated clay and it tends to dilate during shear whereas a normally consolidated clay consolidates. Hence when an overconsolidated clay is sheared under undrained conditions negative pore water pressures are induced, the effective strength is increased, and the undrained strength is much higher than the drained strength, the exact opposite to a normally consolidated clay. When the negative pore pressures gradually dissipate the strength falls as much as 60 or 80% to the drained strength.

In both normally consolidated and overconsolidated clays the shear strength reaches a peak value and then, as displacements increase, decreases to the residual strength. The development of residual strength is therefore a continuous process. Put another way, if at any particular point the soil is stressed beyond its peak strength, its strength decreases and additional stress is transmitted to other points in the soil. These in turn become overstressed and decrease in strength. The failure process continues until the strength at every point along the potential slip surface has been reduced to the residual strength.

3.5.5 Desiccation of clay soils

The volume change which occurs due to evaporation from a clay soil can be conservatively predicted by assuming the lower limit of the soil moisture content to be the shrinkage limit. Desiccation beyond this value cannot bring about further volume change. However, according to Williams and Pidgeon (1983) the effects of evaporation in the highveld of South Africa are not significant because of the presence of an inactive layer at the surface or because vegetation affords shade to the surface of the soil. On the other hand transpiration from vegetative cover is a major cause of water loss from soils in semi-arid regions. Indeed the distribution of consequent soil suction in the soil is primarily controlled by transpiration from vegetation. Very high values of soil suction have been recorded: for example, Williams and Pidgeon referred to the pF at a depth of 6 m being 4.8, whilst that at 15 m was 3.8.

The maximum soil suction which can be developed is governed by the ability of vegetation to extract moisture from the soil. The level at which moisture is no longer available to plants is termed the permanent wilting point and this corresponds to a pF value of about 4.2. The complete depth of active clay profiles usually does not become fully saturated during the wet season in semi-arid regions. For example, in the high veld even during the summer rainfall season the soil suction can still approach the wilting point where the soil is covered with vegetation. Nonetheless, the pF may fall to 3.2 during the wet season and changes in soil suction may be expected over a depth of 2.1 m between the wet and dry seasons.

The moisture characteristic (moisture content *v.* soil suction, Figure 3.15) of a soil provides valuable data concerning the moisture contents corresponding to the

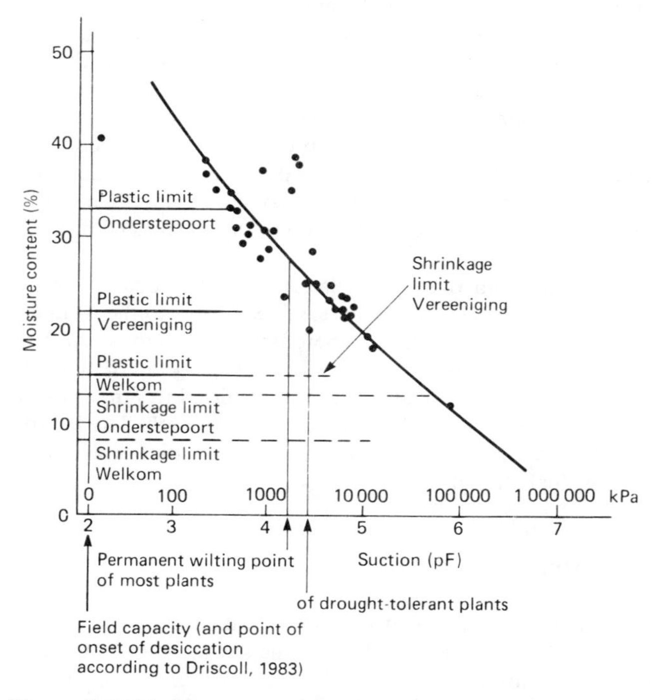

Figure 3.15 *Moisture content-suction relationships for Ondersteport, Vereeniging and Welkom clays (from Williams and Pidgeon, 1983)*

field capacity (defined in terms of soil suction this is a pF value of about 2.0) and the permanent wilting point (pF of 4.2 and above), as well as the rate at which changes in soil suction take place with variations in moisture content. This enables an assessment to be made of the range of soil suction and moisture content which is likely to occur in the zone affected by seasonal changes in climate.

The extent to which the vegetation is able to increase the suction to the level associated with the shrinkage limit is obviously important. In fact the moisture content at the wilting point exceeds that of the shrinkage limit in soils with a high content of clay and is less in those possessing low clay contents. This explains why settlement resulting from the desiccating effects of trees is more notable in low to moderately expansive soils than in expansive ones.

When vegetation is cleared from a site, its desiccating effect is also removed. Hence the subsequent regain of moisture by clay soils leads to them swelling. Williams and Pidgeon (1983) noted that the swelling movements on expansive clays in South Africa, associated with the removal of vegetation and the subsequent erection of buildings, in many areas amounted to about 150 mm. The largest upward movement recorded so far is 374 mm. Similar swelling movement was reported by O'Neill and Poormoayed (1980) from building sites in Texas which were located on areas of expansive clay cleared of vegetation. Uncleared areas did not swell to the same extent. In the latter areas surface movement is cyclical but irregular and may have an amplitude of anything up to 50–60 mm.

The suction pressure associated with the onset of cracking is approximately pF 4.6. The presence of desiccation cracks enhances evaporation from the soil. Mahar and O'Neill (1983) assumed that such cracks led to a variable development of suction pressure, the highest suction occurring nearest the cracks. This, in turn, influences the preconsolidation pressure as well as the shear strength, leading to variability within the soil.

Sridharan and Allam (1982), with reference to arid and semi-arid regions, found that repeated wetting and drying of clay soils can bring about aggregation of soil particles and cementation by compounds of Ca, Mg, Al and Fe. This enhances the permeability of the clays and increases their resistance to compression. Furthermore interparticle desiccation bonding increases the shear strength, the aggregations offering higher resistance to stress. Indeed, depending on the degree of bonding, the expansiveness of an expansive clay soil may be reduced or it may even behave as a non-expansive soil. It has been claimed that the effect of desiccation on clay soils is similar to that of heavy overconsolidation. For example, Williams and Pidgeon (1983) suggested that if the pF in a soil rose to 4.5, then the state of stress therein approximates to a preconsolidation pressure of 3.1 MN/m^2.

Many clay soils in Britain, especially in south-east England, possess a large potential for volume change. However, the mild damp climate means that any significant deficits in soil moisture developed during the summer are confined to the upper 1.0–1.5 m of the soil and the field capacity is re-established during the winter. Even so, deeper permanent deficits can be brought about by large trees. With this in mind, Driscoll (1983) suggested that desiccation could be regarded as commencing when the rate of change in moisture content (and therefore volume) with increasing soil suction, increases significantly. He proposed that this point approximates to a suction of pF about 2 (10 kN/m^2). Similarly notable suction could be assumed to have taken place if, on its disappearance, a low-rise building was uplifted due to the soil swelling. Driscoll maintained that this suction would have a pF value of 3 (100 kN/m^2) He went on to state that the moisture content (m) of the soil at these two values of suction is approximately related to the liquid limit (LL) as follows:

(a) at pF $= 2$
$m = 0.5\,\text{LL}$
(b) at pF $= 3$
$m = 0.4\,\text{LL}$

Hence these values of liquid limit provide crude estimates of the moisture contents at the beginning of desiccation and when it becomes significant.

3.3.6 The strength of clay soils

The value of strength derived by testing a sample of clay in the laboratory depends upon the type of test used, the time which has elapsed between sampling in the field and performing the test, the size of the specimen tested and its orientation in the testing apparatus (Bishop and Little, 1967). It also depends on the type of sample used. For example, Samuels (1975) showed that the strength values of the Gault Clay obtained by testing material from block samples were significantly higher than those obtained from borehole samples, 700–1290 kN/m^2 for the former compared with 270–720 kN/m^2 for the latter.

In relation to applied stress saturated clays behave as purely cohesive materials provided that no change of moisture content occurs. Thus, when a load is applied to a saturated clay it gives rise to excess pore water pressures which are not quickly dissipated. In such instances the angle of shearing resistance (ϕ) is equal to zero. The assumption that $\phi = 0$ forms the basis of normal calculations of the ultimate bearing capacity of clays (Skempton, 1951). The strength may then be taken as the undrained shear strength or one-half the unconfined compressive strength. To the extent which consolidation does occur, the results of analyses based on the premise that $\phi = 0$ are on the safe side. Only in special cases, with prolonged loading periods or with very silty clays, is the assumption sufficiently far from the truth to justify a more elaborate analysis. The undrained shear strength of most British clays ranges from soft (20–40 kN/m^2) to hard (over 300 kN/m^2) and tends to increase with depth below the surface (Table 3.4 and Cripps and Taylor, 1981).

The significance of time-dependent behaviour was highlighted by Graham *et al.* (1983) who considered some engineering properties of lightly overconsolidated clays, notably the Belfast Clay. They showed, for example, that the stress–strain relationship of such clays is especially time-dependent. In other words the undrained shear strengths alter by between 10 and 20% with a tenfold change in the rate of strain. Previously Tavenas *et al.* (1978) had found that the yield behaviour of the Winnepeg Clay was also time-dependent.

Skempton (1964) pointed out that when clay is strained it develops an increasing resistance (strength), but that under a given effective pressure the resistance offered is limited, the maximum value corresponding to the peak strength. If testing is continued beyond the peak strength, than as displacement increases, the resistance decreases, again to a limiting value which is termed the residual strength. Skempton wrote that in moving from peak to residual strength, cohesion falls to almost, or actually, zero and the angle of shearing resistance is reduced to a few degrees (it may be as much as 10° in some clays). He explained the drop in strength which occurred in overconsolidated clays as due to their expansion on passing peak strength and associated increasing water content on the one hand. On the other, he maintained that platey clay minerals became orientated in the direction of shear and thereby offered less resistance. Failure occurs once the stress on a clay exceeds its peak strength and as failure progresses the strength of the clay along the shear surface is reduced to the residual value.

It was suggested that under a given effective pressure, the residual strength of a clay is the same whether it is normally or overconsolidated (Figure 3.16). In other words, the residual shear strength of a clay is independent of its post-depositional history, unlike the peak undrained shear strength which is controlled by the history of consolidation as well as diagenesis. Furthermore, the value of residual shear strength (ϕ_r') decreases as the amount of clay fraction increases in a deposit. In this context not only is the proportion of detrital minerals important but so is that of the diagenetic minerals. The latter influence the degree of induration of a deposit of clay, and the value of ϕ_r' can fall significantly as the ratio of clay minerals to detrital and diagenetic minerals increases.

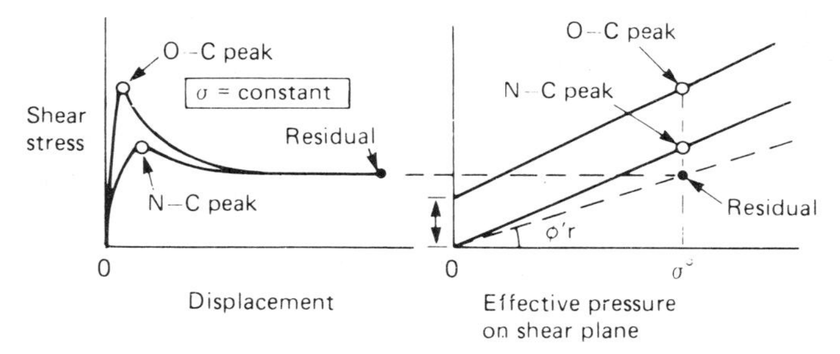

Figure 3.16 *Peak strength and residual strength of normally (N-C) and overconsolidated (O-C) clays (after Skempton, 1964)*

It is generally asserted that the value of residual shear strength depends principally on the method of testing and the effective normal pressure used. For instance, an increase in effective normal pressure leads to a reduction in the residual shear strength. However, Bromhead (1983) found that generally the measured residual strengths obtained in both ring shear and direct shear tests were very similar. He also noted that the results of ring shear tests were much more consistent than those derived from shear box tests and he went on to suggest that the belief that the former test underestimates the value of the residual shear strength of clay was incorrect.

The shear strength of an undisturbed clay is frequently found to be greater than that obtained when it is remoulded and tested under the same conditions and at the same water content. The ratio of the undisturbed to the remoulded strength at the same moisture content was defined by Terzaghi (1936) as the sensitivity of a clay. Subsequently Skempton and Northey (1952) proposed six grades of sensitivity (see Chapter 1). The liquidity index can be used to show increases in sensitivity in clay soils, the general trend being illustrated in Figure 3.17.

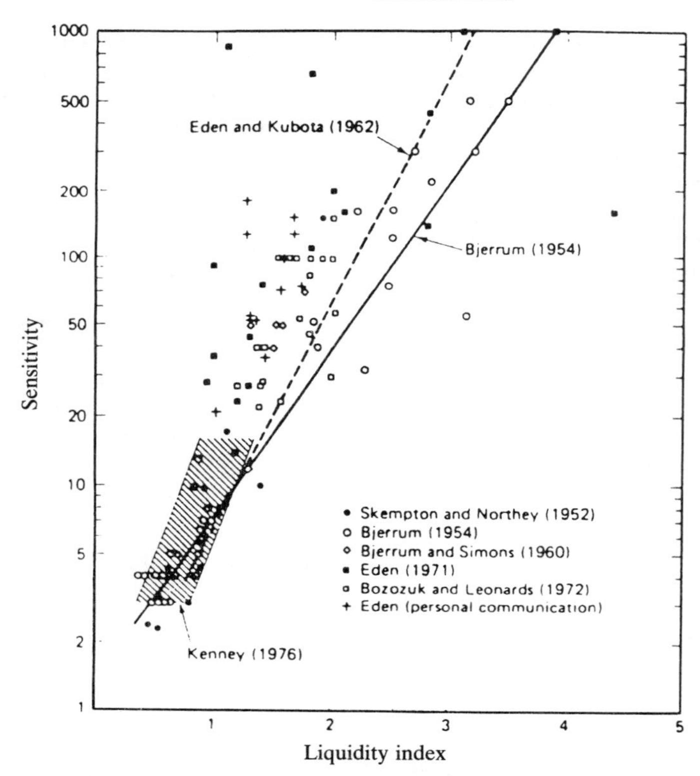

Figure 3.17 *Relationship between sensitivity and liquidity index for clays (from Lutenegger and Hallberg, 1988)*

3.3.7 Weathering of clay soils

The greatest variation in the engineering properties of clays can be attributed to the degree of weathering which they have undergone. For instance, consolidation of a clay deposit gives rise to an anisotropic texture due to the rotation of the platey minerals. Secondly, diagenesis bonds particles together either by the development of cement, the intergrowth of adjacent grains or the action of van der Waals charges which are operative at very small grain separations. Weathering reverses these processes, altering the anisotropic structure and destroying or weakening interparticle bonds.

In addition, removal of overburden leads to vertical expansion of a deposit which facilitates the development of joints and fissures together with softening. The opening of fissures is accompanied by water entrainment and chemical degradation. The process is progressive.

Ultimately weathering, through the destruction of interparticle bonds, leads to a clay deposit reverting to a normally consolidated sensibly remoulded condition. Higher moisture contents are found in the more weathered clay. This progressive degrading and softening is also accompanied by reductions in strength and deformation moduli with a general increase in plasticity.

The reduction in strength has been illustrated by Cripps and Taylor (1981) who quoted the strength parameters shown in Table 3.6 for brown (weathered) and blue (unweathered) London Clay.

Table 3.6

Parameter	*Brown*	*Blue*
τ_u (kN/m^2)	100–175	120–250
c' (kN/m^2)	0–31	35–252
$\phi°$	20–23	25–29

These indicate that the undrained shear strength (s_u) is reduced by approximately half and that the effective cohesion (c') can suffer significant reduction on weathering. The effective angle of shearing resistance is also reduced and at $\phi' = 20$ the value corresponds to a fully softened condition (Skempton, 1970).

An extensive study of how weathering has affected the Lias Clay was undertaken by Chandler (1972). He found that the degree of oxidation provided an obvious basis for classifying the extent of weathering in the Lias Clay and that, in turn, it was closely related to the variation in strength–water content associated with weathering. He chose the ratio Fe_2O_3:FeO as an index of the amount of oxidation which had taken place – the higher the value, the greater the degree of oxidation.

Chandler (1972) recognized four zones of weathering in the Upper Lias, a summary of which is given in Table 3.7. The influence of weathering on the geotechnical properties of this clay is illustrated in Figure 3.18. The strength–moisture content relationship is illustrated in Figure 3.19. This shows that each of the weathered zones can be related to different strength–moisture content curves. The strength–depth relationship is not simply related to variation in moisture content but also takes into account the decreasing number of fissures with depth. Increasing plasticity also is brought about by increases in weathering. In the less weathered zones differences in the water content reflect disturbances in the fabric of the clay and these are not related to increasing plasticity. A similar sequence of weathering was recognized in the Lower Lias by Coulthard and Bell (1992).

Russell and Parker (1979) recognized a similar zoning of the grades of weathering in the Oxford Clay as did Chandler (1972) in the Lias Clay. They found a reasonable correlation between shear strength and moisture content, although this is not as good as in the Lias Clay, probably because of the greater variability of the Oxford Clay.

In general, changes in bulk clay mineralogy with weathering are slight. However, the degree of illite degradation increases in weathered clay and that with the most degraded illite appears to have the lowest strength.

Table 3.7 *Weathering scheme for Upper Lias Clay (after Chandler, 1972)*

Zone	*Oxidation, etc.* *(1) Description of hand specimen* *(2) Range of Fe_2O_3/FeO ratios*	*Fabric, based on examination of thin sections*	*Discontinuities* *(1) Fissure spacing* *(2) Shears*	*Other chemical observations* *(1) %$CaCO_3$* *(2) Selenite*	*Water content* *(1) Range* *(2) Liquidity index*
(Landslip/ solifluction)	(1) Mottled light grey/light brown clay becoming grey with depth; considerable oxidation near surface (with frequent small haematite pellets), minimal oxidation at depth (i.e. >3 m); extensive gleying at shallow depths	*0.2 m+*:– Heterogeneous, with small (<1 mm) rotated lithorelicts; *greater depths*:– rotated lithorects (up to 30 mm); otherwise as Zone III	(1) Apparently intact (except for desiccation cracks) at shallow depths; as Zone III at greater depths (2) As Zone III	(1) 0–4% (2) None observed	(1) 32–36% (2)0 to +0.12
IV	Not observed – probably absent on slopes				
III	(1) Fissured clay with light grey (gleyed) fissure surfaces; centres of lithorelicts seen to be pale brown (oxidized) (2) 2.5–5.0; mean 4.0	Lithorelicts (up to 30 mm) have horizontal bedding; matrix occupies less than 50% of section, is often gleyed, and where limited in extent is usually sheared; larger areas of matrix show oriented bands	(1) 10–30 mm (2) Minor shearing in gleyed fissures is common	(1) 5–6% (2) Common	(1) 27–32% (2) −0.03 to +0.08
IIb	(1) Grey/blue-grey clay with brown (oxidized) areas typically along fissures; no gleying (2) 1.0–6.0; mean 3.0	As Zone I but with brown staining usually in areas parallel to fissures	(1) 20–100 mm (2) Minor shears, 1–2 mm displacement, sometimes associated with oxidation	(1) 0–5% (2) Infrequent	(1) 23–27% (2) −0.15 to −0.05
IIa	(1) Blue-grey clay/soft mudstone, oxidation on surfaces of joints and fissures only (2) 0.3–2.0; mean 0.8	As Zone I but with brown staining only along fissures	(1) 20–100 mm (2) No shears	(1) 1–6 (2) Rare	(1) 19–25% (2) −0.34 to −0.14
I	(1) Blue-grey soft mudstone; no oxidation (2) 0.3–1.5; mean 0.6	Bedding horizontal; fabric variations are those resulting from depositional changes	(1) >100 mm (2) No shears	(1) 1–7% (2) None observed	(1) 14–19% (2) −0.57 to 0.38

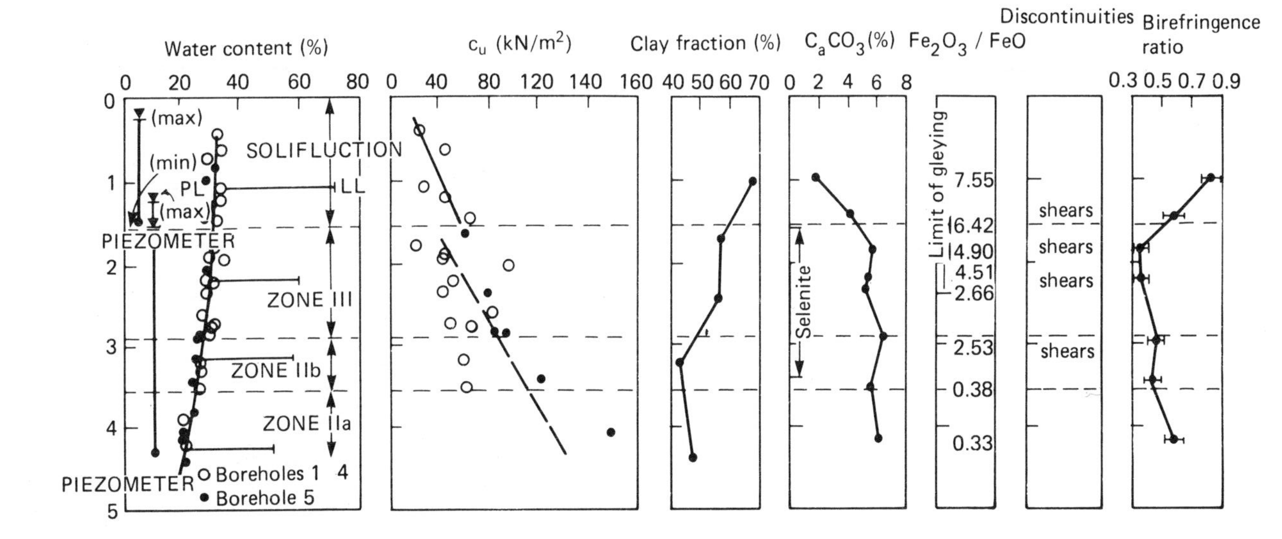

Figure 3.18 *Composite geotechnical profile of the Upper Lias Clay (after Chandler, 1972). Weathering of pyrite produces ferrous sulphate and sulphuric acid. The latter reacts with calcium carbonate to form gypsum, crystals of selenite being formed. Selenite occurs down to the boundary between zones IIa and IIb. Fissures have allowed percolation of acid and therefore selenite formation has occurred at a greater depth than that of general oxidation. There is no general relationship between the degree of weathering and carbonate content, the variability of the carbonate content probably representing depositional or diagenetic variations. A measure of the average orientation of clay particles seen in a thin section of clay beneath the petrological microscope is afforded by the birefringence ratio (the ratio between minimum and maximum light transmitted under crossed polars). This varies from zero, with perfect parallel orientation, to one, with perfect random orientation. With increasing weathering the birefringence ratio increases, indicating that the fabric of the clay becomes more disordered, as the ground surface is approached*

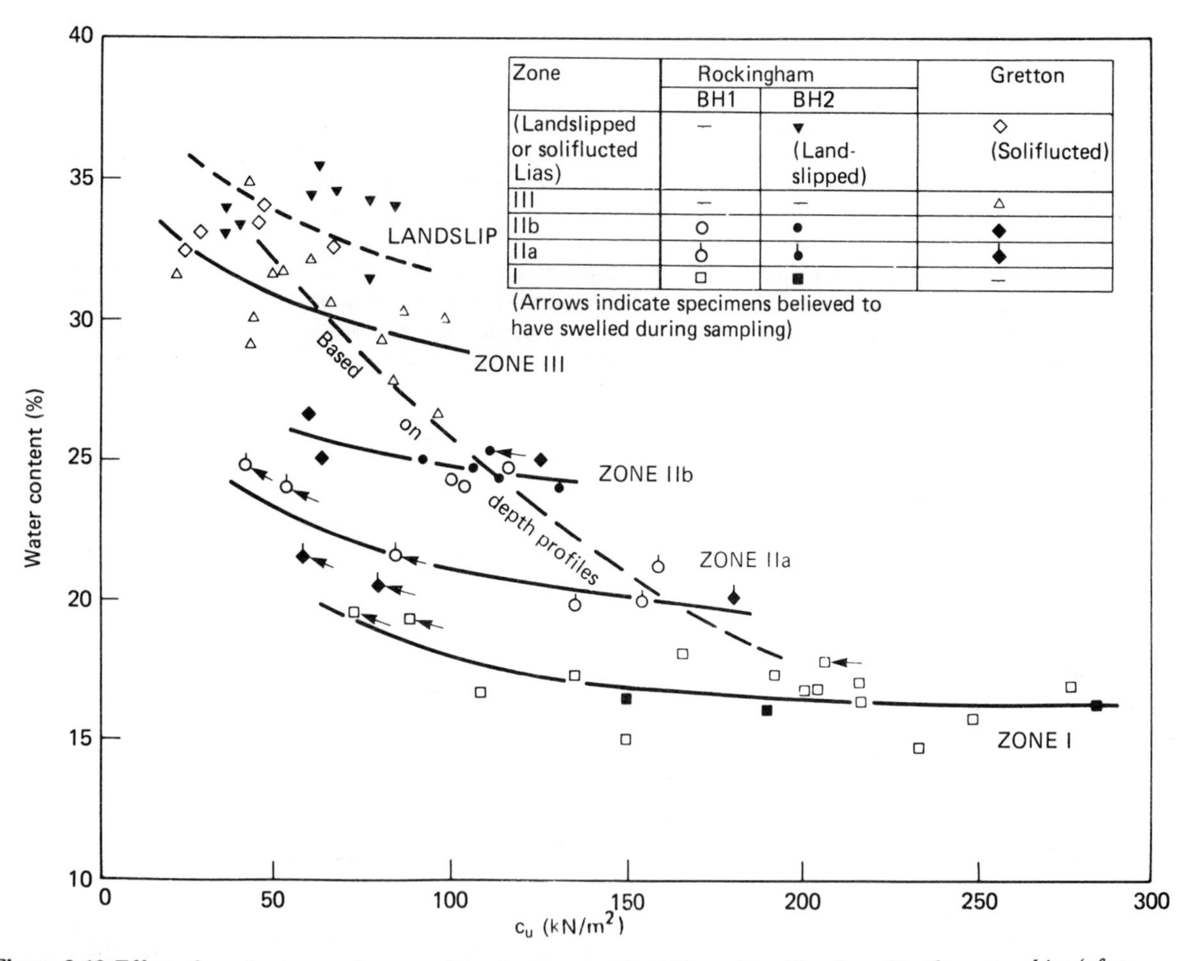

Figure 3.19 *Effect of weathering on the strength and water content of Upper Lias Clay from Northamptonshire (after Chandler, 1972)*

3.4 Fissures in clay

3.4.1 Character of fissures

Ward *et al.* (1959), after noting the presence of fissures in the London Clay, defined them as small fractures confined to a bed or horizons within a bed. Unfortunately bedding is not always readily discernible in some clays. Minor changes in lithology do not appear to influence the attitude of fissuring but do influence the size of fissures and the intensity of fissuring (Fookes and Denness, 1969). Generally with increasing time after exposure the proportion of non-planar to planar fissures increases since the former can develop from the latter by extension from their extremities. Most fissures are probably original features which on exposure tend to open due to stress release and weathering. For instance, Fookes and Parrish (1968) noted that the orientation of many fissures in the London Clay is very closely similar to the orientation of joints in adjacent rock. This suggests that fissures in stiff clay and joints in semi-brittle rocks, when in the same stress field, develop with more or less similar orientations.

In addition to fissures, Skempton *et al.* (1969) recorded joints in the London Clay. These are planar structures, the surfaces of which have a matt texture, although a thin layer of clay gouge occurs along some joints. Joints were predominantly normal to the horizontal bedding, with a pronounced trend in two orthogonal directions. They ranged in height up to 2.6 m and in length up to 6.0 m. Skempton *et al.* also recognized smooth surfaces of moderate size (over 0.03 m^2 in area) which dipped at angles varying between 5 and 25°. These surfaces they termed sheeting. Sheeting usually had a slightly undulating shape but sometimes it was planar. These authors noted that the fissures in the uppermost 13 m of the London Clay may be curved or planar. Fissures usually have a matt surface texture. The mean size of the fissures decreases and the number per unit volume correspondingly increases as the upper surface of the clay is approached. Similarly Ward *et al.* (1965) found that at Ashford the weathered brown London Clay was more fissured than the blue below. At depth the high pressures tend to keep the fissures closed, at least in the sense that they are not visibly open fractures. No appreciable relative movements have taken place along fissures. However, 5% of all fissures in the London Clay are slickensided with polished and striated surfaces. The slickensides are due to shearing consequent upon minor internal distortions of the clay mass. The polished nature of the surface indicates a considerable degree of particle orientation.

3.4.2 Strength of fissured clay

Fissures play an extremely important role in the failure mechanism of fissured clay. Indeed many clays are seriously weakened by the presence of a network of fissures. Terzaghi (1936) provided the first quantitative data relating to the influence of fissures and joints on the strength of clays, pointing out that such features are characteristic of overconsolidated clays. He maintained that fissures in normally consolidated clays have no significant practical consequences. On the other hand fissures can have a decisive influence on the engineering performance of an overconsolidated clay, in that the overall strength of such fissured clay can be as low as one-tenth that of the intact clay. For instance, it appears that the average shearing resistance of stiff fissured clay at the moment of sliding usually ranges between 15 and 30 kN/m^2 whereas the initial shearing resistance of such clays ranges between 100 and 300 kN/m^2. In addition to allowing clay to soften, fissures and joints allow concentrations of shear stress which locally exceed the peak strength of clay thereby giving rise to progressive failure. Under stress the fissures in clay seem to propagate and coalesce in a complex manner.

Skempton (1948) attributed the reduction in strength of the London Clay exposed in cuttings to softening along fissures which opened as a result of small movements consequent upon the removal of lateral support on excavation. The ingress of water into fissures means that the pore water pressure in the clay concerned increases, which in turn means that its strength is reduced. Skempton and La Rochelle (1965) investigated slope failures in the London Clay at Bradwell, Essex. These began to occur on cut slopes a few days after excavation, even though the slopes had been designed on strengths lower than the undrained parameters for the intact clay. The investigation revealed that a large proportion of the reduction in the overall strength below these design parameters was attributed to the fissures. For example, it was noted that blocks of intact clay which were found on the failed slopes showed no signs of deformation or failure.

Skempton and La Rochelle (1965) considered that where unfavourable orientation of fissures exists the major part of a failure plane may follow fissures. As a result the clay may be near its residual strength. The overall strength of the clay could be further reduced by the separation of the walls of closed fissures. As intact clay has a low tensile strength, there is no resistance to the opening of a fissure and once open there is no shear resistance along the fissure itself. They recommended that, if regular fissure patterns with an unfavourable orientation occur at a site, an attempt should be made to estimate the influence of these fissures on the overall strength of the clay mass. It is therefore necessary to determine the average area of a potential failure plane which would pass through open fissures, closed fissures and intact clay. It is also necessary to establish the stress distribution in the clay mass in order to obtain the correct value of the overall residual strength for design purposes.

Skempton *et al.* (1969) summarized the shear strength parameters of the London Clay in terms of effective stress as follows:

(1) peak strength of intact clay: $c' = 31\,kN/m^2$, $\phi' = 20°$,
(2) 'peak' strength on fissure and joint surfaces: $c' = 6.9\,kN/m^2$, $\phi' = 18.5°$,
(3) residual strength of intact clay: $c'_r = 1.4\,kN/m^2$, $\phi'_r = 16°$.

Thus the strength along joints or fissures in clay is only slightly higher than the residual strength of the intact clay. Similar results previously had been obtained by Marsland and Butler (1967) who carried out a series of large-scale shear box tests on stiff fissured Barton Clay. They found that the strength developed along a closed fissure was hardly more than the residual value in the drained or undrained conditions. They also concluded that if a plane of failure develops along fissures which facilitate more effective drainage, then the drained strength parameters should be used in a stability analysis of slopes.

It can be concluded that the upper limit of the strength of fissured clay is represented by its intact strength whilst the lower limit corresponds to the strength along the fissures. The operational strength, which is somewhere between the two, is, however, often significantly higher than the fissure strength.

Ward *et al.* (1959) recommended that tests to determine the mechanical behaviour of fissured clay should be made on as large a scale as possible. The effect of size of the test specimen on the undrained strength of stiff fissured clay subsequently was demonstrated by Bishop and Little (1967). For example, they showed that the strength of the London Clay, when determined by an *in situ* shear box, 600 × 600 mm, was only 55% of that obtained in the laboratory by testing unconsolidated–undrained samples, 37.5 mm in diameter, in triaxial conditions. The greater the curvature and complexity of the fissures, the greater the strengths obtained, particularly those measured on small specimens.

Lo (1970) showed that as the size of the test specimen of fissured clay increased, not only would the number of fissures it contained increase, but the probability of the test specimen possessing larger fissures would also increase, as would the probability of these having a critical orientation. Lastly, there would be a greater likelihood of coalescing adjacent fissures corresponding with the potential plane of failure.

Fissures open fairly rapidly once fissured clay is exposed. This has an important effect upon the properties measured both in the field and in the laboratory. For instance, Ward *et al.* (1965) showed that the strength of the London Clay measured 4–8 h and 2.5 days after excavation was 85% and 75% of that obtained 0.5 h after excavation. This was attributed to the gradual extension of fissures and microcracks in the clay with time.

References

Audric, T. and Bouquier, L. (1976). 'Collapsing behaviour of some loess soils from Normandy', *Q. J. Engg Geol.*, **9**, 265–78.

Barden, L. (1972). 'The relation of soil structure to the engineering geology of clay soil', *Q. J. Engg Geol.*, **5**, 85–102.

Berre, T. and Bjerrum, L. (1973). 'The shear strength of normally consolidated clays', *Proc. 8th Int. Conf. Soil Mech. Foundation Engg, Moscow,* **1**, 39–49.

Bjerrum, L. (1967). 'Progressive failure in slopes of overconsolidated plastic clay and clay shales', *Proc. ASCE Soil Mech. Foundation Engg Div.*, **93**, SM5, 2–49.

Bjerrum, L. (1972). 'Embankments on soft ground', (in *Performance of Earth and Earth-Supported Structures*), Vol. II, *ASCE Proc. Speciality Conf.*, Purdue University, Lafayette, Indiana, 32–3.

Bishop, A. W. and Little, A. L. (1967) 'The influence of the size and orientation of the sample on the apparent strength of the London Clay at Maldon, Essex', *Proc. Geot. Conf., Oslo,* **2**, 89–96.

Boswell, P. G. H. (1961) *Muddy Sediments,* Heffer, Cambridge.

Brackley, I. J. A. (1980) 'Prediction of soil heave from suction measurements', *Proc. 7th Reg. Afr. Conf. Soil Mech., Accra,* **1**, 159–66.

Bromhead, E. N. (1983) 'A comparison of alternative methods of measuring the residual strength of London Clay', *Ground Engineering,* **16**, No. 3, 39–41.

Burland, J. B., Longworth, T. I. and Moore, J. F. A. (1977). 'A study of ground movement and progressive failure caused by a deep excavation in Oxford Clay', *Geotechnique,* **27**, 557–91 (also in *Building Res. Esta., Watford*, 33–78 (1978)).

Burnett, A. D. and Fookes, P. G. (1974). 'A regional engineering geological study of the London Clay in the London and Hampshire Basins', *Q. J. Engg Geol.,* **7**, 257–96.

Chandler, R. I. (1972). 'Lias Clay: weathering processes and their effect on shear strength', *Geotechnique,* **22**, 403–31.

Clevenger, W. A. (1958) 'Experiences with loess as a foundation material', *Trans. Am. Soc. Civil Engrs,* **123**, Paper No. 2961, 151–80.

Coulthard, J. M. and Bell, F. G. (1992). 'The influence of weathering on the engineering behaviour of the Lower Lias Clay'. In *The Engineering Geology of Weak Rocks, Engineering Geology Special Publication No 8*, Cripps, J. C., Culshaw, M. G., Coulthard, J. M., Moon, C. F. and Hencher, S. R. (eds.), A A Balkema, Rotterdam.

Cripps, J. C. and Taylor, R. K. (1981) 'The engineering properties of mudrocks', *Q. J. Engg Geol.,* **14**, 325–46.

Cripps, J. C. and Taylor, R. K. (1986). 'Engineering characteristics of British overconsolidated clays and mudrocks. 1 Tertiary deposits', *Engg Geol.,* **22**, 349–76.

Cripps, J. C. and Taylor, R. K. (1987). 'Engineering characteristics of British overconsolidated clays and mudrocks. II Mesozoic deposits', *Engg Geol.,* **23**, 213–53.

Denisov, N. Y. (1963). 'About the nature and sensitivity of quick clays', *Osnov. Fudami Mekh. Grant.,* **5**, 5–8.

Derbyshire, E. and Mellors, T. W. (1988). 'Geological and geotechnical characteristics of some loess and loessic soils from China and Britain: a comparison', *Engg Geol.,* **25**, 135–75.

Driscoll, R. (1983). 'The influence of vegetation on the swelling and shrinkage of clay soils in Britain', *Geotechnique,* **33**, 93–105.

Feda, J. (1966). 'Structural stability of subsident loess from Praha-Dejvice', *Engg Geol.,* **1**, 201–19.

Feda, J. (1988). 'Collapse of loess on wetting', *Engg Geol.,* **25**, 263–9.

Fookes, P. G. and Denness, B. (1969). 'Observational studies on fissure patterns in Cretaceous sediments of south east England', *Geotechnique,* **19**, 453–77.

Fookes, P. G. and Parrish, D. G. (1968). 'Observations on small scale structural discontinuities in the London Clay and their relationship to regional geology', *Q. J. Engg Geol.,* **1**, 217–40.

Gao, Guorui (1988). 'Formation and development of the structure of collapsing loess in China', *Engg Geol.,* **25**, 235–45.

Gibbs, H. H. and Bara, J. P. (1962). *Predicting Surface Subsidence from Basic Soil Tests.* ASTM Spec. Tech. Publ., No 322, 231–46.

Gibbs, H. H. and Holland, W. Y. (1960). *Petrographic and Engineering Properties of Loess.* US Bureau of Reclamation, Engineering Monograph No 28.

Grabowska-Olszewska, B. (1988). 'Engineering-geological problems of loess in Poland', *Engg Geol.,* **25**, 177–99.

Graham, J., Crooks, J. H. A. and Bell, A. L. (1983). 'Time effects on the stress–strain behaviour of natural soft clays', *Geotechnique,* **33**, 327–40.

Grim, R. E. (1962). *Applied Clay Mineralogy*, McGraw-Hill, New York.

Hardy, R. M. (1950). 'Construction problems on silty soils', *Engg J.,* **33**, 775–9.

Hardy, R. M. (1966). 'Identification and performance of swelling soil types', *Can. Geot. J.*, **2**, 141–53.

Holtz, W. G. and Gibbs, J. H. (1956). 'Engineering properties of expansive clays', *Trans. Am. Soc. Civil Engrs.*, **121**, 641–63.

Houston, S. L., Houston, W. L. and Spadola, D. J. (1988). 'Prediction of field collapse of soils due to wetting', *Proc. ASCE, J. Geot. Engg Div.*, **114**, 40–58.

Jackson, J. O. and Fookes, P. G. (1974). 'The relationship of the estimated former burial depth of the Lower Oxford Clay to some soil properties', *Q. J. Engg, Geol.*, **7**, 137–80.

Jennings, J. E. and Knight, K. (1975). 'A guide to construction on or with materials exhibiting additional settlement due to "collapse" of grain structure', *Proc. 6th African Conf. Soil Mech. Found. Engg*, Durban, 99–105.

Ladd, C. C., Foott, R., Ishihara, K., Schlosser, F. and Poulos, H. G. (1977). 'Stress deformation and stress characteristics'. State-of-the-Art Report. *Proc. 9th Int. Conf. Soil Mech. Found. Engg, Tokyo,* **3**, 421–94.

Larionov, A. K. (1965). 'Structural characteristics of loess soils for evaluating their constructional properties', *Proc. 6th Int. Conf. Soil Mech. Foundation, Engg, Montreal,* **1**, 64–8.

Lin, Z. G. and Wang, S. J. (1988). 'Collapsibility and deformation characteristics of deep-seated loess in China', *Engg Geol.*, **25**, 271–82.

Lo, K. Y. (1970). 'Operational strength of fissured clays', *Geotechnique,* **20**, 57–74.

Lutenegger, A. J. and Hallberg, G. R. (1988). 'Stability of loess', *Engg Geol.*, **25**, 247–61.

Mahar, L. J. and O'Neill, M. W. (1983). 'Geotechnical characterization of desiccated clay', *Proc. ASCE, J. Geot. Engg Div.*, **109**, GT1, 56–71.

Marsland, A. (1973). *Large In Situ Tests to Measure the Properties of Stiff Fissured Clays.* Current Paper 1/73, Building Research Establishment, Watford.

Marsland, A. and Butler, M. E. (1967). 'Strength measurements on stiff fissured Barton Clay from Fawley, Hampshire', *Proc. Geot. Conf., Oslo,* **1**, 139–46.

Mielenz, R. C. and King, M. E. (1955). 'Physical-chemical properties and engineering performance of clays'. In *Clays and Clay Technology*, Pask J. A. and Turner, M. D. (eds.). California Division of Mines, Bull. No. 169, 196–254.

Obermeier, S. F. (1974). 'Evaluation of laboratory techniques for measurement of swell potential of clays', *Bull. Ass. Engg Geologists,* **11**, 293–314.

Olson, R. E. (1974). 'Shearing strength of kaolinite, illite and montmorillonite', *Proc. ASCE, J. Geot. Engg Div.*, **100**, Paper No. 10947, 1215–29.

O'Neill, M. W. and Poormoayed, A. M. (1980). 'Methodology for foundations on expansive clays', *Proc. ASCE. J. Geot. Engg Div.*, **106**, GT12, 1345–67.

Padfield, C. J. and Sharrock, M. J. (1983). *Settlement of Structures on Clay Soils*, CIRIA, Spec. Publ. 27, London.

Penman, A. D. M. (1953). 'Shear characteristics of a saturated silt in triaxial compression', *Geotechnique,* **3**, 312–15.

Popescu, M. E. (1979). 'Engineering problems associated with expansive clays from Romania', *Engg Geol.*, **14**, 43–53.

Popescu, M. E. (1986). 'A comparison between the behaviour of swelling and collapsing soils', *Engg Geol.,* **23**, 145–63.

Russell, D. J. and Parker, A. (1979). 'Geotechnical, mineralogical and chemical inter-relationships in weathering profiles of an overconsolidated clay', *Q. J. Engg Geol.*, **12**, 197–216.

Samuels, S. G. (1975). 'Some properties of the Gault Clay from the Ely-Ouse Essex water tunnel', *Geotechnique,* **25**, 239–64.

Schmertmann, J. H. (1969). 'Swell sensitivity', *Geotechnique,* **19**, 530–3.

Schultze, E. and Horn, A. (1961). 'The shear strength of silt', *Proc. 5th Int. Conf. Soil Mech. Foundation Engg, Paris,* **1**, 350–3.
Schultze, E. and Kotzias, A. B. (1961). 'Geotechnical properties of lower Rhine silt', *Proc. 5th Int. Conf. Soil Mech. Foundation Engg, Paris,* **1**, 329–33.
Skempton, A. W. (1948). 'The rate of softening in stiff fissured clay with special reference to the London Clay', *Proc. 2nd Int. Conf. Soil Mech. Found Engg*, Rotterdam, **2**, 50–3.
Skempton, A. W. (1951). 'The bearing capacity of clays', *Building Research Congress*, Div. 1, 180–9.
Skempton, A. W. (1964). 'Long-term stability of clay slopes', *Geotechnique,* **14**, 77–102.
Skempton, A. W. (1970). 'First time slides in overconsolidated clays', *Geotechnique,* **20**, 320–4.
Skempton, A. W. and La Rochelle, P. (1965). 'The Bradwell slip: a short term failure in London Clay', *Geotechnique,* **15**, 221–41.
Skempton, A. W. and Northey, R. D. (1952). 'The sensitivity of clays', *Geotechnique,* **2**, 30–53.
Skempton, A. W., Schuster, R. L. and Petley, D. J. (1969). 'Joints and fissures in the London Clay at Wraysbury and Edgeware', *Geotechnique,* **19**, 205–17.
Sridharan, A. and Allam, M. M. (1982). 'Volume change behaviour of desiccated soils', *Proc. ASCE. J. Geot. Engg Div.,* **108**, GT8, 1057–71.
Tan, Tjong Kie (1988). 'Fundamental properties of loess from northwestern China', *Engg Geol.,* **25**, 103–22.
Tavenas, F., Leroueil, S., La Rochelle, P. and Roy, M. (1978). 'Creep behaviour of an undisturbed lightly overconsolidated clay', *Can. Geot. J.,* **15,** 402–23.
Terzaghi, K. (1936). 'Stability of slopes of natural clay', *Proc. 1st Int. Conf. Soil Mech. Found. Engg, Cambridge, Mass.,* **1**, 161–5.
Van Olphen, H. (1963). *An Introduction to Clay Colloid Chemistry*, Wiley, New York.
Vijayvergiya, V. N. and Ghazzaly, O. I. (1973). 'Prediction of swelling potential for natural clays', *Proc. 3rd Int. Conf. on Expansive Soils*, Haifa, 227–32.
Wahls, H. E. (1962). 'Analysis of primary and secondary consolidation', *Proc. ASCE, J. Soil Mech. Foundation Engg Div.,* **88**, SM6, 207–31.
Ward, W. H., Samuels, S. G. and Butler, M. E. (1959). 'Further studies of the properties of London Clay', *Geotechnique,* **9**, 33–8.
Ward, W. H., Marsland, A. and Samuels, S. G. (1965). 'Properties of the London Clay at the Ashford Common Shaft: *in situ* and undrained strength tests', *Geotechnique,* **15**, 321–44.
Williams, A. A. B. and Donaldson, G. (1980) 'Building on expansive soils in South Africa 1973–1980', *Proc. 4th Int. Conf. on Expansive Soils*, Denver, **2**, 834–8.
Williams, A. A. B. and Jennings, J. E. (1977). 'The *in situ* shear behaviour of fissured soils', *Proc. 9th Int. Conf. Soil Mech. Foundation Engg, Tokyo,* **2**, 169–76.
Williams, A. A. B. and Pidgeon, J. T. (1983). 'Evapotranspiration and heaving clays in South Africa', *Geotechnique,* **33**, 141–50.

Chapter 4

Tills and other deposits associated with cold climates

4.1 Character and types of tills

Till is regarded as being synonymous with boulder clay. It is deposited directly by ice whilst stratified drift or tillite is deposited in melt waters associated with glaciers. An extensive review of the various types of glacial deposits and their engineering properties has been provided by Fookes *et al.* (1975).

The character of a till deposit depends on the lithology of the material from which it was derived, on the position in which it was transported in the glacier, and on the mode of deposition (Boulton and Paul, 1976). The underlying bedrock usually constitutes up to about 80% of basal tills, depending on its resistance to abrasion. Argillaceous rocks such as shales and mudstones are more easily abraded and produce fine-grained tills which are presumably richer in clay minerals and therefore more plastic than other tills. Mineral composition also influences the natural moisture content which is slightly higher in tills containing appreciable quantities of clay minerals or mica. The uppermost tills in a sequence contain a high proportion of far-travelled material and may not contain any of the local bedrock.

Deposits of till consist of a variable assortment of rock debris ranging from fine rock flour to boulders. The shape of the rock fragments found in till varies but is largely conditioned by the initial shape of the fragment at the moment of incorporation into the ice. Angular boulders are common, their irregular sharp edges resulting from crushing. Crushing or grinding of a rock fragment occurs when it comes in contact with another fragment or the rock floor. The random occurrence in a till of boulders has been referred to as *raisin cake* structure, however, more frequently fragments are crudely aligned in rows. The larger elongated fragments generally possess a preferred orientation in the path of ice movement. On the one hand tills may consist essentially of sand and gravel with very little binder, alternatively they may have an excess of clay. Lenses and pockets of sand, gravel and highly plastic slickensided clay are frequently encountered in some tills. Some of the masses of sand and gravel are interconnected, due to the action of meltwater, but many are isolated. Small distorted pockets of sand and silt have been termed flame structures. Most tills contain a significant amount of quartz in their silt–clay fractions.

Distinction has been made between tills derived from rock debris which was carried along at the base of a glacier and those deposits which were transported within and on the ice. The former is referred to as lodgement till whereas the latter is known as ablation till. Lodgement till is thought to be plastered onto the ground beneath the moving glacier in small increments as the basal ice melts. Because of the overlying weight of ice such deposits are overconsolidated. Ablation till accumulates on the surface of the ice when englacial debris melts out, and as the glacier decays the ablation till is slowly lowered to the ground. It is therefore normally consolidated. Lodgement till contains fewer, smaller stones (they generally possess a preferred orientation) than ablation till and they are rounded and striated. Clast orientation in ablation till varies from almost random to broadly parallel to the ice flow direction.

Due to abrasion and grinding the proportion of silt and clay size material is relatively high in lodgement till (e.g. the clay fraction varies from 15 to 40%). Lodgement till is commonly stiff, dense and relatively incompressible. Hence it is practically impermeable. Oxidation of lodgement till takes place very slowly so that it is usually grey.

Fissures are frequently present in lodgement till, especially if it is clay matrix dominated. Sub-horizontal fissures have been developed as a result of incremental loading and periodic unloading whilst sub-vertical fissures owe their formation to the overriding effects of ice and stress relief. McGown *et al.* (1974) noted a very definite preferred orientation of fissures in till and that their intensity increased as the surface of the till was approached. This was attributed to greater stresses caused by ice movement and to the effects of weathering.

Because it has not been subjected to much abrasion ablation till is characterized by abundant large stones that are angular and not striated, the proportion of sand and gravel is high and clay is present only in small amounts (usually less than 10%). Because the texture is loose, ablation till oxidizes rapidly and commonly is brown or yellowish brown. It also may have an extremely low *in situ* density. Since ablation till consists of the load carried at the time of ablation it usually forms a thinner deposit than lodgement till.

Till sheets can comprise one or more layers of different material, not all of which are likely to be found at any one locality. Shrinking and reconstituting of an ice sheet can complicate the sequence.

McGown and Derbyshire (1977) devised an elaborate system for the classification of tills. Their classification is based upon the mode of formation, transportation and deposition of glacial material and provides a general basis for the prediction of the engineering behaviour of tills (Figure 4.2; Table 4.1). However, a glacial deposit can undergo considerable changes after deposition due to the influence mainly of wind and water.

The particle size distribution and fabric (stone orientation, layering, fissuring and jointing) are among the most significant features as far as the engineering behaviour of a till is concerned. McGown and Derbyshire (1977) used the percentage of fines to distinguish granular, well graded and matrix dominated tills, the boundaries being placed at 15 and 45% respectively. The fabric of tills includes features of primary and secondary origins such as folds, thrusts, fissures (macrofabric), disposition of clasts (macro and mesofabric) and the organization of the matrix. It would seem that distinctive macro and mesofabric patterns characterize flow till, lodgement till and till deformed by thrusting and loading both in terminal moraines and till flutes.

4.2 Basic properties of tills

According to McGown (1971) tills are frequently gap graded, the gap generally occurring in the sand fraction (Figure 4.1). He also noted that large, often very

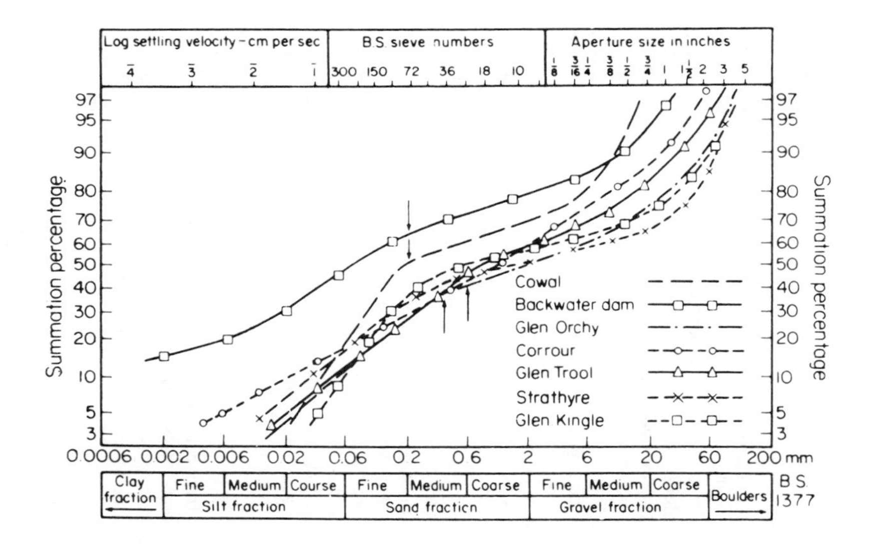

Figure 4.1 *Typical gradings of some Scottish morainic soils (after McGown, 1971)*

local, variations can occur in the gradings of till which reflect local variations in the formation processes, particularly the comminution processes. The clast size consists principally of rock fragments and composite grains, and presumably was formed by frost action and crushing by ice. Single grains predominate in the matrix. The range in the proportions of coarse and fine fractions in tills dictates the degree to which the properties of the fine fraction influence the properties of the composite soil. The variation in the engineering properties of the fine soil fraction is greater than that of the coarse fraction, and this often tends to dominate the behaviour of the till.

In a survey of tills which occur in north Norfolk, Bell (1991) noted that their clast fraction accounts for less than 40% of the deposits and usually less than 20%. Hence they are matrix-dominated tills, the approximate proportions of sand, silt and clay varying as shown in Table 4.2.

Table 4.1 *Till types and processes (Modified after McGown and Derbyshire, 1977)*

Formative processes	*Transportation processes*	*Depositional processes*
Comminution till: Produced by abrasion and interaction between particles in the basal zone of a glacier. It is a common element in most tills. Pebbles show a preferred orientation and are surrounded by a compact matrix in which there is a high concentration of fines.	Superglacial till: Derived from frost riving of adjacent rocks or by differential melting out of glacier dirt beds. It may or may not become incorporated in the glacier and may suffer frost shattering and washing by meltwaters as it is transported on the top of the glacier.	*Ablation till: Accumulated by melting out on the surface of a glacier or as a coating on inert ice.
Deformation till: Produced by plucking, thrusting, folding and brecciation of the glacier bed. Varies from mixtures of local material and erratic debris to disrupted, but little transported, masses of material derived from local bedrock. The materials are porous and some initially contain more water than required for compaction to maximum density. Hence these are soft tills which tend to be deformed rather than crushed.	Englacial till: Derived from superglacial till subsequently buried by accumulating snow or entrained in shear zones. It is transported within the ice mass and is more abundant in Polar regions than in temperate zones. Englacial debris occurs mainly in the lower 30–60 m of a glacier where rock detritus may comprise as much as 10–20% of its volume. Consequently an appreciable thickness of englacial till can be melted out from the base of a glacier, although this suffers a reduction in volume of anything up to 90%. Such till may be precompressed by overlying ice and may be sliced by thrust planes. Elongate stones may possess preferred orientation. The average grain size is much smaller than that of superglacial till.	*Meltout till: Accumulated as the ice of an ice–debris mixture melts out. Meltout tills exhibit a relatively low bulk density but show some variation depending on the particle size distribution. Generally the material is poorly sorted although occasionally thin layers or lenses of washed sediment occur. The microfabric is normally rather open. The clast fabrics of meltout tills show a wide variation depending on the disposition and debris concentration of the melting out mass.
	Basal till: Derived from comminution products in the ice-rock contact zones, particularly the lowermost regions of a glacier. It is generally transported in concentrated bands in the bottom metre or so of a glacier.	Lodgement till: Accumulated subglacially by accretion from debris-rich basal ice. Lodgement tills generally possess a wide range of particle sizes and are frequently anisotropic, fissured or jointed, especially when well graded or clay matrix dominated. They are usually stiff, dense, relatively incompressible soils. Macro, meso and microfabric patterns show high consistency. Sub-horizonal fissuring is due to incremental lodgement and periodic unloading, while subvertical fissures are evidence of ice over-riding and stress relief. Low-angle shear failure planes also occur.

Flow till: Consists essentially of melted out superglacial comminution debris but also occurs due to flow of subglacial meltout tills in subglacial cavities and at the ice margin. It usually contains a wide range of particle sizes. Flow till is frequently interbedded with fluvioglacial material. Orientation of clasts is broadly parallel to the deposition plane and imbricated up flow. The fines also reflect the flow mechanism in the parallelism of clay–silt fractions which produces a locally dense micro-fabric.

Waterlain till: Accumulates on subaqueous surface under a variety of depositional processes and may thus show a wide variety of characters. They vary from rather soft lodgement tills to subaqueous mudflows and to crudely stratified lacustrine clay silts. Stratification, deformation and very diffuse to random clast fabrics are common.

* This distinction between superglacial meltout and ablation tills is based on the degree of disturbance of their ice-inherited fabric, including loss of fines. The term ablation till is thus best avoided.

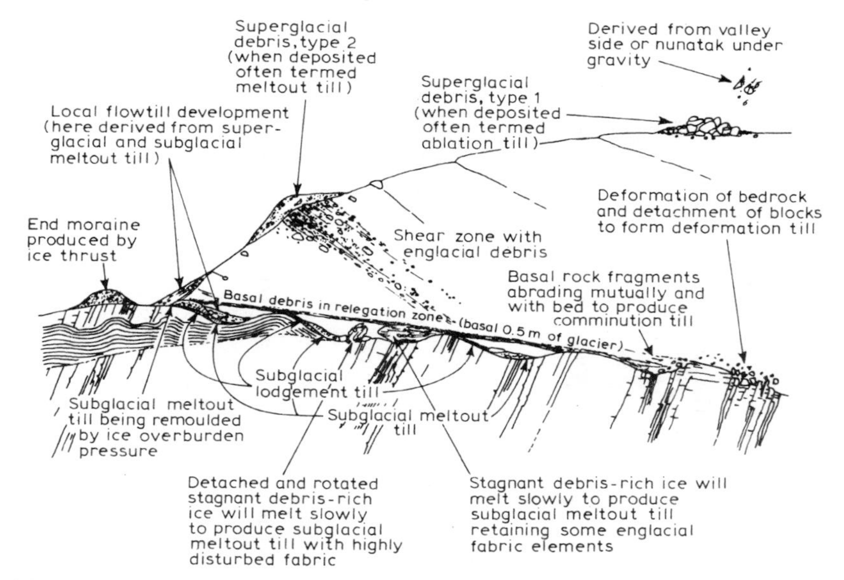

Figure 4.2 *Acquisition, transportation and deposition of tills by a glacier (after McGown and Derbyshire, 1977)*

The till with the most clay size material in the matrix is the Marly Drift. In fact a large proportion of the fine material in the Marly Drift often consists of chalky matter. The least clay size material occurs in the Contorted Drift, most of this consisting of silt. Sand tends to form the major component of the matrix in the Cromer and Hunstanton Tills.

A similar situation was found in the tills of Holderness in Humberside by Bell and Forster (1991). There the fine fraction of the tills generally accounts for over 60% and frequently over 80% of the individual deposits. Hence they similarly are all matrix-dominated tills. The Basement Till is the finest, containing the largest amount of clay size material, usually between 22 and 40%. The proportion of silt tends to vary from 27 to 35% and fine sand from 15 to 20%. The remaining proportion of sand is always less than 15%. In the Skipsea till the fine sand fraction can range between 20 and 30% whilst medium and coarse sands together do not account for more than 15% and frequently for less than 10%. The silt fraction varies between 30 and 40% whilst that of clay constitutes between 15 and 35%. The Withernsea Till contains between 20 and 45% silt and between 10 and 27% clay. The fine sand fraction is between 15 and 20% and the remaining sand amounts to between 12 and 18%. A wider range of particle size distribution occurs in the Hessle Till than the other three tills. The proportion of clay varies from 6 to 37% of silt from 28 to 32% and of fine sand from 17 to 21%. Medium and coarse sand may comprise from 9 to 18%. This wide spread of particle size distribution can be attributed to the fact that the Hessle Till is the weathered product of the Skipsea and Withernsea Tills. Like all tills, these are characteristically unsorted.

Table 4.2 *Sand, silt and clay fractions of till from north Norfolk*

	Hunstanton Till	*Marly Drift*	*Contorted Drift*	*Cromer Till*
Sand	34–58%	15–45%	26–40%	38–64%
Silt	36–27%	30–23%	54–48%	32–18%
Clay	30–15%	55–32%	20–12%	30–18%

The consistency limits of tills are dependent upon water content, grain size distribution and the properties of the fine grained fraction. Generally, however, their plasticity index is small and the liquid limit of tills decreases with increasing grain size.

The plastic and liquid limits of the tills of north Norfolk and Holderness are set out in Table 4.3, from which it can be seen that in the Cromer, Hunstanton,

Table 4.3 *Natural moisture content (m), plastic limit (PL), liquid limit (LL), plasticity index (PI), liquidity index (LI), consistency index (CI), and activity (A) of tills from north Norfolk and Holderness. (after Bell, 1991; and Bell and Forster, 1991)*

A. North Norfolk

	m	*PL* (%)	*LL* (%)	*PI* (%)	*LI*	*CI*	*A*
1. Hunstanton Till (Holkham)							
Max	18.6	23	40 (IP)	23	0.07	0.97 (S)	1.00 (N)
Min	16.8	15	34 (L)	15	−0.19	0.89 (S)	0.75(N)
Mean	17.8	18	37 (IP)	20	−0.02	0.92 (S)	0.85 (N)
2. Marly Drift (Weybourne)							
Max	25.2	21	45 (IP)	26	0.48	0.85 (S)	0.50 (I)
Min	22.4	18	32 (L)	14	0.15	0.50 (F)	0.40 (I)
Mean	23.6	20	37 (IP)	18	0.32	0.68 (F)	0.45 (I)
3. Contorted Drift (Trimingham)							
Max	18.9	18	29 (L)	13	0.33	0.86 (S)	0.80 (N)
Min	13.2	9	19 (L)	8	0.07	0.72 (F)	0.65 (I)
Mean	15.6	14	25 (L)	11	0.16	0.78 (S)	0.75 (N)
4. Cromer Till (Happisburgh)							
Max	15.8	20	40 (IP)	24	−0.16	1.16 (VS)	0.95 (N)
Min	11.9	14	27 (L)	13	−0.18	0.98 (S)	0.65 (I)
Mean	13.2	17	35 (IP)	19	−0.17	1.09 (VS)	0.80 (N)

B. Holderness

	m	*PL* (%)	*LL* (%)	*PI* (%)	*LI*	*CI*	*A*
1. Hessle Till (Dimlington, Hornsea)							
Max	26.6	26	53 (H)	32	0.072	1.147 (VS)	2.10 (A)
Min	18.5	20	38 (I)	17	−0.019	0.794 (S)	0.96 (N)
Mean	22.6	22	47 (I)	25	0.044	0.972 (S)	1.24 (N)
2. Withernsea Till (Dimlington)							
Max	19.3	21	39 (I)	20	−0.276	1.016 (VS)	1.21 (N)
Min	12.3	15	22 (L)	12	−0.095	0.828 (S)	0.72 (I)
Mean	16.9	18	34 (L)	17	−0.164	0.986 (S)	0.93 (N)
3. Skipsea Till (Dimlington)							
Max	18.2	19	36 (I)	18	−0.294	1.288 (VS)	0.67 (I)
Min	13.5	14	20 (L)	9	−0.044	0.978 (S)	0.51 (I)
Mean	15.5	16	30 (L)	14	−0.188	1.108 (VS)	0.56 (I)
4. Basement Till (Dimlington)							
Max	20.4	23	42 (I)	22	−0.158	1.081 (VS)	0.59 (I)
Min	15.6	16	28 (L)	12	−0.032	0.984 (S)	0.53 (I)
Mean	17.0	20	36 (I)	19	−0.127	1.009 (VS)	0.55 (I)

Liquid limit
LP = low plasticity, less than 35%.; IP = intermediate plasticity, 35 to 50%.; H = high plasticity, 50–70%.

Consistency index
VS = very stiff, above 1.; S = stiff, 0.75 to 1.; F = firm, 0.5 to 0.75.

Activity
I = inactive, less than 0.75.; N = normal, 0.75 to 1.25.; A = active, over 1.25.

Basement, Skipsea and Withernsea Tills the natural moisture content is generally slightly below that of the plastic limit and hence they commonly have negative liquidity indices. In heavily overconsolidated unweathered lodgement tills the natural moisture content is generally rather low and slightly below that of the plastic limit. Furthermore the liquidity indices of such tills typically lie within the range −0.1 to −0.35.

The reverse situation exists in the case of the Drifts and Hessle Till. As far as the Marly Drift is concerned the high content of chalky matter may help explain the fact that this material possesses a high natural moisture content, which exceeds the plastic limit. On the other hand this situation in the Contorted Drift may be attributable to its low to intermediate plasticity. The frequent positive liquidity index value, together with higher liquid and plastic limits of the Hessle Till compared with the other tills of Holderness, help to confirm the view that this till is a weathered derivative of the two tills below it.

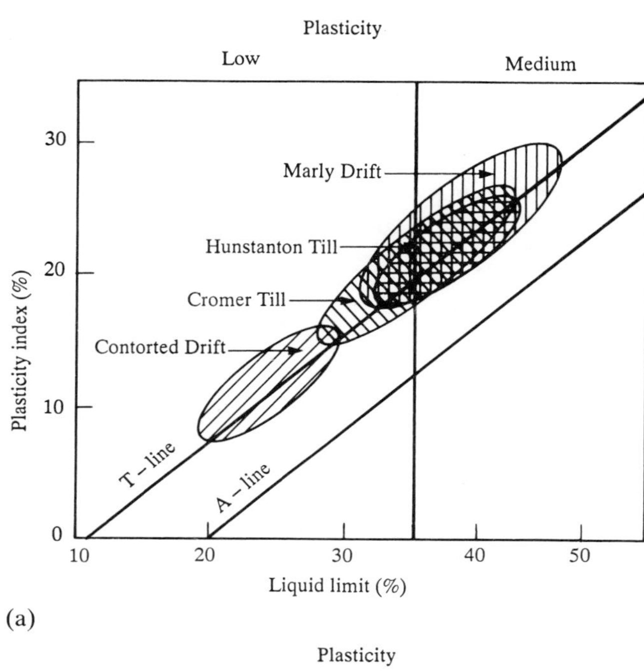

(a)

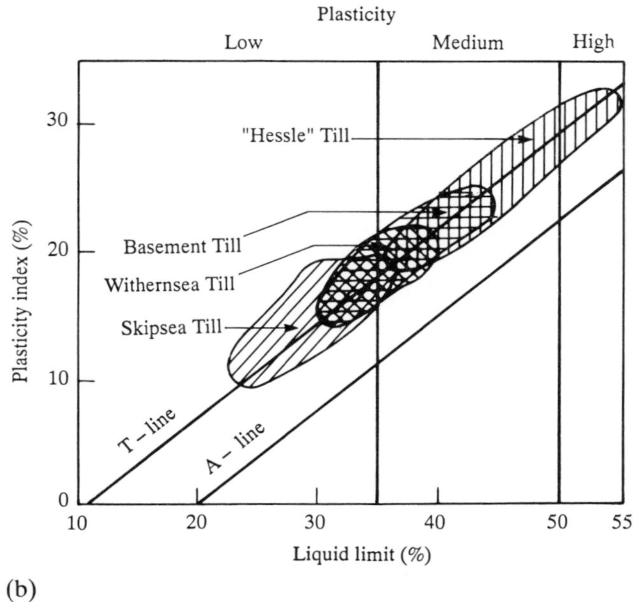

(b)

Figure 4.3 *Plasticity data for tills of* (a) *north Norfolk (after Bell, 1991);* (b) *Holderness (after Bell and Forster, 1991)*

Table 4.4 *Strength of tills of north Norfolk and Holderness (after Bell, 1991; Bell and Forster, 1991)*

A. North Norfolk

	Unconfined compressive strength (kN/m^2)			*Cohesion* (kN/m^2)			*Angle of friction* (°)		
	Intact	*Remoulded*	*Sensitivity*	c_u	c'	c_r	ϕ_u	ϕ'	ϕ_r
1. Hunstanton Till (Holkham)									
Max	184	164	1.22 (L)	43	18	4	9	34	28
Min	152	128	1.18 (L)	22	8	0	3	26	21
Mean	158	134	1.19 (L)	29	12	2	5	29	23
2. Marly Drift (Weybourne)									
Max	120	94	1.49 (L)	49	16	0	3	28	25
Min	104	70	1.28 (L)	16	7	0	0	21	16
Mean	110	81	1.34 (L)	27	11	0	1	24	21
3. Contorted Drift (Trimingham)									
Max	180	168	1.67 (L)	46	20	3	10	33	25
Min	124	76	1.08 (L)	20	6	0	3	27	20
Mean	160	136	1.23 (L)	26	11	1	6	30	22
4. Cromer Till (Happisburgh)									
Max	224	188	1.19 (L)	48	19	5	6	32	29
Min	154	140	1.10 (L)	26	12	0	2	26	18
Mean	176	156	1.13 (L)	35	14	3	4	29	23

B. Holderness

	Unconfined compressive strength (kN/m²)			*Direct shear*				*Triaxial*			
	Intact	*Remoulded*	*Sensitivity*	c	ϕ°	c_r	ϕ_r°	c_u	ϕ_u°	c′	ϕ'
1. Hessle Till (Dimlington, Hornsea)											
Max	138	116	1.31 (L)	30	25	3	23	98	8	80	24
Min	96	74	1.10 (L)	16	16	0	13	22	5	10	13
Mean	106	96	1.19 (L)	20	24	1	20	35	7	26	25
2. Withernsea Till (Dimlington)											
Max	172	148	1.18 (L)	38	30	2	27	62	19	42	34
Min	140	122	1.15 (L)	21	20	0	18	17	5	17	16
Mean	160	136	1.16 (L)	26	24	1	21	30	9	23	25
3. Skipsea Till (Dimlington)											
Max	194	168	1.15 (L)	45	38	5	35	50	21	35	36
Min	182	154	1.08 (L)	25	20	0	19	17	10	22	24
Mean	186	164	1.13 (L)	27	26	1	25	29	12	28	30
4. Basement Till (Dimlington)											
Max	212	168	1.27 (L)	47	34	2	30	59	17	42	36
Min	163	140	1.19 (L)	23	20	0	18	22	6	19	20
Mean	186	156	1.21 (L)	29	24	1	23	38	9	34	29

c = cohesion in kN/m² ϕ = angle of friction. *L* = low sensitivity

When the plasticitiy indices of these tills are plotted against their liquid limits they all fall well above the A-line on the plasticity chart (Figure 4.3). In fact they tend to fall along the T-line of Boulton (1976), indicating the unsorted nature of the tills, their somewhat different locations reflecting the composition and particle size distribution of their matrix material. Again the position of the 'Hessle' Till along the T-line suggests that this is weathered till (Sladen and Wrigley, 1983). Generally these tills are inactive, the Hessle Till once more proving the exception.

4.3 Compressibility and strength of tills

The compressibility and consolidation characteristics of tills are principally determined by the clay content. For example, the value of compressibility index tends to increase linearly with increasing clay content whilst for moraines of very low clay content, less than 2%, this index remains about constant ($C_c = 0.01$).

Klohn (1965) noted that dense, heavily overconsolidated till is relatively incompressible and that when loaded undergoes very little settlement, most of which is elastic. For the average structure such elastic compressions are too small to consider and can therefore be ignored. However, for certain structures they are critical and their magnitude must be estimated prior to construction. Klohn carried out a series of plate loading tests which indicated that the modulus of elasticity of the till deposit concerned was very high, being of the order 1500 MN/m^2. Observations subsequently, taken on all major structures, indicated that settlement occurred almost instantaneously on application of load.

In another survey of dense till, Radhakrishna and Klym (1974) found the undrained shear strength, as obtained by pressuremeter and plate loading tests, to average around 1.6 MN/m^2, while the values from triaxial tests ranged between 0.75 and 1.3 MN/m^2. The average values of the initial modulus of deformation were around 215 MN/m^2 which was approximately twice the laboratory value. These differences between field and laboratory results were attributed to stress relief of material on sampling and sampling disturbance.

Significantly lower values of shear strength were obtained by Bell (1991) and Bell and Forster (1991) for tills from north Norfolk and from Holderness when tested in unconfined compression. When remoulded and tested again all tills experienced only small losses in strength on remoulding and therefore are of low sensitivity. These authors also subjected these tills to a series of shear box and triaxial tests, a summary of the results being provided in Table 4.4.

The presence of fissures influences the shear strength of till and therefore its stability. For example, McGown *et al.* (1977) showed that the opening of fissures sympathetically oriented to cut slopes, and softening of till along fissures as a result of weathering, were responsible for small slip failures. In other words, these two factors gave rise to a rapid reduction of undrained shear strength along the fissures. In fact McGown *et al.* (1977) found that the undrained shear strength of fissures in till may be as little as one-sixth that of the intact soil. They emphasized that the distinction between the nature of the various fissure coatings (sand, silt or clay-size material) is of critical importance in determining the shear strength behaviour of the fissured soil mass. Deformation and permeability also are controlled by the nature of the fissure surface and coatings.

Table 4.5 *A weathering scheme for Northumberland lodgement tills (after Eyles and Sladen, 1981)*

Weathering state	*Zone*	*Description*	*Maximum depth* (m)
Highly weathered	IV	Oxidized till and surficial material Strong oxidation colours High rotten boulder content Leaching of most primary carbonate Prismatic gleyed jointing Pedological profile usually leached brown earth	3
Moderately weathered	III	Oxidized till Increased clay content Low rotten boulder content Little leaching of primary carbonate Usually dark brown or dark red brown Base commonly defined by fluvioglacial sediments	8
Slightly weathered	II	Selective oxidation along fissure surfaces where present, otherwise as Zone I	10
Unweathered	I	Unweathered till No post-depositionally rotted boulders No oxidation No leaching of primary carbonate Usually dark grey	

4.4 Character of weathered tills

Eyles and Sladen (1981) recognized four zones of weathering within the soil profile of lodgement till in the coastal area of Northumberland (Table 4.5). As the degree of weathering of the till increases, so does the clay fraction and moisture content. This, in turn, leads to changes in the liquid and plastic limits and in the shear strength (Table 4.6 and Figure 4.4).

4.5 Fluvio-glacial deposits; stratified drift

Stratified deposits of drift are often subdivided into two categories, namely, those deposits which accumulate beyond the limits of the ice, forming in streams, lakes or seas and those deposits which develop in contact with the ice. The former type are referred to as pro-glacial deposits whilst the latter are termed ice-contact deposits.

Most meltwater streams which deposit outwash fans do not originate at the snout of a glacier but from within or upon the ice. Many of the streams which flow through a glacier have steep gradients and are therefore efficient transporting agents, but when they emerge at the snout, they do so on to a shallower incline and deposition results. Outwash deposits are typically cross bedded and range in size

Table 4.6 *Typical geotechnical properties for Northumberland lodgement tills (after Eyles and Sladen, 1981)*

Property	*Weathering zones*	
	I	*III & IV*
Bulk density (Mg/m^3)	2.15–2.30	1.90–2.20
Natural moisture content (%)	10–15	12–25
Liquid limit (%)	25–40	35–60
Plastic limit (%)	12–20	15–25
Plasticity index	0–20	15–40
Liquidity index	−0.20 to −0.05	III −0.15 to +0.05 IV 0 to +30
Grading of fine (<2 mm) fraction		
% clay	20–35	30–50
% silt	30–40	30–50
% sand	30–50	10–25
Average activity	0.64	0.68
c' (kN/m^2)	0–15	0–25
ϕ' (degrees)	32–37	27–35
ϕ'_r (degrees)	30–32	15–32

from boulders to coarse sand. When first deposited the porosity of these sediments varies from 25 to 50%. They are therefore very permeable and hence can resist erosion by local run-off. The finer silt–clay fraction is transported further downstream. Also in this direction an increasing amount of stream alluvium is contributed by tributaries so that eventually the fluvio-glacial deposits cannot be distinguished. Most outwash masses are terraced.

Valley trains are outwash deposits which are confined within long narrow valleys. Since deposition occurs more rapidly at the centres of valleys than at the sides, they are thickest there. If outwash quickly accumulates in a valley it may eventually dam tributary streams so that small lakes form along the sides of the main valley. Such deposition also may bury small watersheds and divert pro-glacial streams.

Deposition which takes place at the contact of a body of ice is frequently sporadic and irregular. Locally the sediments possess a wide range of grain size, shape and sorting. Most are granular and variations in their engineering properties reflect differences in particle-size distribution and shape. Deposits often display abrupt changes in lithology and consequently in relative density. They are characteristically defomed since they sag, slump or collapse as the ice supporting them melts.

Kame terraces are deposits by meltwater streams which flow along the contact between the ice and the valley side. The drift is principally derived from the glacier although some is supplied by tributary streams. They occur in pairs, one each side of the valley.

Kames are mounds of stratified drift which originate as small deltas or fans built against the snout of a glacier where a tunnel in the ice, along which meltwater travels, emerges. Many kames do not survive deglaciation for any appreciable period of time.

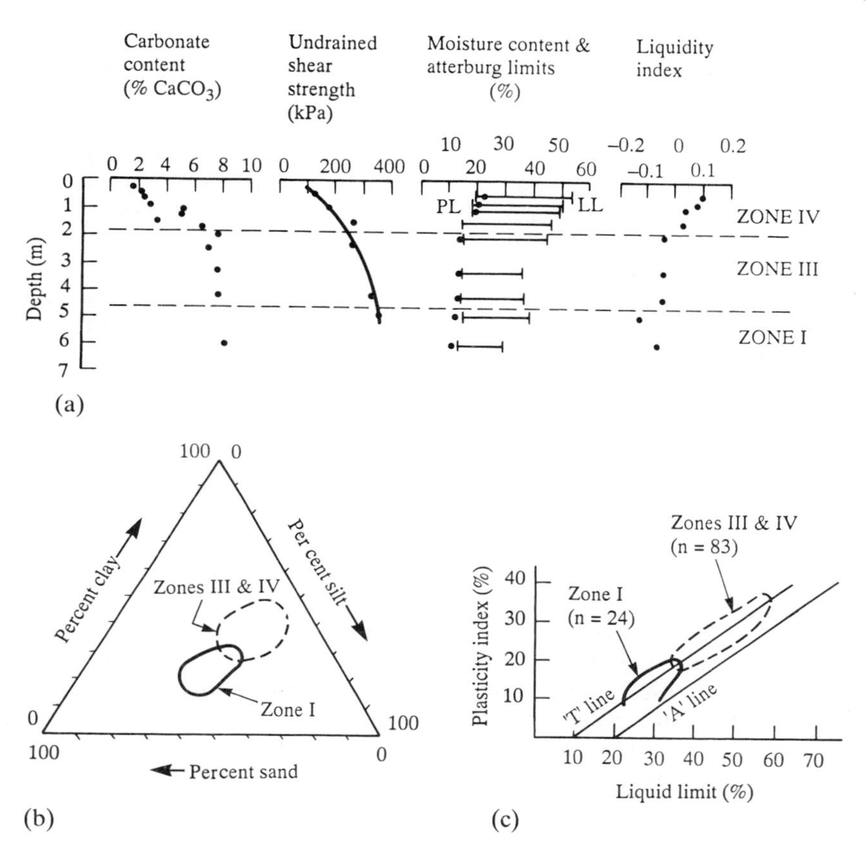

Figure 4.4 (a) *Northumberland lodgement tills: carbonate content, undrained strength, moisture content, Atterberg indices and liquidity index versus depth at a single representative site (Sandy Bay).* (b) *Particle size distribution envelopes for weathered and unweathered Northumberland lodgement tills.* (c) *A plasticity chart showing envelopes for weathered and unweathered Northumberland lodgement tills (n* = number of determinations) (after Eyles and Sladen, 1981)

Eskers are long, narrow, sinuous, ridge-like masses of stratified drift which are unrelated to surface topography. They represent sediments deposited by streams which flowed within channels in a glacier. Although eskers may be interrupted their general continuity is easily discernible and indeed some may extend lengthwise for several hundred kilometres. Eskers may reach up to 50 m in height, whilst they range up to 200 m wide. Their sides are often steep. Eskers are composed principally of sands and gravels, although silts and boulders are found within them. These deposits are generally cross bedded.

Lenses of openwork gravel may mean that the permeability of a fluvio-glacial deposit is higher than expected. The bearing capacity of openwork gravel is generally less than surrounding sandy gravel because of the higher void space. On loading this may give rise to differential settlement.

Lacustrine clays and silts generally occur as layers interbedded with fluvio-glacial sand and gravel deposits, but may also occur as pockets and lenses. Such inclusions may lead to differential settlement or even bearing capacity failure. These interbedded deposits may reduce vertical permeability significantly. Inclusions of till may have similar effects.

The most familiar pro-glacial deposits are varved clays. These deposits accumulated in pro-glacial lakes and are generally characterized by alternating laminae of finer and coarser grain size, each couplet being termed a varve. The thickness of the individual varve is frequently less than 2 mm although much thicker layers have been noted. Generally the coarser layer is of silt size and the finer of clay size. In a survey of the varved clays of the Elk Valley, British Columbia, George (1986) noted that the clay layers invariably contained very fine silty to sandy laminations and/or lenses. Piping occurred in some of the thicker layers (up to 0.5 m in thickness) of silt and fine sand. According to Bell and Coulthard (1991) clay is the dominant particle size in the Tees Laminated Clay comprising between 44% and 76% with an average of 61%. The average silt content is 37%, varying from 27% to 43%, and the fine sand fraction ranges up to 10%.

Taylor *et al.* (1976) showed that in varved clays from Gale Common in Yorkshire, the clay minerals were well orientated around silt grains such that at boundaries between silty partings and matrix the clay minerals tended to show a high degree of orientation parallel to the laminae. A similar situation occurs in the Tees Laminated Clay and in the varved clay of the Elk Valley, British Columbia. In the latter the fabric of the clay material typically possesses a preferred orientation parallel to bedding and consists of elongated clay platelets and microaggregates. Voids are present due to the edge-to-face contacts betweeen particles.

Usually very finely comminuted quartz, feldspar and mica form the major part of varved clays rather than clay minerals. For example, the clay mineral content may be as low as 10%, although instances where it has been as high as 70% have been recorded. What is more montmorillonitic clay has also been found in varved clays.

The results of X-ray diffraction analysis of specimens of Tees Laminated Clay obtained by Bell and Coulthard (1991) indicate that its bulk mineralogical composition is dominated by illite and kaolinite with lesser amounts of chlorite. The quartz content of the bulk specimens ranged from 4% to 26%, reflecting the differing thicknesses of the silt–fine sand layers, and the carbonate content was always less than 15%. Traces of potash feldspar, muscovite and expandable mixed-layer clays were also recorded. Similarly George (1986) found that kaolinite and illite are the dominant minerals in the clay material of the varved clay of the Elk Valley, British Columbia, with minor amounts of mixed-layer clays. Calcite, dolomite and quartz also are present in the clay fraction.

Varved clays tend to be normally consolidated or lightly overconsolidated, although it is usually difficult to make the distinction. In many cases the precompression may have been due to ice loading. However, Saxena *et al.* (1978) reported that the upper part of the varved clay of Hackensack Valley, New Jersey was highly overconsolidated. This they attributed to the effects of fluctuating water levels or to desiccation.

The two normally discrete layers formed during the deposition of the varve present an unusual problem in that it may invalidate the normal soil mechanics analyses, based on homogeneous soils, from being used. As far as the Atterberg limits are concerned assessment of the liquid and plastic limits of a bulk sample may not yield a representative result. However, Metcalf and Townsend (1961) suggested that the maximum possible liquid limit obtained for any particular varved deposit must be that of the clayey portion, whereas the minimum value must be that of the silty portion. Hence they assumed that the maximum and minimum values recorded for any one deposit approximate to the properties of the individual layers. The range of liquid limits for varved clays tends to vary between 30 and 80% whilst that of plastic limit often varies between 15 and 30%. These limits, obtained from varved clays in Ontario, allow the material to be classified as inorganic silty clay of medium to high plasticity. In some varved clays in Ontario the natural moisture content would appear to be near the liquid limit. They are consequently soft and have sensitivities generally of the order of 4. Since triaxial and unconfined compression tests tend to give very low strains at failure, around 3%, Metcalf and Townsend presumed that this indicated a structural effect in the varved clays. The average strength reported was about 40 kN/m^2, with a range of 24 to 49 kN/m^2. The effective stress parameters of apparent cohesion and angle of shearing resistance range from 5 to 19 kN/m^2, and 22 to 25° respectively.

Some of the geotechnical properties of the varved clays of the Elk Valley, British Columbia, are given in Table 4.7. The failure envelopes produced by consolidated undrained triaxial tests possessed both over- and normally consolidated segments which respectively gave values of cohesion and angle of friction as $c' = 53.2$ kN/m^2, $\phi' = 4.7°$; and $c' = 0$, $\phi = 14.8°$. The overconsolidated effect was attributed to cementation bonding. This is supported by the values of preconsolidation pressure that were obtained (Table 4.7). These values, according to George (1986), are significantly greater than any overburden stress the site has undergone throughout its geological history.

Turning to the Tees Laminated Clay Bell and Coulthard (1991) showed that it displays a wide range of plastic and liquid limit values (Table 4.8). The plastic limits are usually less than the natural moisture content and so the liquidity indices normally are positive. From the plasticity chart (Figure 4.5) it can be seen that these clays range from low to very high plasticity. The wide spread reflects the different proportions of clay and silt in the specimens tested, obviously the higher the content of the latter, the lower the plasticity. However, most of the results show that the Tees Laminated Clay is of high plasticity. Indeed the average value of liquid limit is 56%. All the specimens tested fell above the A-line and straddled the T-line (Boulton, 1976). In fact the distribution of results for this clay bears a close similarity to those of other laminated clays in northern England (Taylor *et al.*, 1976)

Table 4.7 *Some properties of the varved clays of the Elk Valley, British Columbia (After George, 1986)*

(a) Basic soil properties

Varved clay description (typical)	yellowish brown; typical thicknesses clay: 33–37 mm silt: 2–8 mm usually fine silty to sandy laminations and lenses (<0.5 mm thick) in clay layers, 2–14 mm apart	
Soil properties	Clay	Bulk
USCS Classification	CL	CL
Natural moisture content (%)	41.0	35.4
Liquid limit (%)	39.7	34.3
Plastic limit (%)	24.2	22.4
Plasticity index (%)	15.5	11.9
Liquidity index	1.08	1.09
% Clay (<0.002 mm)	43	–
Activity	0.36	–

(b) Results of oedometer tests

Parameter	*Value*
Compression index	0.405–0.587* (0.496)†
Recompression index	0.030–0.060 (0.045)
Effective preconsolidation pressure (kN/m^2)	
–Schmertmann	250–330 (290)
–Casagrande	350–400 (375)
Initial void ratio	1.09–1.14 (1.02)
Unit weight (kN/m^3)	17.46–18.23 (17.85)
Natural water content (%)	37.9–41.0 (39.5)

* Observed range for two tests
† Average value

and Norfolk (Kazi and Knill, 1969). The Tees Laminated Clay is inactive, it having an average value of 0.54 and ranging from 0.47 to 0.65.

Quick undrained triaxial tests carried out on the Tees Laminated Clay indicated that it tended to fail at less than 10% strain (Bell and Coulthard, 1991). In some of those specimens where the test was continued beyond failure the stress, after attaining a peak, tended to fall away quickly. In others, however, this was not the case (Figure 4.5). The average shear strength of the Tees Laminated Clay tends to be about 60 kN/m^2, although there is a wide variation – from the upper 20s up to 102 kN/m^2. The residual strength is between 20 and 30 kN/m^2 lower than the peak strength. Some specimens of laminated clay were remoulded and retested to determine their sensitivity. This ranged between 3.5 and 4.2 with an average value of 4.0. Hence these clays are medium sensitive to sensitive and undergo a notable

Table 4.8 *Index properties of Tees Laminated Clay (after Bell and Coulthard, 1991)*

	Moisture content (%)	*Plastic limit (%)*	*Liquid* limit (%)*	*Plasticity index*	*Liquidity index*	*Linear shrinkage (%)*	*Consistency index†*
Max	35	31	78 (VH)	49	0.35	14	0.89 (S)
Min	22	18	29 (L)	16	−0.12	9	0.42 (SO)
Mean	30	26	56 (H)	33	0.15	11	0.70 (F)

* L = low plasticity (<35%); I = Intermediate plasticity (35–50%)
H = High plasticity (50–75%); VH = Very high plasticity (70–90%)
† So = Soft (<0.5); F = Firm (0.5–0.75); S = Stiff (0.75–1.0).

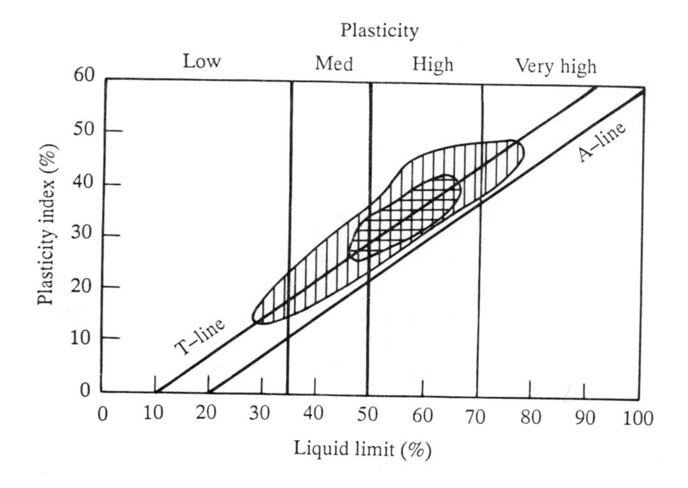

Figure 4.5 *Plasticity chart showing distribution of over 100 specimens of Tees Laminated Clay. Inner shaded area contains the results form over two thirds of the tests (after Bell and Coulthard, 1991)*

reduction in strength when remoulded. However, remoulded strength increases with time (Figure 4.5).

Values of the coefficient of consolidation obtained for Tees Laminated Clay ranged from 3.2 to 7.9 m^2/year, with an average of 5.54 m^2/year. These values are characteristic of clay of medium plasticity. Those of volume compressibility fall between 0.13 and 0.41 m^2/MN with an average of 0.21 m^2/MN, which suggests a slightly overconsolidated clay. The swell index and compression index for the Tees Laminated Clay vary between 0.11 and 0.68, and 0.24 and 1.55 respectively, which

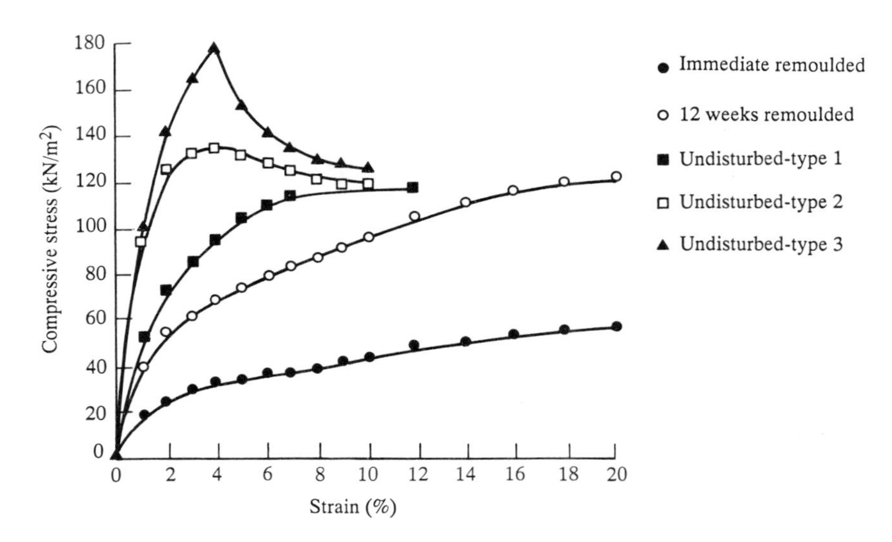

Figure 4.6 *Examples of stress–strain curves of undisturbed and remoulded Tees Laminated Clay tested in quick undrained triaxial conditions (after Bell and Coulthard, 1991)*

are again characteristic of a medium to high plasticity clay. In addition the ratio of the swell index and the compression index provides an indication of the swelling properties of a clay soil. If the ratio is less than 0.2, then the soil is non-swelling, whereas if it exceeds 0.2 the soil is medium to high swelling. The actual ratios range from 0.35 to 0.45, suggesting that this clay has a potential for swelling. The average preconsolidation pressure for this clay was found to be 253 kN/m^2.

4.6 Frozen ground phenomena in periglacial environments

Frozen ground phenomena are found in regions which experience a tundra climate, that is, in those regions where the winter temperatures rarely rise above freezing point and the summer temperatures are only warm enough to cause thawing in the upper metre or so of the soil. Beneath the upper or active zone the subsoil is permanently frozen and so is known as the permafrost layer. Because of this summer meltwater cannot seep into the ground, the active zone then becomes waterlogged and the soils on gentle slopes are liable to flow. Layers or lenses of unfrozen ground termed taliks may occur, often temporarily, in the permafrost.

Permafrost is an important characteristic, although it is not essential to the definition of periglacial conditions. It covers 20% of the Earth's land surface and during Pleistocene times it was developed over an even larger area (Figure 4.7). Ground cover, surface water, topography and surface materials all influence the distribution of permafrost. The temperature of perennially frozen ground below

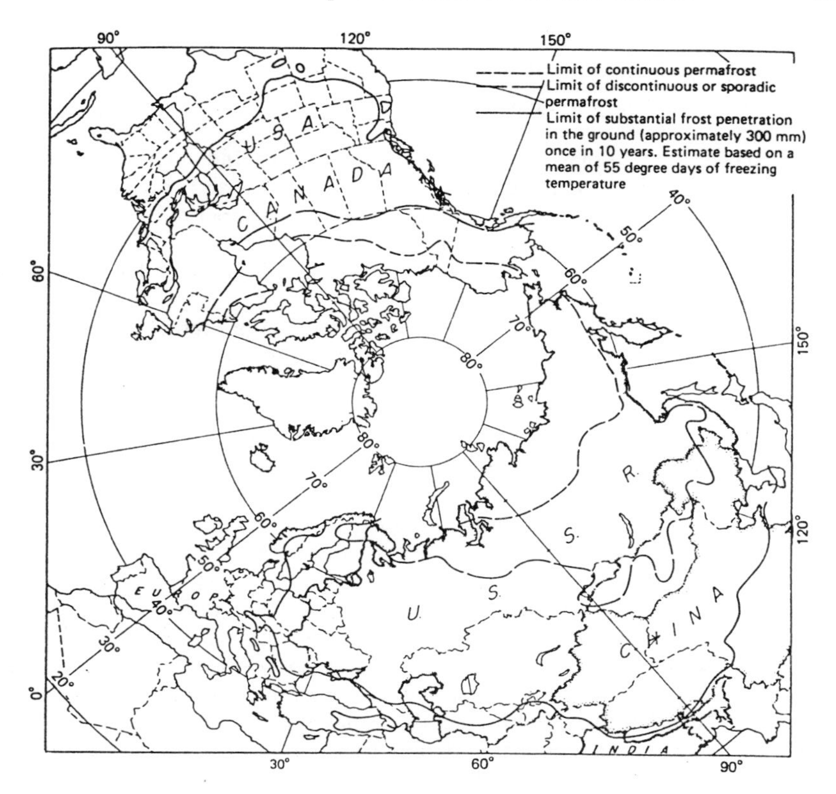

Figure 4.7 *Distribution of seasonally and perennially frozen ground (permafrost) in the northern hemisphere (from Andersland, 1987)*

the depth of seasonal change ranges from slightly less than 0°C to −12°C. Generally the depth of thaw is less, the higher the latitude. It is at a minimum in peat or highly organic sediments and increases in clay, silt and sand to a maximum in gravel where it may extend to 2 m in depth.

Prolonged freezing gives rise to shattering in the frozen layer, fracturing taking place along joints and cracks. For example, frost shattering, due to ice action in Pleistocene times, has been found to extend to depths of 30 m in the Chalk.

The rock concerned suffers a reduction in bulk density and an increase in deformability and permeability. Fretting and spalling are particularly rapid where the rock is closely fractured. Frost shattering may be concentrated along certain preferred planes, if joint patterns are suitably oriented. Preferential opening takes place most frequently in those joints which run more or less parallel with the ground surface. Silt and clay frequently occupy the cracks in frost shattered ground, down to appreciable depth, having been deposited by meltwater. Their presence may cause stability problems. If the material possesses a certain range of grain size and permeability, the freezing of intergranular water causes expansion and disruption of previously intact material and frost shattering may be very pronounced.

Frozen soils often display a polygonal pattern of cracks. Individual cracks may be 1.2 m wide at their top, may penetrate to depths of 10 m and may be some 12 m apart. They form when, because of exceptionally low temperatures, shrinkage of the ground occurs. Ice wedges occupy these cracks and cause them to expand. When the ice disappears an ice wedge pseudomorph is formed by sediment, frequently sand, filling the crack.

Ground may undergo notable disturbance as a result of mutual interference of growing bodies of ice or from excess pore pressures developed in confined water-bearing lenses. Involutions are plugs, pockets or tongues of highly disturbed material, generally possessing inferior geotechnical properties, which have been intruded into overlying layers. They are formed as a result of hydrostatic uplift in water trapped under a refreezing surface layer. They usually are confined to the active layer.

Ice wedges and involutions usually mean that one material suddenly replaces another. This can cause problems in shallow excavations.

The movement downslope as a viscous flow of saturated rock waste is referred to as solifluction. It is probably the most significant process of mass wastage in tundra regions. Solifluction deposits commonly consist of gravels, which are characteristically poorly sorted, sometimes gap-graded, and poorly bedded. These gravels consist of fresh, poorly worn, locally derived material. Individual deposits are rarely more than 3 m thick and frequently display flow structures.

Sheets and lobes and solifluction debris, transported by mudflow activity, are commonly found at the foot of slopes. These materials may be reactivated by changes in drainage, by stream erosion, by sediment overloading or during construction operations. Solifluction sheets may be underlain by slip surfaces, the residual strength of which controls their stability.

4.7 Quick clays

The material of which quick clays are composed is predominantly smaller than 0.002 mm but many deposits seem to be very poor in clay minerals, containing a high proportion of ground-down, fine quartz. For instance, it has been shown that quick clay from St Jean Vienney consists of very fine quartz and plagioclase. Indeed examination of quick clays with the scanning electron microscope has revealed that they do not possess clay based structures, although such work has not lent unequivocal support to the view that non-clay particles govern the physical properties.

Cabrera and Smalley (1973) suggested that such deposits owe their distinctive properties to the predominance of short-range interparticle bonding forces which they maintained were characteristic of deposits in which there was an abundance of glacially produced, fine non-clay minerals. In other words they contended that the ice sheets supplied abundant ground quartz in the form of rock flour for the formation of quick clays. Certainly quick clays have a restricted geographical distribution, occurring in certain parts of the northern hemisphere which were subjected to glaciation during Pleistocene times.

Gillott (1979) has shown that the fabric and mineralogical composition of sensitive soils from Canada, Alaska and Norway are qualitatively similar. He pointed out that they possess an open fabric, high moisture content and similar index properties (Table 4.9).

An examination of the fabric of these soils revealed the presence of aggregations. Granular particles, whether aggregations or primary minerals, are rarely in direct contact, being linked generally by bridges of fine particles. Clay minerals usually are non-oriented and clay coatings on primary minerals tend to be uncommon, as are cemented junctions. Networks of platelets occur in some soils. Primary minerals, particularly quartz and feldspar, form a higher than normal proportion of the clay size fraction and illite and chlorite are the dominant phyllosilicate minerals. Gillott noted that the presence of swelling clay minerals varies from almost zero to significant amounts.

The open fabric which is characteristic of quick clays has been attributed to their initial deposition, during which time colloidal particles interacted to form loose aggregations by gelation and flocculation. Clay minerals exhibit strongly marked colloidal properties and other inorganic materials such as silica behave as colloids when sufficiently fine grained. Gillott (1979) suggested that the open fabric may have been retained during very early consolidation because it remained a near equilibrium arrangement. Its subsequent retention to the present day may be due to mutual interference between particles and buttressing of junctions between granules by clay and other fine constituents, precipitation of cement at particle contacts, low rates of loading, and low load increment ratio.

Quick clays often exhibit little plasticity, their plasticity indices at times varying between 8 and 12%. Their liquidity index normally exceeds 1, and their liquid limit is often less than 40%. Quick clays are usually inactive, their activity frequently being less than 0.5. The most extraordinary property possessed by quick clays is their very high sensitivity. In other words a large proportion of their undisturbed strength is permanently lost following shear. The small fraction of the original strength regained after remoulding may be attributable to the development of some different form of interparticle bonding. The reason why only a small fraction of the original strength can ever be recovered is because the rate at which it develops is so slow. As an example the Leda Clay is characterized by exceptionally high sensitivity, commonly between 20 and 50, and a high natural moisture content and void ratio, the latter is commonly about 2. It has a low permeability, being around 10^{-10} m/s. The plastic limit is around 25%, with a liquid limit about 60%, and undrained shear strength of 700 kN/m^2. When subjected to sustained load an undrained triaxial specimen of Leda Clay exhibits a steady time-dependent increase in both pore pressure and axial strain. Continuing undrained creep may often result in a collapse of the sample after long periods of time have elapsed.

In a review of the mineralogy and physical properties of sensitive clays from eastern Canada, Locat *et al.* (1984) found that plagioclase was the dominant

Table 4.9 *Engineering properties of sensitive soils (after Gillott 1979)*

*Location**	*Depth (m)*	*Natural moisture content (%)*	*Preconsolidation pressure* (kN/m^2)	*Undrained strength* (kN/m^2)	*Sensitivity*	*Liquid limit* (%)	*Plastic limit* (%)	*Liquidity index*	*Activity*
O	13.7	60	450	160	–	49	23	1.4	0.35
Q	5.2	75	150	50	–	70	26	1.1	0.64
Q	14.3	81	150	50	–	65	28	1.4	0.45
O	2.6	65	60	20	100	55	22	1.3	0.73
Q	12.2	28	590	230	–	23	16	1.7	0.18
O	5.2	78	320	120	–	65	28	1.3	0.44
BC	20.1	38	–	20	30	28	22	2.7	0.22
BC	14.0	29	–	–	4	23	16	1.9	0.33
BC	35.4	37	–	60	5	28	23	2.8	0.17
A	61.3	17	–	–	–	26	21	0.8	–
A	60.7	–	–	–	–	23	20	–	–

* O = Ontario, Q = Quebec, BC = British Columbia, A = Alaska

Table 4.10 *Geotechnical properties of sensitive clays from eastern Canada (After Locat et al., 1984) Strength (C_u, C_{ur}), sensitivity (S_t), pore water salinity (S) index properties (LL, PL, PI, LI), specific surface area (SS), cation exchange capacity (G.E.C.), amount of phyllosilicates and amorphous minerals (P)*

Site	c_u kN/m^2	c_{ur} kN/m^2	S_t	*S* (g/L)	*w* (%)	*LL* (%)	*PL* (%)	*PI* (%)	*LI*	<2 μm (%)	*SS* (m^2/g)	*G.E.C.* (meq./100 g)	*P* (%)	<2 μm − *P* (%)
Grande-Baleine	32	0.1	282	0.7	58	36.0	24.0	12.0	2.8	74	26	14.4	8.5	65.5
Grande-Baleine	24	0.2	120	–	46	35.8	23.0	12.8	1.8	66	27	–	15	51.0
Grande-Baleine	23	0.2	114	0.7	58	36.9	25.0	11.9	2.7	70	24	7.5	–	–
Olga	23	1.0	19	0.3	83	73.4	26.0	47.4	1.2	94	85	44.4	34.5	59.5
Olga	23	0.9	25	0.3	85	66.7	28.0	38.7	1.5	82	48	10.0	–	–
Olga	28	1.19	24	–	88	67.6	27.8	39.8	1.5	88	55	–	31.3	56.7
St. Marcel	81	0.8	24	<2	81	64.0	28.0	36.0	1.5	85	67	21.9	26.7	58.3
St. Marcel	–	1.21	–	<2	82	61.8	25.8	36.0	1.6	85	58	–	24.9	60.0
St. Léon	80	3.4	24	12	58	61.0	23.5	37.5	0.9	74	56	14.1	32.2	41.8
St. Alban	80	2.2	37	2.0	40	36.0	21.0	15.0	1.3	43	46	11.3	33.2	9.8
St. Barnabé	157	3.1	50	4	48	43.0	20.0	23.0	1.2	50	40	8.5	19.4	30.6
Shawinigan	109	1.8	62	0.6	33	34.0	18.0	16.0	0.9	36	29	6.1	12.8	23.2
Chicoutimi	245	0.5	532	–	33	28.0	18.0	10.0	1.5	55	25	22.5	14.0	41.0
Outardes	109	0.6	181	–	64	34.0	21.5	12.5	3.4	50	30	7.3	–	–
Outardes	130	<0.07	>10^3	–	35	25.0	17.0	8.0	2.3	47	23	–	–	–

c_u = undrained shear strength, c_{ur} = remoulded undrained shear stength, w = natural moisture content, LL = liquid limit, PL = plastic limit, PI = plasticity index, LI = liquidity index.

mineral (from 25 to 48%) followed by quartz, microcline and hornblende. Small quantities of dolomite and calcite were also present. The phyllosilicates (including the clay minerals of which illite is the most common) together with the amorphous materials never represented more than one third of the minerals present in the samples analysed.

As far as the physical properties of these sensitive clays are concerned, with one exception, all the samples fall above the 'A' line on the plasticity chart indicating inorganic material. The nature moisture contents, plastic limits, liquid limits and plasticity indices of the soils are given in Table 4.10 from which it can be seen that their natural moisture contents always exceed the plastic limits and commonly exceed the liquid limits. In such cases their liquidity indices are greater than 1. The relationship between liquidity index, undrained shear strength and sensitivity for silty and clay soils has been referred to in Chapter 3 (see Figures 3.3 and 3.17). Strength decreases and sensitivity increases dramatically as liquidity index increases. This is illustrated by the fall cone strengths quoted by Locat *et al.* (1984), the strength more or less disappearing on remoulding, giving sensitivity values varying from 19 to over 1000. Values of specific surface, cation exchange capacity and the salinity of the pore water also are provided in Table 4.10. The variation in the geotechnical properties of these soils was primarily attributed to their differences in mineralogy and texture.

Quick clays are associated with several serious engineering problems. Their bearing capacity is low, settlement is high and prediction of consolidation of quick clays by the standard methods is unsatisfactory. Slides in quick clays sometimes have proved disastrous and unfortunately the results of slope stability analyses are often unreliable. Dewatering leads to irreversible shrinkage.

Quick clays can liquefy on sudden shock. This has been explained by the fact that if quartz particles are small enough, having a very low settling velocity, and if the soil has a high water content, then the solid–liquid transition can be achieved.

References

Bell, F. G. (1991). 'A survey of the geotechnical properties of some till deposits occurring on the north coast of Norfolk', in *Quaternary Engineering Geology*, Special Publication No. 7, Forster, A., Culshaw, M. G., Cripps, J. C. and Little J. L. (eds.). Geological Society, London, 103–110.

Bell, F. G. and Coulthard, M. C. (1991). 'The Tees Laminated Clay', in *Quaternary Engineering Geology*, Engineering Geology Special Publication No. 7, Forster, A., Culshaw, M. G., Cripps, J. C., Little, J. and Moon, C. F. (eds.). Geological Society, London, 111–180.

Bell, F. G. and Forster, A. (1991). 'The till deposits of Holderness', in *Quaternary Engineering Geology*, Special Publication No. 7, Forster, A., Culshaw, M. G., Cripps, J. C., Little, J. and Moon, C. F. (eds.). Geological Society, London, 339–48.

Boulton, G. S. (1976). 'The development of geotechnical properties in glacial tills', in *Glacial Till – An Interdisciplinary Study*. Legget, R. F. (ed.). Special Publication No. 12, Royal Society of Canada, Ottawa, 292–303.

Boulton, G. S. and Paul, M. A. (1976). 'The influence of genetic processes on some geotechnical properties of glacial tills', *Q. J. Engg Geol.,* **9**, 159–94.

Cabrera, J. G. and Smalley, I. J. (1973) 'Quick clays as products of glacial action: a new approach to their nature, geology, distribution and geotechnical properties', *Engg Geol.*, **7**, 115–33.

Eyles, N. and Sladen, J. A. (1981) 'Stratigraphy and geotechnical properties of weathered lodgement till in Northumberland England', *Q. J. Engg Geol.*, **14**, 129–142.

Fookes, P. G., Gordon, D. L. and Higginbotton, I. E. (1975). 'Glacial landforms, their deposits and engineering characteristics', in *The Engineering Behaviour of Glacial Materials, Proc. Symp. Midland Soil Mechanics and Foundation Engineering Society*, Birmingham University, 18–51.

George, H. (1986). 'Characteristics of varved clays of the Elk Valley, British Columbia, Canada', *Engg Geol.*, **23**, 59–74.

Gillott, J. E. (1979). 'Fabric, composition and properties of sensitive soils from Canada, Alaska and Norway', *Engg Geol.*, **14**, 149–72.

Kazi, A. and Knill, J. L. (1969) 'The sedimentation and geotechnical properties of the Cromer Till between Happisburgh and Cromer, Norfolk', *Q. J. Engg Geol.*, **2**, 63–86.

Klohn, E. J. (1965). 'The elastic properties of dense glacial till deposit', *Can. Geot. J.*, **2**, 115–28.

Locat, J., Lefebvre, G. and Ballivy, G. (1984). 'Mineralogy, chemistry and physical properties interrelationships of some sensitive clays from eastern Canada', *Can. Geot. J.*, **21**, 530–40.

McGown, A. (1971). 'The classification for engineering purposes of tills from moraines and associated landforms', *Q. J. Engg Geol.*, **4**, 115–30.

McGown, A. and Derbyshire, E. (1977). 'Genetic influences on the properties of tills', *Q. J. Engg Geol.*, **10**, 389–410.

McGown, A., Saldivar-Sali, A. and Radwan, A. M. (1974). 'Fissure patterns and slope failures in till at Hurlford, Ayreshire', *Q. J. Engg Geol.*, **7**, 1–26.

McGown, A., Radwan, A. M. and Gabr, A. W. A. (1977). 'Laboratory testing of fissured and laminated soils, *Proc. 9th Int. Conf. Soil Mech. Foundation Engg*, Tokyo, **1**, 205–10.

Metcalf, J. B. and Townsend, D. L. (1961). 'A preliminary study of the geotechnical properties of varved clays as reported in Canadian case records', *Proc. 14th Can. Conf. Soil Mech.*, Section 13, 203–25.

Radhakrishna, H. S. and Klym, T. W. (1974) 'Geotechnical properties of a very dense glacial till', *Can. Geot. J.*, **11**, 396–408.

Saxena, S. K., Helberg, J. and Ladd, C. C. (1978) 'Geotechnical properties of Hachensack Valley varved clays of New Jersey', *Geot. Testing J.*, **1**, No. 3, 148–61.

Sladen, J. A. and Wrigley, W. (1983) 'Geotechnical properties of lodgement till – a review', in *Glacial Geology: An Introduction for Engineers and Earth Scientists*', Eyles, N. (ed.), Pergamon Press, Oxford, 184–212.

Taylor, R. K., Barton, R., Mitchell, J. E. and Cobb, A. E. (1976). 'The engineering geology of Devensian deposits underlying P.F.A. lagoons at Gale Common, Yorkshire', *Q. J. Engg Geol.*, **9**, 195–218.

Chapter 5

Tropical soils

5.1 Introduction

In humid tropical regions weathering of rock is more intense and extends to greater depths than in other parts of the world. Residual soils develop in place as a consequence of weathering, primarily chemical weathering. Consequently climate (temperature and rainfall), parent rock, water movement (drainage and topography), age and vegetation cover are responsible for the development of the soil profile.

The influence of temperature and moisture content on rock weathering in tropical areas has been related by Weinert (1973) to the following climatic index:

$$N = \frac{12E_j}{P_a} \tag{5.1}$$

where E_j is evaporation during the hottest month and P_a is the annual precipitation. The transition from tropical humid conditions in which chemical weathering predominates to hot semi-arid and arid conditions in which physical weathering is important was placed at a value of 5. Deep weathering and chemical decomposition occur in areas where the N value is less than 5. Where it drops below 2 montmorillonite in the residual soil changes to kaolinite. Silica is often lost when the N value is below 1, hydrated oxides of iron and aluminium being significant soil components. Above an N value of 5 the residual cover is relatively thin.

The mineralogy of residual soils is partly inherited from the parent rock from which they were derived and partly produced by the processes of weathering. Hence the mineralogy varies widely, as does grain size and unit weight. The particles and their arrangement evolve gradually as weathering proceeds. The coarser grains present in residual soils normally are inherited from the parent rock and may consist of unweathered quartz or partially weathered feldspars. The latter may degrade during shear. In addition, weathering of parent rock *in situ* may leave behind relict structures which may offer weak bonding even in extremely weathered material. Such bonding can influence engineering behaviour and may make a contribution towards strength and stiffness. The effect of parent rock is more notable in the initial stages of weathering as other pedological factors dominate during later stages of development. For example, residual soils may inherit anisotropic structures from their parent rock. This is common in soils formed from metamorphic rocks, notably schists and gneisses. Relict jointing and bedding, as well as the development of fissures, affect the behaviour of residual soil. Low strength along relict discontinuities may be attributable to particles being coated with low-friction iron/manganese organic compounds. Angles of shearing resistance along such surfaces may be around 15–20°, falling to around 10° when they are slickensided (Anon, 1990).

On the other hand lateritic soils ultimately may be developed above most types of rock. Organic matter is rapidly broken down and rarely occurs beneath a thin surface layer. Weathering takes place mainly by hydrolysis in near-neutral conditions at depths well below the influence of acidic organic decomposition products. Although the influence of weathering can extend to great depth its degree does decrease with depth. As rocks are weathered the bases (Na, K, Ca, Mg) are removed in solution or become part of 2:1 clay minerals such as montmorillonite or illite but rarely of kaolinite (a 1:1 clay mineral). Silica also is removed in solution or combines with other weathered products to form montmorillonite or kaolinite, to some extent depending on the environment. In fact montmorillonite ultimately may give rise to kaolinite. Montmorillonite and illite tend to form in alkaline environments, kaolinite in acidic, whilst free-draining conditions favour the formation of the two latter minerals, montmorillonite developing in poorly drained areas. The 2:1 minerals may move down the profile as dispersed clay particles in suspension (lessivage or illuviation) to form clay-depleted upper horizons and clay-enriched lower horizons, but kaolinite is less susceptible to this process.

Smectites are often present in tropical black clays (vertisols) and if they are present in quantity they can cause large volume changes in response to small seasonal changes in effective stress. Furthermore when surface evaporation is impeded these changes can produce large strains in the surface soils.

Kaolinites are present in fersiallitic, ferruginous and ferralitic soils. Like smectites, kaolinites have a platey form with a low coefficient of interparticle friction but the particles are much larger and less active. Hence, residual soil containing kaolinite has higher strength and a lower compressibility than one with a similar clay fraction containing smectite.

Allophane and halloysite often are present in residual soils. These minerals, according to Vaughan (1990), do not possess a platey shape, neither do they give peak or residual strengths, or form low strength discontinuities after shear at large strains. They give rise to small volume changes on wetting and drying. In addition they can contain water within their structure but this water has no influence on their mechanical behaviour.

Platey clay minerals of low coefficient of friction can be orientated when shearing occurs, thus giving rise to low residual frictional strength and to the development of polished shear surfaces. Such surfaces may be present in residual soils having formed as a result of strains accompanying soil formation, shrinkage and swelling (Lupini *et al.*, 1981).

Iron and aluminium oxides and hydrated oxides released on weathering tend to remain in place. Iron oxides crystallize as hematite when the soil is seasonally desiccated, or as goethite in a constantly humid environment. Hematite gives the soil a red colour, goethite a brown or ochreous colour. Gibbsite is the main aluminium oxide to form.

Usually some interparticle bonding occurs in residual soils. Under moderate weathering, some bonding may be inherited from the parent rock, but in a mature residual soil it is probably due to the effects of crystallization during weathering and mineral alteration, and to the precipitation of cementitious material. In the formation of duricrusts the quantities of cement precipitated may be high enough to cause rock-like material to be developed.

Soil profiles formed by weathering in tropical areas can be represented by a gradational series of zones which pass generally upwards from fresh rock to residual soils (see Chapter 9). The slightly and moderately weathered zones occur above the parent rock and in engineering terms behave in a similar manner to the fresh rock. Anon (1990) referred to these two zones collectively as weathered bedrock. In the highly and completely weathered zones the textural and structural features of the parent rock can still be recognized and this part of the weathered mantle has been termed saprolite. Saprolite, in the engineering context, is regarded as a soil in that its behaviour is controlled by mass deformation. Above this the original textures and structures are lost and this soil zone has been termed solum. It must be emphasized that the profile described is very much idealized and that considerable variations exist in the field.

5.2 Classification of tropical residual soils

The classification of tropical residual soils adopted here is that proposed by the Geological Society (Anon, 1990). This is a pedogenetic classification which considers the breakdown of parent material in humid, semi-arid and arid tropical climates. Three phases of mature residual soil development have been recognized in tropical areas depending on the degree of weathering and the formation of new minerals, namely, fersiallitization, ferrugination and ferrallitization. Their global distribution is shown in Figure 5.1. According to Duchaufour (1982) in subtropical climates with a marked dry season the stage of fersiallitization is rarely exceeded; in semi-arid tropical climates development ceases with ferrugination, and it is only in humid tropical climates that ferrallitization occurs. Another important factor which is taken into account is whether the soil is free or poorly drained. In certain situations combinations of these conditions can give rise to the formation of hardened horizons in the soil profile which are the result of residual accumulation of iron or aluminium oxides, or the precipitation of calcite, dolomite, gypsum or silica. Such hardened horizons tend to occur at or near the ground surface and are referred to as duricrusts. The occurrence of a dry season is essential for the full development of duricrusts. Accordingly the classification formulated by Anon (1990) consists of two basic divisions, namely, the mature soils and the duricrusts, each of which are further subdivided (Table 5.1).

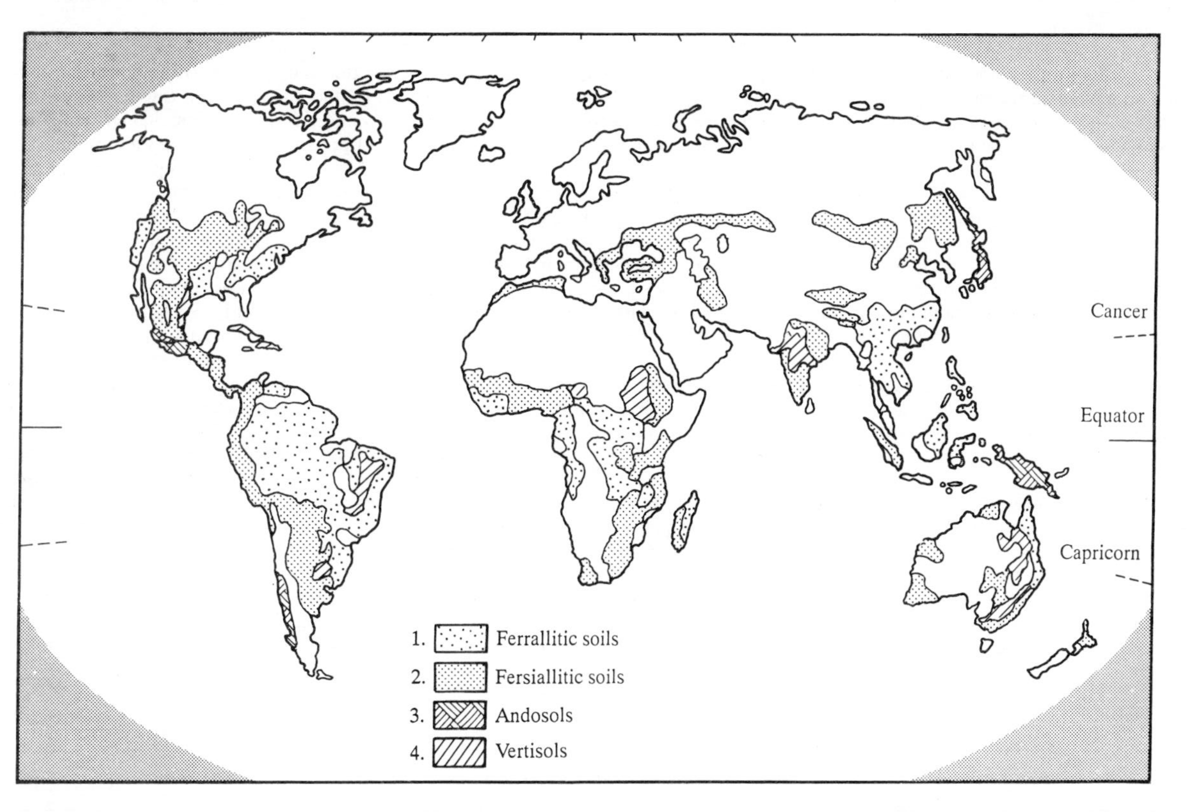

Figure 5.1 *Simplified world distribution of the principal types of tropical residual soils (based on FAO World Soil Map). These broad classes of soils extend beyond the tropics in favourable circumstances, which include high rainfall sub-tropical, continental east coasts (ferrallitic soils), and the west coast/Mediterranean and continental interiors in mid-latitudes (fersiallitic soils)*

Table 5.1 *Classification of residual soils (From Anon, 1990)*

Subdivision	*Definition*	*Classification criteria groups*	*General characteristics*

A: Mature Soils

Ferrallitic soils

Ferrallitic soils form in the hot, humid tropics (annual rainfall greater than 1500 mm, mean temperature 25°C, with little or no dry season), and have profiles several metres thick. All primary minerals except quartz are weathered by hydrolysis in the neutral conditions, and much of the silica and bases are removed in solution. Any remaining silica combines with alumina to form kaolinite, but usually there is an excess of alumina, which forms gibbsite. The exchange capacity of the clay fraction is less than 16 meq/100 g, and usually there is no clay lessivage. Upper horizons of the profile are weakly acidified by organic decomposition products which cause dissolution, chelation and mobilization of iron and aluminium oxides, and decompose any kaolinite present to produce more gibbsite. Depending on the balance between iron and aluminium oxides, ferrallitic soils may be divided into ferrites, in which iron oxides dominate and which occur mainly over rocks low in aluminium, and allites in which aluminium oxides (usually gibbsite) predominate.

Over permeable quartz-rich substrates, high groundwater tables may induce podzolization leading to the removal of iron and aluminium in acid environments and the formation of residual 'white sand' (albic) horizons from 1 to 3 m thick. Typically the white sands are found on coastal plains where sandy sediments occur under high rainfall conditions. Similar materials are also present over acidic crystalline rocks and sandstones on the ancient plateaux of former Gondwanaland. They also occur under savanna conditions where the precipitation is from 600 to 1000 mm, but here they may be relict features of previous (Pleistocene) periods of wetter climate.

Although most ferrallitic soils probably take 10 000 or more years to form, development is more rapid on silica-poor rocks such as basalt than on silica-rich parent materials like granite or quartz-rich sediments. The greater silica content of quartzose parent materials is often reflected also in the presence of kaolinite in the subsoil horizons formed by neutral hydrolysis. This zone of kaolinite formation is often poorly drained and coarsely mottled with white, red and ochreous patches; it may be overlain by a'lateritic' horizon enriched with iron mobilized from acidic near-surface horizons or by changes in the position of the water table. Formation of kaolinite is encouraged by the poor drainage, whereas free drainage removes dissolved silica more rapidly and favours development of gibbsite. The iron-enriched horizon may be moderately or strongly indurated, pisolitic from welding of concretions, or vesicular from precipitation of iron in a polyhedral network of fissures and subsequent removal of the softer material between the fissures.

The only climate suitable for formation of ferrallitic soils is the very hot, humid environment of the tropical evergreen rainforest. Where similar soils extend into drier climatic zones, it is probably the result of climatic shifts during the Quaternary. Conversely, some residual soils in areas currently favourable to ferrallitization are only at the fersiallitic or ferruginous stage; this is either because they have been rejuvenated by erosion on slopes, or because they are formed on recent deposits which have not been exposed to the tropical rainforest environment for long.

Ferralitic (*senso stricto*)	Final phase of development of thick soil profiles in hot, humid climates in which most primary minerals, even quartz, are affected by total hydrolysis. Oxides of iron, aluminium silica and bases are liberated but iron and aluminium are retained in the profiles while the bases and some silica are removed in solution; neoformed kaolinites are poor in silica. Main characteristics are quartz, kaolinite, gibbsite, hematite or goethite. An argillic horizon is generally absent.	(a) Ferrallitic	Profiles retain most of iron and aluminium. Some silica and all bases removed. Profile acidifies rapidly. Neoformed clays are kaolinites. Some free aluminium as gibbsite. Usually reduced to quartz, kaolinite, gibbsite and ferric oxides.
		(1) Ferrallitic (kaolinitic)	Gibbsite absent.
		(2) Ferrallitic (gibbsitic)	Gibbsite dominant.
		(b) Ferrallites	
		(1) Ferrites	On ultrabasic rocks. Poor in aluminium. Silica and magnesium removed. Iron left as goethite.
		(2) Allites	Hydromorphic ferrallites, well drained, permeable; but humid. Iron mobilized by reduction and removed. All that remains is gibbsite – uniformly white.
		(c) Ferrallitic (with hydromorphic segregation of iron)	Zones of plastic regolith, mottled zone is waterlogged. Iron poorly mobilized.
		(1) Hydromorphic	Very humid but well drained.
		(2) Plinthitic	Downslope forms on badly drained areas.
		(3) Indurated	Iron transported over great distances. Acid water tables. At springs.

Table 5.1 *continued*

Subdivision	*Definition*	*Classification criteria groups*	*General characteristics*
Ferruginous soils			
These soils form in climatic zones which are either more humid (without a dry season) or slightly hotter than the areas where most fersiallitic soils originate. They tend to be somewhat more strongly weathered than the fersiallitic soils, but orthoclase and muscovite typically remain unaltered. Kaolinite is the dominant clay mineral; 2:1 minerals are subordinate and gibbsite is usually absent. The exchange capacity of the clay fraction is 16 to 25 meq/100 g and is greatest in the clay enriched horizons because of preferential lessivage of 2:1 minerals. On older land surfaces and the more permeable and base-rich parent materials, ferrisols transitional to ferrallitic soils often occur. These have thicker profiles (often greater than 3 m) than typical ferruginous soils, the lower horizons being kaolinitic saprolite. At high altitudes under forests with tree ferns, their surface horizon is often humus-rich and very acid, leading to partial alteration of kaolinite to gibbsite in lower parts of the humus-rich horizon.			
Ferruginous (*sensu stricto*)	Soils intermediate between those formed by fersiallitization and ferrallitization. Weathering of primary materials is stronger than in fersiallitic soils but not as pronounced as in ferrallitized soils. There is some removal of soluble silica by drainage. Neoformed clays are usually kaolinitic but some 2:1 clays persist; lessivied gibbsite is absent. Horizons are often in the form of a kaolinitic saprolite, and the development is strongly influenced by age	(a) Argillic (Bt horizon present)	Kaolinite dominant. 2:1 clays subordinate. Gibbsite absent. Lower horizons kaolinitic saprolite. Most clays by neoformation. Development strongly influenced by age. Kaolinization by gradual degradation of montmorillonite, illite, interstratified clays and by kaolinization of feldspar.
		(1) Eutrophic Ferruginous	Base saturation >50%
		(2) Oligotrophic Ferruginous	Base saturation <50%.
		(3) Hydromorphic Ferruginous	
		(b) Ferrisols (Bt horizon not essential)	Diffuse accumulation of clay.
		(1) Ferrisol, weak Bt	Without diffuse accumulation of clay.
		(2) Ferrisol, weathered Bt	With segregation of iron.
		(3) Hydromorphic	At high altitudes may have gibbsite.
		(4) Humic	

Fersiallitic soils

Such soils probably form mainly in subtropical or Mediterranean climates, with mean temperatures of 13 to 20°C, rainfall 500–1000 mm, and a hot dry season; tropical sub-types are also known. Under subtropical and Mediterranean conditions, the upper soil horizons undergo decalcification and weathering of primary minerals during the wet season. The elements freed by these processes are largely retained in the profile as a result of capillary rise during the dry season and effective bioturbation of the soil (e.g. by termites). With limestones much of the dissolved calcium carbonate is re-precipitated in this way to form a thin discontinuous calcrete horizon in the subsoil. In regions with a dry season too long to allow a dense forest to develop, the calcareous crust becomes thick and continuous, especially on footslopes which periodically receive bicarbonate-rich water from upslope.

Although weathering of primary minerals is more intense in tropical than temperate (siallitic) soils, it does not affect quartz, alkali feldspars or muscovite. Because of the more intense weathering, fersiallitic soils contain more iron oxide than siallitic soils and the free iron is usually greater than 60% of the total iron content. The main new clay mineral formed is smectite, especially where drainage is impeded so that much of the silica and bases released by weathering are retained in the profile. However, kaolinite may appear on older well-drained land surfaces and silica-poor parent materials, such as basalt. Where the parent material is clay-rich, the composition of the soil clays may be determined mainly by minerals derived with little alteration from the sediment. The 2:1 clays often undergo lessivage to form clay-enriched subsurface horizons, and some iron oxide may be carried down with the clay to form a red or red-mottled, clay-enriched B horizon. The clay fraction (<2 μm) usually has an exchange capacity of about 50 meq/100 g, but it may be as small as 25 meq/100 g. Extremely quartzose rocks without iron or weatherable minerals do not produce fersiallitic soils in any topographic or climatic situation.

Silica and bases lost in solution may also move laterally and accumulate where drainage is impeded, for example on footslopes, valley floors and enclosed hollows. In such situations recombination with other weathering products to form swelling 2:1 clays often results in patches of clay-rich deposits. These crack in dry seasons and humic, weathered topsoil is incorporated into subsoil horizons by self-mulching to form vertisols (black cotton soils).

With recent volcanic ashes, fersiallitic soils are often associated on slopes with uniformly dark-coloured, very porous soils of low bulk density, known as andosols. These immature thixotropic soils owe their characteristics to the formation of complexes between humus and imperfectly crystallized aluminosilicates (allophanes) produced by rapid weathering of volcanic glass. As silica is lost during development of andosols, allophanes are replaced by disordered fibrous clay minerals (imogolites) and eventually by the globular or tubular 1:1 clay mineral halloysite.

Table 5.1 *continued*

Subdivision	*Definition*	*Classification criteria groups*	*General characteristics*
Fersiallitic (sensu stricto)	Soils with some reddening. 2:1 clays are dominant by both transformation and neoformation and the cation exchange capacity of the clays is greater than 25 meq/100 g. Where vertical development is incomplete they form brown fersiallitic soils; where complete, saturated or almost saturated complex, red fersiallitic soils and when complex, desaturated and degraded, acid fersiallitic soils.	(a) Fersiallitic brown soils (1) Tropical brown eutrophic (2) Subtropical and mediterranean	Young, therefore different from eutrophic brown soils. Not as deep as many tropical soils. Organic. Not rubified. 2:1 clays plus montmorillonites.
		(b) Modal fersiallitic red soil (1) Tropical fersiallitic red soil (2) Subtropical and mediterranean	Common on basic rocks. Older soils rubified. 2:1 clays degrading to kaolinite. Loss of silica. Kaolinite not dominant. May have a calcic horizon.
Fersiallitic Andosols	Soils, often clays, derived partly or wholly from volcanic deposits and often dark coloured, essentially amorphous allophane-humus complexes. They have an enormous water-holding capacity which exceeds 100% and can reach 200% in hydromorphic tropical andosols, but prolonged desiccation can lower this capacity, often irreversibly. They have high exchange capacity and are very clay and iron rich; they indurate on drying.	(a) Vitrisols	Young andosols rich in volcanic glass, <10% organo-mineral complexes.
		(b) Andosols (*sensu stricto*) (1) Humic andosols	Little profile differentiation. Strongly developed with >10% organo-mineral complexes. Slight decrease of organic material with depth. All horizons grey-black. Allophanic.
		(2) Differentiated andosols	Marked downward decrease in organic material. Brown B horizon. Little or no silicate clay. Occurs on massive indurated rocks.
		(3) Hydromorphic intergrade	Constantly soaked, decomposition of organics slow. Peaty, porous, but a lot of water retained. Complete reduction of iron.
		(c) Andic soils	Rapid mineralization of organic material concentrated at surface. Kaolinization limited, silica poor. Hydroxide, gibbsite–goethite.

Vertisols	Dark coloured mature soils, rich in swelling clays strongly bonded to humic compounds. Typically show deep mixing by vertical movement due to clay volume change and possess large contraction cracks and slickensides.	(a) Vertisols (*sensu stricto*)	Swelling clays overwhelming.
		(1) Developed (include black cotton soils)	Ratio of ferric iron to total iron high. Clays of neoformation and aggradation dominant.
		(2) Slightly developed intergrade	Parents often crystalline and volcanic. Deep cracking and compaction with desiccation.
		(b) Coloured vertic soils	Characteristic of humid tropical climates with short dry season. Often on crystalline basic or volcanic rocks. Poor drainage.
		(1) Intergrades and ferruginous	Differences occur with degree of maturity of organic and iron content.
		(2) Vertical tropical eutrophic brown soils	High structure, clay content.

B: Duricrusts

Duricrusts are indurated products of surficial and pene-surficial processes formed by cementation or replacement of bedrock, weathering deposits, unconsolidated sediments, soil or other materials produced by low-temperature physico-chemical processes.

Calcrete	An indurated deposit mainly consisting of Ca, Mg carbonates. The term includes non-pedogenetic forms produced by fluvial or groundwater action: they may be pedogenetic by lateral or vertical transfer. Subdivisions are usually made on the basis of degree and type of cementation (e.g. powder, nodular, concretionary).	(a) Calcified soils	Usually loose or soft soil weakly cemented by $CaCO_3$.
		(b) Powder calcrete	Fine, usually loose powder, $CaCO_3$ with few visible particles. Little nodular development.
		(c) Nodular calcrete (1) Nodular (2) Concretionary/concentric	Nodules or concentrations in a loose structureless matrix.
		(d) Honeycomb (1) Coalesced nodules	Stiff to very hard open-textured calcrete with voids usually filled with soil.
		(2) Cemented	Cemented pebbles and fragments coalesced by laminar rinds.
		(e) Hardpan calcrete (1) Cemented honeycomb (2) Cemented powder (3) Recemented (4) Coalesced horizontal nodules (5) Case hardened calcic	A firm to very hard sheet-like layer always underlain by softer or looser material and seldom less than 45 cm thick. May be pseudo-laminated.

Table 5.1 *continued*

Subdivision	*Definition*	*Classification criteria groups*	*General characteristics*
		(f) Laminar	Firm to hard finely laminated undulose sheet layers, frequently capped by hardpan.
		(g) Boulder	Varies from discrete to coalesced hard to very hard boulders in a usually red sand matrix.
Gypcrete	Gypcrete: an indurated deposit mainly consisting of calcium sulphate dihydrate. The main subdivisions are based on the degree and type of induration (e.g. powder, laminated) but also include desert roses in dune sand. It may include non-pedogenetic forms such as blown gypsum dune sand but is normally pedogenetic by vertical transfers.	(a) Desert rose	Individual, twinned and ingrown crystal clusters in loose sand matrix at capillary level.
		(b) Mesocrystalline	Subsurface crusts, euhedral or lenticular.
		(c) Indurated (1) Powder (2) Indurated alabastine (3) Alabastine cobbles	Surface crusts – microcrystalline.
		(d) Evaporitic (1) Laminated (2) Bedded	Horizontal laminae or occasionally layer bedding forms.
		(e) Gypsum rich dune sand	Loose or lightly cemented, often in dune form.
Ferricrete	A form of indurated deposit consisting of accumulations of sesquioxides, mainly iron, within one or several ferruginous or ferrallitic soil horizons. It may form by deposition from solution, moving laterally or vertically, or as a residue after removal of silica, alkalis, etc. The term 'carapace' is sometimes used for moderate induration and 'cuirrasse' for high induration. It may be pedogenetic by retention or accumulation of minerals and by segregation within vadose profiles. Groundwater forms are pisolitic. Subdivisions are based on degree and type of induration.	(a) Water table cuirasses	Occur at breaks of slope and basin margins. Iron carried by lateral circulation.
		(1) Local	Accumulates in topographic lows as goethite or hematite. Localized segregation.
		(2) Plinthite	Red patches of hematite.
		(3) Petroplinthite	Irreversibly indurated by water table fall.
		(b) Plateau cuirasses	Very thick on erosion surfaces. Concentrated in climatic zones with marked seasonal contrasts. Thickens laterally. Goethite replaced by hematite until dominant.

		or	
		(a) Pisolitic	Welded concretions, unbanded to pseudo-rounded.
		(b) Scoraceous and vesicular	Accumulated in old network of fissures of a polyhedral or prismatic horizon.
		(c) Petroplinthite	Induration of plinthites often vesicular.
Alcrete (Alucrete)	A form of indurated deposit often called 'bauxite (alucrete) crust', containing Al and Fe in residual laterite deposits. The Al is in sufficient quantity to be of commercial use. It may be subdivided on the basis of degree and type of induration. It may also be detrital, reworked or pedogenetic.	(a) Pisolitic in plinthites (b) Scoraceous and vesicular (c) Petroplinthite	Varieties as for ferricretes
Silcrete	An indurated deposit mainly consisting of silica, which may be formed by lateral or vertical transfer and which may include pedogenetic, groundwater or leached varieties. Subdivisions may be made on the basis of the dominant fabric type. Commonly they are poor in Fe, Al, Ca, K, Mg and P. Subdivisions may also be made on the basis of the dominant fabric type.	(a) Grain supported fabric	Skeletal grains in a self-supporting matrix. Optically continuous overgrowths chalcedony, micro-quartz cryptocrystalline and opaline alternatives.
		(b) Floating fabric	Skeletal grains floating in matrix, not self-supporting. Massive globules sometimes present.
		(c) Matrix fabric	Skeletal grains, massive globules absent in some forms, common in others.
		(d) Conglomerate fabric	Detrital content.

5.2.1 Laterite and laterization

Nevertheless since particular terms, notably laterite and to a lesser extent red clays and latosols, have been used for many years to describe certain tropical residual soils, even though they have been ill-defined, they are retained here because of the frequency of their use. Much confusion exists in engineering circles regarding the term laterite since there has been a tendency to use the term in connection with any red soil found in the tropics. Laterite was originally defined by Buchanan (1807) as a tropical red soil which hardens on exposure to air. Recently Charman (1988) redefined the term in an attempt to overcome the existing confusion. He suggested that laterite should be regarded as a highly weathered material which forms as a result of the concentration of hydrated oxides of iron and aluminium in such a way that the character of the deposit in which they occur is affected. These oxides may be present in an unhardened soil, as a hardened layer, as concretionary nodules in a soil matrix or in a cemented matrix enclosing other materials. In fact Charman proposed a cycle of concretionary development from an original plinthite layer (a layer in which concentration of oxides occurs but with no concretionary development) to a mature hardpan laterite. He recognized intermediate duricrust forms such as nodular laterite and honeycomb laterite. Laterites therefore fall within the ferruginous and ferralitic classes referred to in Table 5.1. They tend to occur in areas of gentle topography which are not subject to significant erosion.

Laterization is rapid in tropical regions experiencing periods of heavy rainfall alternating with drier periods (Gidigasu, 1976). Decomposition of the parent rock involves the removal of silica, lime, magnesia, soda and potash, leading to an enrichment in iron and aluminium oxides. Repeated wetting and drying can result in aggregation of soil particles and cementation by compounds of Ca, Mg, Al and Fe. Kaolinite also is formed. During drier periods the water table is lowered. The small amount of iron which has been mobilized in the ferrous state by the groundwater then is oxidized, forming hematite, or if hydrated – goethite. Oxides and hydroxides of iron accumulate, since rapid oxidation does not allow the organic compounds in solution to dissolve iron and thereby for it to be carried away. Ola (1978a) maintained that in laterites the ratios of silica (SIO_2) to sesquioxides (Fe_2O_3; Al_2O_3) usually are less than 1.33 to 1, that those ratios between 1.33 and 2.0 to 1 are indicative of lateritic soils and that those greater than 2.0 are indicative of non-lateritic types. The movement of the water table leds to the gradual accumulation of iron oxides at a given horizon in the soil profile. A cemented layer of laterite ultimately forms which may be a continuous or honeycombed mass (cellular laterite) or nodules may be formed, as in pisolitic laterite or laterite gravel. Concretionary layers are often developed near the surface in lowland areas because of the high water table.

Scanning electron microscope studies carried out by Malomo (1989) on laterite soils showed strongly cemented surfaces covered with iron oxides. These oxides initially exist as a semi-gelatinous coating which can become denser on loss of moisture but still remaining non-crystalline. Crystallization takes place slowly with the formation of hetatite or goethite if hydrated. The microfabric consists largely of surfaces at various stages of crystallization. The combined action of leaching and cementation can lead to the development of highly porous soil fabrics.

If laterization proceeds further, as a result of prolonged leaching, then kaolinite is decomposed, the silica being removed in solution and gibbsite remains. Iron compounds also may be removed so that the soil becomes enriched in alumina, a bauxitic soil being developed. Bauxitic clays often have a pisolitic structure.

5.2.2 Red clays and black clays

Red clays and latosols are residual ferruginous soils in which oxidation readily occurs. Such soils tend to develop in undulating country and most of them appear to have been derived from the first cycle of weathering of the parent material. They differ from laterite in that they behave as a clay and do not possess strong concretions. They do, however, grade into laterite.

The residual red clays of Kenya have been produced by weathering, leaching having removed the more soluble bases and silica, leaving the soil rich in iron oxide (hematite) and hydroxide (goethite), and in aluminium. The latter usually occurs in the form of kaolinitic clay minerals or sometimes as gibbsite. Dumbleton (1967) found halloysite in these red clays, as did Dixon and Robertson (1970). These soils contain a high percentage of clay size material.

Black clays are typically developed on poorly drained plains in regions with well-defined wet and dry seasons, where the annual rainfall is not less than 1250 mm. Generally the clay fraction in these soils exceeds 50%, silty material varying between 20 and 40% and sand forming the remainder. The organic content commonly is less than 2%. Montmorillonite normally is present in the clay fraction and is the principal factor determining the behaviour of these clays, in particular the appreciable volume changes which take place on wetting and drying. Changes in stress and strain when expansive soils are wetted and dried by seasonal changes may be high enough to destroy bonding and structure. Calcium carbonate sometimes occurs in these black clays, frequently taking the form of concretions (kankar), but it is not an essential characteristic. Normally it constitutes less than 1%. Black cotton soils are highly plastic silty clays, frequently formed from basaltic rocks, shale or clayey sediments. The soil may contain up to 70% of montmorillonite, kaolinite and quartz comprising most of the remainder.

5.3 Physical properties and engineering behaviour

The science of soil mechanics was largely developed in relation to transported soils found in temperate latitudes. Hence reproducible results from some standard tests may be difficult to obtain from tropical soils formed in place by weathering.

The specific gravity of residual soils may be unusually low or unusually high since that of the particles present may vary more than it does in transported soils. The values of dry unit weight for lateritic soils range from as low as 13 kN/m^3 to 23 kN/m^3, with most of the values falling between 17 and 22 kN/m^3. The dry unit weight decreases with increasing content of clay size particles. The dry unit weights of andisols tend to be much lower than those of typical lateritic soils and values around 12 kN/m^3 seem common, but values have been reported as low as 9 kN/m^3 (Mitchell and Sitar, 1982).

Water of crystallization may be present within some minerals in many tropical residual soils, as well as free water. Some of the former type of water may be lost during conventional testing for moisture content. In order to avoid this, Anon (1990) recommended that comprative tests should be carried out on duplicate samples, taking the measurement of moisture content by drying to constant mass between 105°C and 110°C on one sample, and at a temperature not exceeding 50°C on the other. An appreciable difference in the two values indicates the presence of structured water. Some values of natural moisture content for laterite soils are given in Table 5.2, values frequently falling within the range 6–22%. Natural moisture contents for andosols may be in the range 60–80% and values as high as 150% have been recorded (Mitchell and Sitar, 1982). Residual soils typically have natural moisture contents greater than optimum.

Drying brings about changes in the properties of residual soils not only during sampling and testing, but also *in situ*. The latter occurs as a result of local climatic conditions, drainage, and position within the soil profile. Drying initiates two important effects, namely, cementation by the sesquioxides and aggregate formation on the one hand, and loss of water from hydrated clay minerals on the other. In the case of halloysite, the latter causes an irreversible transformation to metahalloysite. Some consequences of these changes are illustrated in Table 5.3 and Figure 5.2. Drying can cause almost total aggregation of clay size particles into silt and sand size ranges, and a reduction or loss of plasticity (Table 5.3). Unit

Table 5.2 *Some engineering properties of tropical residual soils*

Soil	*Location*	*Natural moisture content (%)*	*Plastic limit (%)*	*Liquid limit (%)*	*Clay content (%)*	*Activity*	*Void ratio*	*Unit weight* kN/m^3)
1. Laterite	Hawaii		135	245	36			
2. Laterite	Nigeria	18–27	13–20	41–51				15.2–17.3
3. Laterite	Sri Lanka	16–49	28–31	33–90	15–45			
4. Lateritic gravel	Cameroon							
5. Red clay	Brazil		26	42	35	0.46		
6. Red clay	Ghana		24	48	31	0.77		
7. Red clay	Brazil		22	28	27	0.3		
8. Red clay	Hong Kong							
9. Latosol	Brazil		30	47	35	0.51		
10. Latosol	Ghana		29	52	23	1.0		
11. Latosol	Java	36–38			80		1.7–1.8	
12. Black clay	Cameroon		21	75				
13. Black clay	Kenya		36	103				
14. Black clay	India		41	·132	48			
15. Andosol	Kenya	62	73	107				11.5
16. Andosol	Java	58–63			29		2.1–2.2	

1. Tuncer and Lohnes (1977); 2. Ola (1978); 3. Nixon and Skipp (1957); 4. Townsend (1985); 5, 6, 7, 9 and 10. Morin (1982); 8. Lumb (1962); 11 and 16. Hobbs *et al* (1990); 12 and 13. Horn (1982); 14. Sabba Rao (1990); 15. Foss (1973).

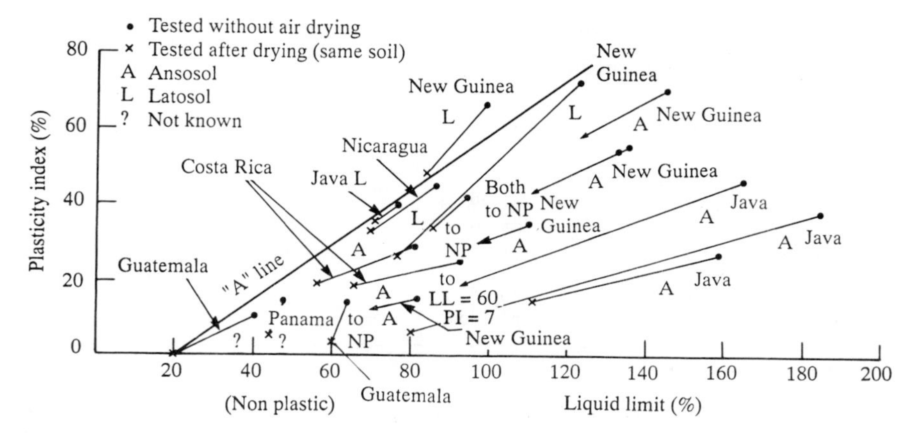

Figure 5.2 *Effect of drying on the Atterberg limits of some tropical soils (from Morin and Todor, 1975)*

weight, shrinkage, compressibility and shear strength also can be affected. Hence classification tests should be applied to the soil with as little drying as possible, at least until it can be established from comparative tests that drying has no effect on the results.

As far as the plasticity of tropical residual soils is concerned disaggregation

Dry density (Mg/m³)	*Specific gravity*	*Shear stength*		*Compressibility*		c_v (m²/yr)	*Permeability (m/s)*
		φ′	c'(kN/m²)	c_c	m_v(m³/MN)		
		27–57	48–345				1.5×10^{-6}–5.4×10^{-9}
		28–35	466–782	0.0186	13	262	
	2.89–3.14	37	0–40				
	2.73						
	2.8						
	2.75				0.466–0.075	2.32–232	5×10^{-6}–2×10^{-6}
	2.86						
	2.83						
	2.74	21	26	0.03–0.06	0.995–0.01		4.7–5×10^{-8}
	2.64	16–20	47–58				
1.27–1.75							
	2.72						
		37	25				
		18	46	0.01–0.04	0.358–0.084		4×10^{-8}
							2.8×10^{-8}–5.6×10^{-9}

Table 5.3 *The effect of drying on index properties*

(a) Liquid (LL) and plastic limits (PL) (from Anon, 1990)

Soil type and location	*Natural LL:PL*	*Air dried LL:PL*	*Oven dried LL:PL*
Laterite (Costa Rica)	81:29		56:19
Laterite (Hawaii)	245:35	NP	
Red clay (Kenya)	101:70	77:61	65:47
Latosol (Dominica)	93:56	71:43	
Andosol (New Guinea)	145:75		NP
Andosol (Java)	184:146		80:74

NP indicates non-plastic

(b) Effect of air drying on particle size distribution of a hydrated laterite clay from Hawaii (From Gidigasu, 1974)

Index properties	*Wet (at natural moisture content)*	*Moist (partial air drying)*	*Dry (complete air drying)*
Sand content (%)	30	42	86
Silt content (%) (0.05–0.005 mm)	34	17	11
Clay content (%) (<0.005 mm)	36	41	3

should be carried out with care so that individual particles are separated but not fragmented. In fact disaggregation may have to be brought about in some cases by soaking in distilled water (Anon, 1990). Moreover the sensitivity of the soil to working with water should be checked by using a range of mixing times before testing, in order to determine the shortest time for thorough mixing.

The heterogeneity in soil characteristics for well-developed laterite soils from Ghana is illustrated in Figure 5.3 by the range of scatter of plasticity data (Gidigasu, 1974). De Graft-Johnson *et al.* (1972) concluded that plasticity data alone could not be used to differentiate between residual soils formed in different climatic regimes and from different parent rock types. It is possible, however, to produce a broad grouping of the plasticity properties of lateritic soils and andisols (Figure 5.4). The lateritic soils straddle the A line and range from low to high plasticity. By contrast, andisols and halloysitic soils generally fall below the A line and exhibit high plasticity. Their plasticity increases as a result of manipulation but decreases, as noted, as a result of drying.

Usually at or near the surface the liquid limits of laterites do not exceed 60% and the plasticity indices are less than 30% (Table 5.2). Consequently such laterites are

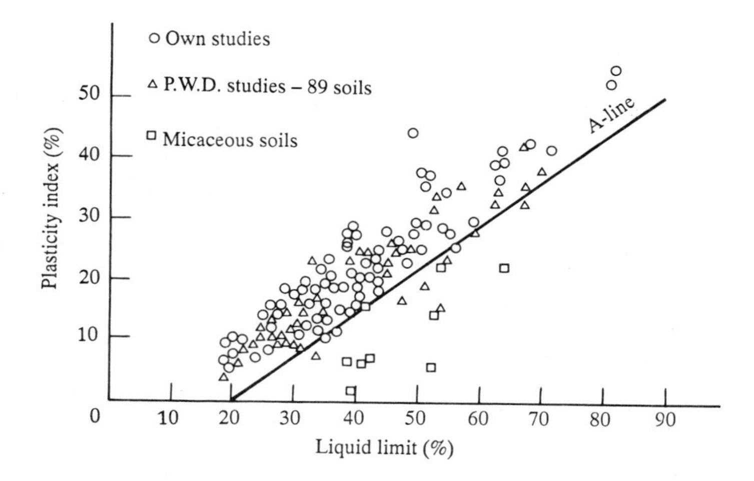

Figure 5.3 *Variability of Atterberg limits for lateritic soils for Ghana (from Gidigasu, 1974)*

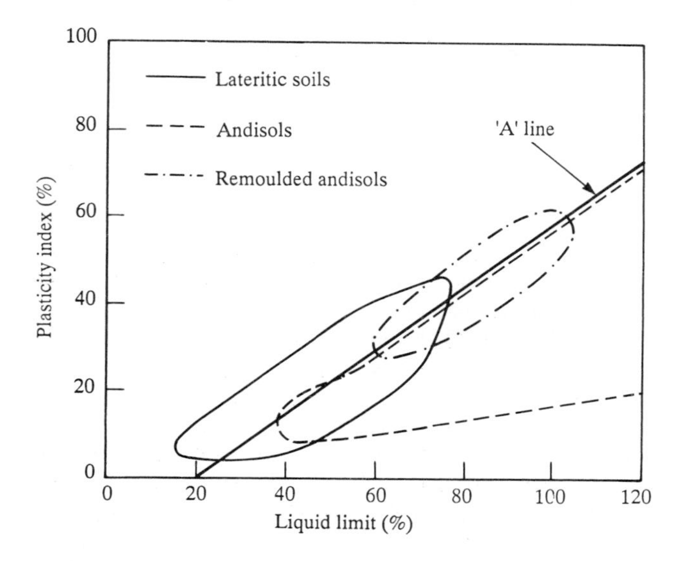

Figure 5.4 *Approximate ranges of Atterberg limits for lateritic soils and andisols (from Mitchell and Sitar, 1982)*

of low-to-medium plasticity. The activity of laterites may vary between 0.5 and 1.75. Red clays and latosols have high plastic and liquid limits (Table 5.2). For example, Clare (1957a) mentioned plastic limits ranging between 34 and 56% and liquid limits between 76 and 104%. The clay content in these soils varied between 63 and 88%, hence the activity of the soils was around 1.5–1.8. The liquid limits of black clays (vertisols) may vary between 50 and 100% with plasticity indices between 25 and 70%. The shrinkage limit frequently is around 10–12%.

Laterite commonly contains all size fractions from clay to gravel and sometimes even larger material (Figure 5.5). The procedures required to obtain particle-size distributions of residual soils have been outlined by Anon (1990).

The fabric of *in situ* residual soils, particularly of the lateritic type, involves a wide range of void ratios and pore sizes. For example, the range of void ratio for lateritic soils in Hawaii, as found by Tuncer and Lohnes (1977), was 0.8–1.91. However, the variability of void ratio does not vary systematically with soil type, parent rock, type of weathering and state of stress. It commonly is due to differential leaching removing varying quantities of material from the soil. The void ratio at a particular state of stress may be in a metastable, stable–contractive or stable–dilatant state (Figure 5.6). The strains which a residual soil with a bonded structure experiences when it yields depend on its void ratio and degree of bonding. When the yield stress is exceeded the strains undergone are determined by the soil state. It is subjected to large contraction if it is metastable whereas it will undergo only small strains if it is stable dilatant. The void ratio and pore size do not change significantly with stress if this does not exceed the yield stress. Although larger pores allow the soil to de-saturate at small suctions, many small pores can retain water at high suctions.

Vaughan *et al.* (1988) introduced the concept of relative void ratio (e_R) for residual soils, defining it as

$$e_R = \frac{e - e_{opt}}{e_L - e_{opt}} \tag{5.2}$$

where e is the natural void ratio, e_L is the void at the liquid limit and e_{opt} is the void ratio at the optimum moisture content. They suggested relating engineering properties to relative void ratio rather than *in situ* void ratio.

According to Mitchell and Sitar (1982) the main sources of data on strength characteristics of both lateritic soils and andisols are from tests performed on remoulded specimens. The reported range of the friction angle is from 18 to 41°, with the majority of the results falling in the range 28–38°. The values of cohesion range from 0 to in excess of 48 kN/m^2. Data on the undisturbed shear strength characteristics of residual soils are fairly limited. Andisols are characterized by fairly high friction angles, 27–57°, as opposed to 23–33° for lateritic soils. The cohesion for laterites can vary from 0 to over 210 kN/m^2 whilst that quoted for andisols ranged from 22 to 345 kN/m^2. Moisture content does affect the strength of andisols significantly as the degree of saturation can have an appreciable effect on cementation.

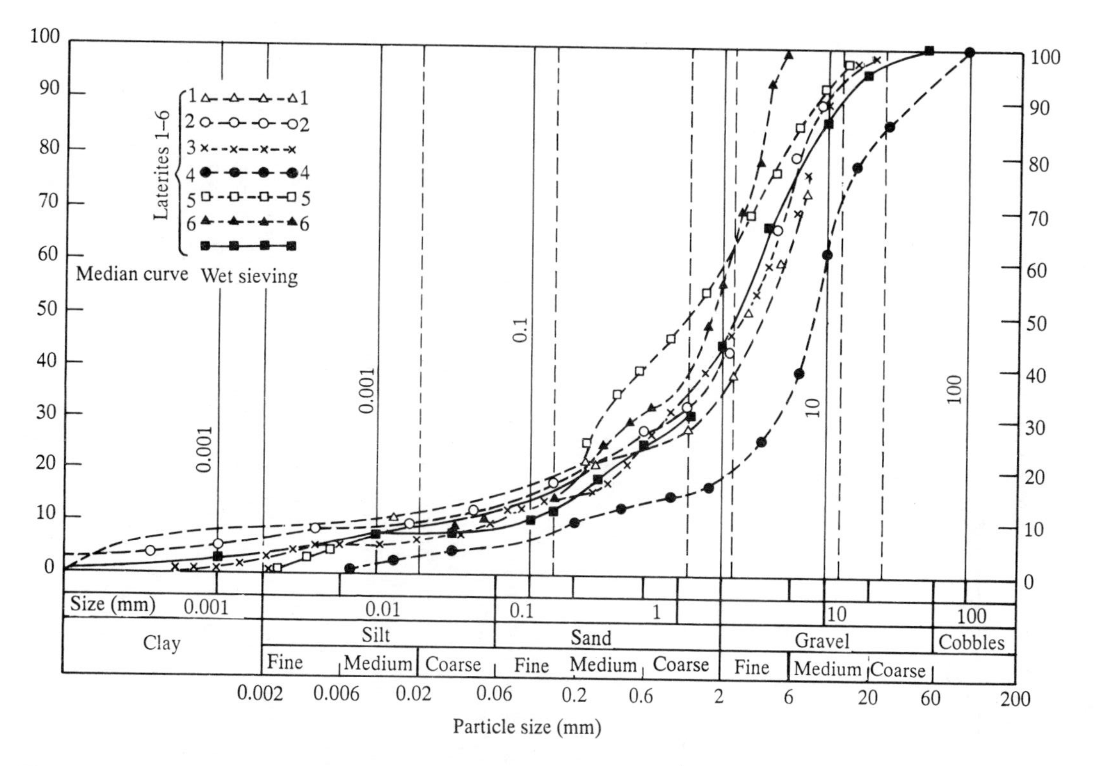

Figure 5.5 *Grading curves of laterites (after Madu, 1977)*

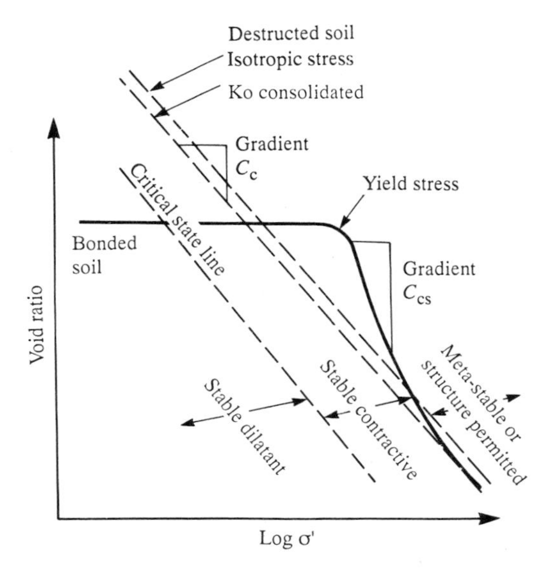

Figure 5.6 *Stress–void ratio of a residual soil related to the states possible for the de-structured soil. The void ratio at a particular stress may be in one of three states.*

(1) Meta-stable in which the soil exists at a void ratio which is impossible for the same soil in the de-structured state at the same stress level. It can exist in this state only due to the strength and stability provided by its inter-particle bonding.
(2) Stable-contractive, in which the soil could exist in the de-structured state, but would contract during shear towards the constant volume 'critical state'.
(3) Stable-dilatant, in which the soil could exist in the de-structured state at the same stress level, but where it would expand during shear towards the 'critical state'.

Bonding and void ratio combine to determine the strains which the soil undergoes when it yields at a particular stress state. If an in situ *residual soil exceeds its yield stress, then the strains which it subsequently is subjected to depend largely on its state: if meta-stable it will be subject to large contractions, if stable-dilatant it will suffer only relatively small strains (unless shear failure occurs). Thus recognition of state allows the consequence of exceeding yield stress to be established. The slope of a void ratio/log stress plot, c_{cs}, after yield is approximately linear and is a function of the yield stress and the initial void ratio rather than the grading and mineralogy of the soil (after Vaughan, 1990)*

Because of the presence of a hardened crust near the surface, the strength of laterite may decrease with increasing depth. For example, Nixon and Skipp (1957) quoted values of shear strength of 90 and 25 kN/m^2 derived from undrained triaxial tests, for samples of laterite taken from the surface crust and from a depth of 6 m respectively. In addition the variation of shear strength with depth is influenced by the mode of formation, type of parent rock, depth of water table and its movement, degree of laterization and mineral content, as well as the amount of cement precipitated. De Graft-Johnson and Bhatia (1969) gave the changes in engineering properties with depth for a laterite occurring over granite in Kumasi, Ghana (Table 5.4).

Table 5.4

Depth (m)	*Dry density* (Mg/m^3)	*Moisture content* (%)	*Plastic limit* (%)	*Liquid limit* (%)	*Stength*	
					ϕ'	c' (kN/m^2)
3.0–3.6	1.63	28	31	44	6°	84
4.2–4.9	1.51	30	26	54	15°	138
7.9–8.5	1.55	26	49	65	5°	63
11.0–11.6	1.71	18	NP	–	23°	216

NP = non-plastic

Obviously the cementation between particles of residual soils has a significant influence on their shear strength, as does the widely variable nature of the void ratio and partial saturation which, as noted, can occur to appreciable depth. A bonded soil structure exhibits a peak shear strength, unrelated to density and dilation, which is destroyed by yield as large strains develop. Examples of residual strength for different materials are shown in Figure 5.7. Deformation of residual soils is small unless the yield stress is exceeded. Post-yield deformation depends on yield stress and initial porosity.

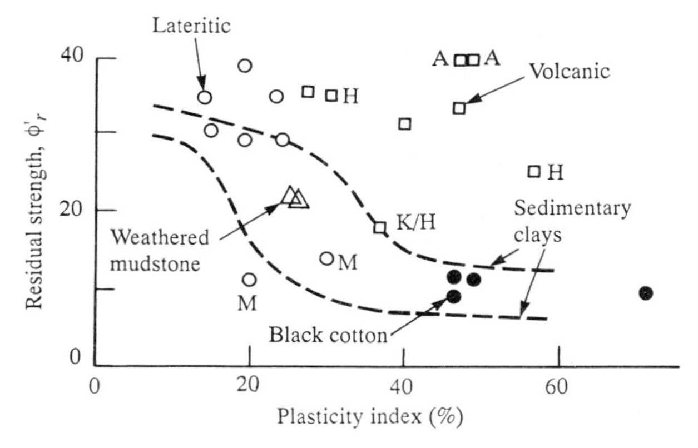

Figure 5.7 *Residual strength, ϕ'_r, of residual tropical soils.* $c' = 0$, $\sigma'_n = 400\ kN/m^2$ *(after Boyce, 1985). A, Allophane; H, Halloysite; K, Kaolinite; M, Mica*

The shear strength of a bonded soil may be underestimated if the bonded structure yields during a test. This can be minimized by using a low value of effective confining pressure or by the selection of a suitable stress path, as explained by Anon (1990). The application of an excessive rate of loading should also be avoided in soils with a high void ratio and sensitive grain structure.

Allam and Sridharan (1981) subjected red clay to a number of cycles of wetting and drying in an attempt to simulate the notable changes in natural moisture content which these soils experience as the seasons change. They found that such changes increase the stiffness of the soil fabric, which gives rise to an increase in shear strength and a decrease in compressibility. In particular repeated wetting and drying can increase the strength of some saprolitic soils. On the other hand wetting and drying may weaken the shearing resistance. For instance, the cohesion component is significantly affected by the degree of saturation. The amount of reduction in cohesion is influenced by the origin of the parent material. Foss (1973) reported a reduction of up to 50% in the cohesion and about 30% in the angle of friction of red clay from Kenya due to saturation. Subsequently Sridharan (1990) presented some results of undisturbed and remoulded shear strength tests on desiccated lateritic soils (Figure 5.8; Table 5.5). The undisturbed soils exhibited larger shear strength and a stiffer stress–strain response. Furthermore the strain at failure was greater for remoulded soils and this was more pronounced in the unconsolidated undrained test. A residual soil of high porosity can show sensitivity provided that it has a high degree of saturation.

It frequently is assumed that lateritic soils are relatively incompressible due to the presence of cementation. The compression index for a bonded residual soil is a

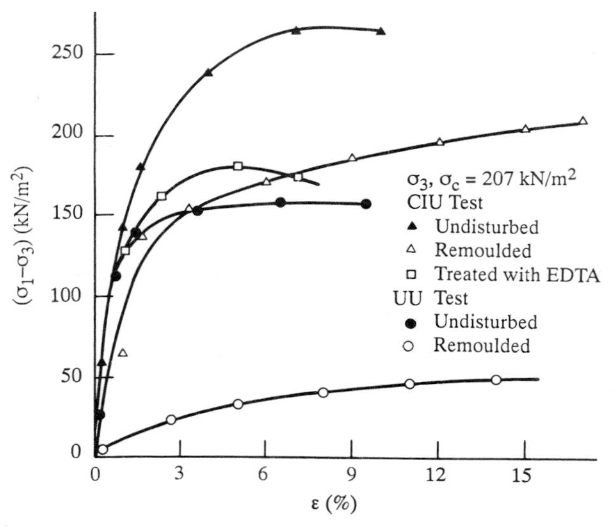

Figure 5.8 *Typical stress–strain curve for a desiccated lateritic soil (after Sridharan, 1990)*

Table 5.5 *Test results of lateritic desiccated soils (After Sridharan, 1990)*

Soil	*1*		*2*		*3*		*4*	
Soil state	*U*	*R*	*U*	*R*	*U*	*R*	*U*	*R*
Sand (%)	53.0		51.9		48.3		70.0	
Clay and silt (%)	47.0		48.1		51.7		30.0	
Liquid limit (%)	49.0		53.2		50.6		38.0	
Plastic limit (%)	25.0		32.6		31.0		18.0	
Specific gravity, G	2.63		2.63		2.68		2.60	
Void ratio, e	0.80	0.75	0.63	0.65	0.68	0.62	0.69	0.62
UU Shear tests:								
c (kN/m^2)	38.9	15.5	39.8	29.3	32.8	27.6	24.5	17.2
ϕ (deg)	13.4	1.7	5.5	2.5	4.5	3.0	11.0	2.5
c' (kN/m^2)	8.0	0.0	0.0	0.0	8.3	0.0	22.8	3.9
ϕ' (deg)	30.6	35.2	31.4	31.4	25.8	35.2	25.1	29.9
CIU Shear tests								
c (kN/m^2)	36.9	28.6	57.9	28.5	39.3	31.8	32.8	14.5
ϕ (deg)	21.0	15.5	17.8	15.3	15.0	13.4	19.3	16.7
c' (kN/m^2)	23.4	0.0	47.5	0.0	32.8	0.0	22.8	3.9
ϕ' (deg)	27.8	31.3	19.6	27.8	19.0	23.8	25.1	27.5
Average failure UU	11.7	15.6	13.2	18.3	13.6	18.0	11.3	23.5
Strain (per cent) CIU	11.8	14.8	11.8	14.8	13.1	16.0	14.6	16.3

U = Undisturbed; R = Remoulded.

function of the initial void ratio rather than soil grading and plasticity. Vargas (1974) presented the following relationship for lateritic soils from Brazil:

$$C_c = 0.005\,(\mathrm{LL} + 22) \pm 0.1 \qquad (5.3)$$

Many tropical residual soils behave as if they are overconsolidated in that they exhibit a yield stress at which there is a discontinuity in the stress–strain behaviour and a decrease in stiffness. This is the apparent preconsolidation pressure. The degree of overconsolidation depends on the amount of weathering. The cementation of soils formed in regions with a distinct dry season can be weakened by saturation and this leads to a collapse of the soil structure. As a consequence the apparent preconsolidation pressure decreases and the compressibility of the soil increases (Figure 5.9).

Collapsible soils have an open textured fabric which can withstand reasonably large stresses when partly saturated, but undergo a decrease in volume due to

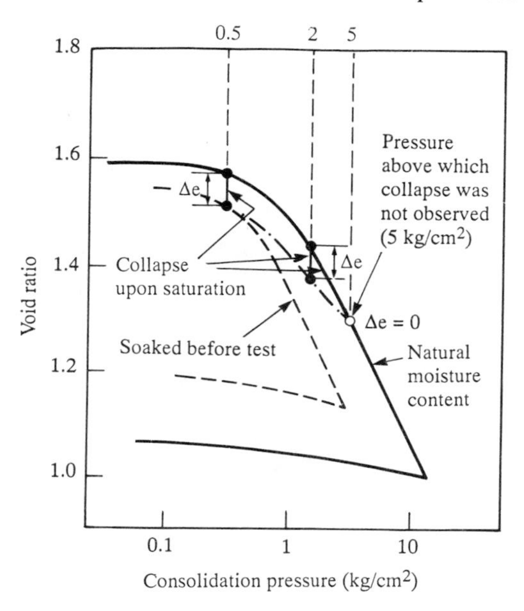

Figure 5.9 *Typical consolidation curves for collapsible lateritic soils showing the effects of saturation (after Vargas, 1974)*

collapse of the soil structure on wetting, even under low stresses (Vargas, 1990). Many partially saturated tropical residual soils are of this nature. Collapse and associated settlement normally is due to loss or reduction of bonding between soil particles because of the presence of water. Intensively leached residual soils formed from quartz-rich rocks tend to be prone to collapse. Other ways in which collapse is brought about include loss of the stabilizing effect of surface tension in water menisci at particle contacts in partially saturated soil and loss of strength of 'dry' clay bridges between particles. Methods of assessing collapsible soils have been referred to in Chapter 3.

Expandable clay minerals frequently are present in tropical residual clay soils of medium to high plasticity. When these clays occur above the water table, they can undergo a high degree of shrinkage on drying. By contrast, when wetted they swell, exerting high pressures. Indeed, highly expansive clay minerals may develop suficient swelling pressure to break the bonded structure of the soil and thereby give rise to large heave movements. Large heave movements can occur when desiccated black clay soils are wetted. Alternating wet and dry seasons may produce significant vertical movements as soil suction changes. The amount of heave which takes place depends on the effective stress changes and on the swelling modulus of the soil $((\delta V/V)/\delta\sigma')$. Swelling modulus increases with the proportion of clay in the soil and the amount of expansive clay minerals, notably montmorillonite, in the clay fraction. Methods of estimating the expansivity of clay soils are outlined in Chapter 3.

The black cotton soil of Nigeria, described by Ola (1978b), is a highly plastic silty clay in which shrinkage and swelling is a problem in many regions that experience alternating wet and dry seasons. These volume changes, however, are confined to an upper critical zone of the soil, which is frequently less than 1.5 m thick. Below this the moisture content remains more or less the same, for instance, around 25%. Ola (1980) noted an average linear shrinkage of 8% for some of these soils, with an average swelling pressure of 120 kN/m^2 and a maximum of about 240 kN/m^2. He went on to state that in such situations the dead load of a building should be at least 80 kN/m^2 to counteract the swelling pressure.

Collapse and swelling characteristics depend on void ratio and applied stress. Some soils can exhibit either collapse or swelling under different conditions, depending on the stress level at which the decrease in effective stress occurs.

The effects of leaching on lateritic soils has been investigated by Ola (1978a). As mentioned, the cementing agents in lateritic soils help to bond the finer particles together to form larger aggregates. However, as a result of leaching these aggregates break down, which is shown by the increase in liquid limit after leaching. Moreover removal of cement by leaching gives rise to an increase in compressibility of more than 50%. Again this is mainly due to the destruction of the aggregate structure. Conversely there is a decrease in the coefficient of consolidation by some 20% after leaching (Table 5.6). The change in effective angle of shearing resistance and effective cohesion before and after leaching, is similarly explained. Prior to leaching the larger aggregates in the soil cause it to behave as a coarse-grained, weakly bonded particulate material. The strongly curvilinear form of the Mohr failure envelopes also can be explained as a result of the breakdown of the larger aggregates.

As referred to above, some residual soils have high void ratios and large macro-pores and therefore are associated with high permeability. For instance, *in situ* permeabilities of clay-type lateritic soils may be as high as 10^{-5} to 10^{-4} m/s.

Table 5.6 *Engineering properties of a lateritic soil before and after leaching (After Ola, 1987a)*

Property	*Before leaching*	*After leaching*
Natural moisture content %	14	–
Liquid limit %	42	53
Plastic limit %	25	21
Relative density	2.7	2.5
Angle of shearing resistance, ϕ'	26.5°	18.4°
Cohesion, c′, kN/m^2	24.1	45.5
Coefficient of compressibility, m^2/MN*	12	15
Coefficient of consolidation, m^2/year*	599	464

* For a pressure of 215 kN/m^2

Some typical values of permeability for saprolites and laterites are as shown in Table 5.7.

Table 5.7

Saprolite (m/s)	*Laterite* (m/s)
4×10^{-3}–5×10^{-9}	4×10^{-6}–5×10^{-9}
5×10^{-6}–1×10^{-7}	5×10^{-5}–1×10^{-6}
–	3×10^{-6}–1×10^{-9}

Both in the saprolite horizon and the underlying rock the permeability is governed by discontinuities. Brand (1984) considered that the saprolitic soils in Hong Kong were generally relatively permeable and that therefore drained conditions normally occurred. Water tables are often quite low in such soils. Blight (1990) noted that water tables are often deeper than 5–10 m. If evapo-transpiration exceeds infiltration, then deep desiccation of the soil profile is likely. As a result residual soils may be cracked and fissured, which further increases permeability.

5.4 Soils of arid and semi-arid regions

In arid and semi-arid regions the evaporation of moisture from the surface of the soil may lead to the precipitation of salts in its upper layers. The most commonly precipitated material is calcium carbonate. These caliche deposits are referred to as calcrete (Braithwaite, 1983; Netterberg, 1980). Calcrete occurs where soil drainage is reduced due to long and frequent periods of deficient precipitation and high evapotranspiration. Nevertheless, the development of calcrete is inhibited beyond a certain aridity since the low precipitation is unable to dissolve and drain calcium carbonate towards the water table. Consequently in arid climates gypcrete may take the place of calcrete. Climatic fluctuations which, for example, took place in north Africa during Pleistocene times, therefore led to alternating calcification and gypsification of soils. Certain calcretes were partially gypsified and elsewhere gypsum formations were covered with calcrete hardpans (Horta, 1980).

The hardened calcrete crust may contain nodules of limestone or be more or less completely cemented (this cement may, of course, have been subjected to differential leaching). In the initial stages of formation calcrete contains less than 40% calcium carbonate and the latter is distributed throughout the soil in a discontinuous manner. At around 40% carbonate content the original colour of the soil is masked by a transition to a whitish colour. As the carbonate content increases it first occurs as scattered concentrations of flaky habit, then as hard concretions. Once it exceeds 60% the concentration becomes continuous (Table 5.1). The calcium carbonate in calcrete profiles decreases from top to base, as generally does the hardness.

Gypcrete is developed in arid zones, that is, where there is less than 100 mm precipitation annually. In the Sahara aeolian sands and gravels often are encrusted with gypsum deposited from selenitic groundwaters. A gypcrete profile may contain three horizons. The upper horizon is rich in gypsified roots and has a banded and/or nodular structure. Beneath this occurs massive gypcrete-gypsum cemented sands. Massive gypcrete forms above the water table during evaporation from the capillary fringe (newly formed gypcrete is hard but it softens with age). At the water table gypsum develops as aggregates of crystals, this is the sand–rose horizon.

Very occasionally in arid areas enrichment of iron or silica gives rise to ferricrete or silcrete deposits respectively.

Clare (1957b) described two types of tropical arid soils rich in sodium salts, namely, kabbas and saltmarsh. Both are characterized by their water retentive properties. Kabbas consists of a mixture of partly decomposed coral with sand, clay, organic matter and salt. The salt content in saltmarsh varies up to 40 or 50%, the soil basically consisting of silt with variable amounts of sand and organic material. It occurs in low-lying areas that either have a very high water table or are periodically inundated by the sea. For a review of ground conditions in arid regions such as the Middle East see Fookes (1978) and Epps (1980).

5.5 Dispersive soils

Potentially dispersive soils are red-brown, yellow or grey in colour, and can contain a moderate to high content of clay, illite being the dominant clay mineral. Their pH value ranges between 6 and 8 and they may contain up to 12% sodium. They occur on old aggradation surfaces in areas where the annual rainfall is less than 860 mm.

Dispersive clay soils deflocculate in the presence of relatively pure water to form colloidal suspensions and therefore are highly susceptible to erosion and piping. Such soils contain a higher content of dissolved sodium in their pore water than ordinary soils. There are no significant differences in the clay contents of dispersive and non-dispersive soils, except that soils with less than 10% clay particles may not have enough colloids to support dispersive piping. In non-dispersive soil there is a definite threshold velocity below which flowing water causes no erosion. The individual particles cling to each other and are only removed by water flowing with a certain erosive energy. By contrast, there is no threshold velocity for dispersive clay, the colloidal clay particles go into suspension even in quiet water. Hence retrogressive erosion can occur at very low pore water flow velocities.

Damage due to internal erosion of dispersive clay leads to the formation of pipes and internal cavities on slopes. Piping is initiated by dispersion of clay particles along desiccation cracks, fissures and rootholes. Colloidal erosion damage also has led to the failure of earth dams (McDaniel and Decker, 1979; Tadanier and Ingles, 1984). Indications of piping take the form of small leakages of muddy coloured water after initial filling of the reservoir. In addition severe erosion damage forming deep gullies occurs on embankments after rainfall. Fortunately, when dispersive soils are treated with lime they are transformed to a non-dispersive state.

Dispersive erosion depends on the mineralogy and chemistry of the clay on the one hand, and the dissolved salts in the pore and eroding water on the other. The presence of exchangeable sodium is the main chemical factor contributing towards dispersive clay behaviour. This is expressed in terms of the exchangeable sodium percentage (ESP):

$$\text{ESP} = \frac{\text{exchangeable sodium}}{\text{cation exchange capacity}} \times 100 \tag{5.4}$$

where the units are given in meq/100 g of dry soil. A threshold value of ESP of 10 has been recommended, above which soils that have their free salts leached by seepage of relatively pure water are prone to dispersion (Elges, 1985). Soils with ESP values above 15%, according to Gerber and Harmse (1987), are highly dispersive (Figure 5.10). Those with low cation exchange values (15 meq/100 mg of clay) were found to be completely non-dispersive at ESP values of 6% or below. Similarly soils with high cation exchange capacity values and a plasticity index greater than 35% swell to such an extent that dispersion is not significant. High ESP values and piping potential generally exist in soils in which the clay fraction is composed largely of smectitic and other 2:1 clays. Some illites, however, are highly dispersive. High values of ESP and high dispersibility are rare in clays composed largely of kaolinites.

The main property of clay soils governing their susceptibility to dispersion is the total content of dissolved salts in the water. The lower the content of dissolved salts in the water, the greater the susceptibility of sodium saturated clay to dispersion. Dispersion occurs when repulsive forces between clay particles exceed the

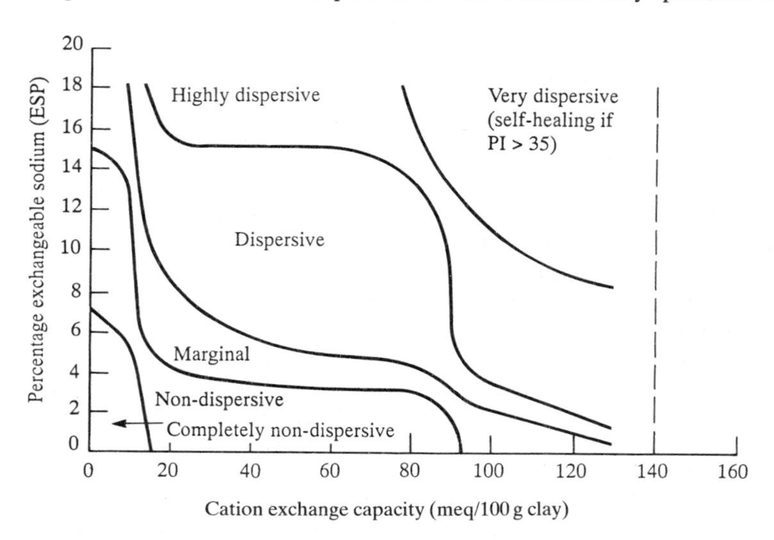

Figure 5.10 *Diagram of determination of dispersion potential as a function of ESP and CEC 100 g of clay (after Gerber and Harmse, 1987)*

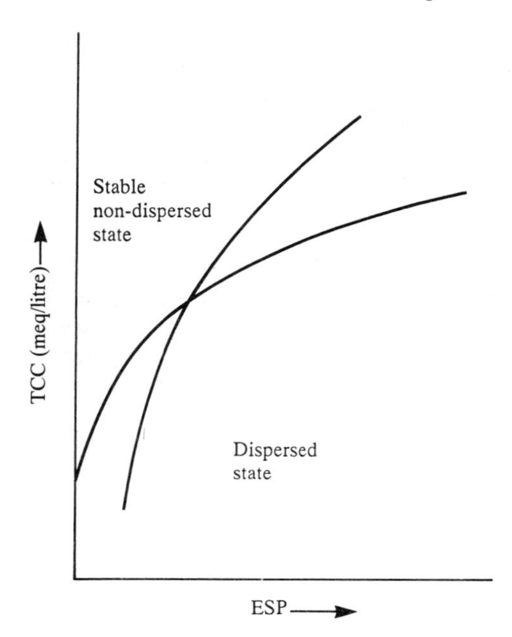

Figure 5.11 *Threshold for total cation concentration (TCC) for given exchangeable sodium percentage (ESP) above which soils remain flocculated*

attractive forces. Thus when the clay mass is in contact with water individual clay particles deflocculate to form a suspension.

There is a threshold value for total cation concentration (TCC) in the pore water (for a given ESP) above which the soil remains flocculated. Figure 5.11 shows the zones in which a soil of a given ESP can exist in either a dispersed or flocculated state, dependent on the concentration of the salts in the pore water. A soil with a high ESP and initially in a flocculated state (due to high salt content of the pore water) can be rendered dispersive by leaching of salts from the pore water.

The sodium absorption ratio (SAR) is used to quantify the role of sodium where free salts are present in the pore water and is defined as

$$\text{SAR} = \text{Na}/\frac{(\text{Ca} + \text{Mg})}{2} \tag{5.5}$$

with units expressed in meq/litre. If no free salts are present, then the use of SAR to help define dispersive soils is not applicable.

With only few exceptions dispersive clays are found in regions lying between Weinert's (1973) climatic N values 2 and 10. Dispersive clays develop in low-lying areas where the rainfall is such that seepage water has a high SAR, especially in regions where the N values are higher than 2. Soils formed on granites are especially prone to high ESP values. In more arid regions where the *N* values exceed 10 the development of dispersive soils is generally inhibited by the presence of free salts, despite high SAR values. Highly dispersive soils, however, can develop if the free salts with high SAR values are removed by leaching.

Although a number of tests have been used to identify dispersive clay soils no single test can be relied on completely to identify them. One of the simplest tests is the crumb test. This involves preparing a specimen by lightly compressing a moist soil sample into a cube 15 mm per side. The specimen then is placed in 250 ml of distilled water. The tendency for colloidal sized particles to deflocculate and to go into suspension is observed as the specimen begins to hydrate. Four grades can be distinguished, namely, no reaction, slight reaction, moderate reaction and strong reaction. Although the crumb test generally gives a good indication of the potential erodibility of clay soils, a dispersive soil sometimes may give a non-dispersive reaction.

In the dispersion or double hydrometer test the particle size distribution is first measured using the standard hydrometer test in which the soil sample is dispersed by a chemical dispersant and mechanical agitation. Then a second test is carried out without the soil being dispersed. Hence two particle size distribution curves are derived and the difference between the two provides a measure of the tendency of the clay to disperse naturally. The content of 5 μm size which goes into suspension in the latter test is expressed as a percentage of 5 μm size measured in the standard test. For ratios greater than 50% the soil is considered highly dispersive, whereas if it is less than 30% it is non-dispersive.

In the pinhole test distilled water is caused to flow through a 1.0 mm diameter hole punched in a specimen of clay 50 mm thick (Sherard *et al.*, 1976a). Water is passed through the pinhole under heads of 50, 180 and 380 mm and the flow rate and effluent turbidity are recorded. The flow emerging from a specimen of dispersive clay is coloured with a colloidal cloud and does not clear with time. Within 10 min the hole enlarges to about 3 mm or larger and the test is completed. The flow of water for erosion resistant clay is clear or becomes so in a few seconds and the hole does not erode.

The results of physical methods, namely, the crumb test, the double hydrometer test and the pinhole test will be invalid if the water quality is not considered or if the soil solution contains soluble free salts. Gerber and Harmse (1987), for example, found that these methods were unable to identify dispersive soils when free salts were present which is frequently the case with sodium saturated soils.

The standard test used by soil scientists is the soluble salts in pore water test whereby the soil is mixed with distilled water to a consistency near its liquid limit. Then a sample of pore water (the saturation extract) is drawn from the soil and is tested to determine the amount of calcium, magnesium, sodium and potassium in milli-equivalents per litre. The total dissolved salts (TDS) is regarded as the total of these four cations. The sodium content is expressed as a percentage of TDS. Sherard *et al.* (1976b) used the relationship between percentage sodium and TDS to derive Figure 5.12. The relationships shown therein are only valid for relatively pure eroding waters (such as TDS less than 0.5 meq/l or about 300 mg/l). In fact Sherard *et al.* found that soils with high pore water TDS and 50% sodium, or more,

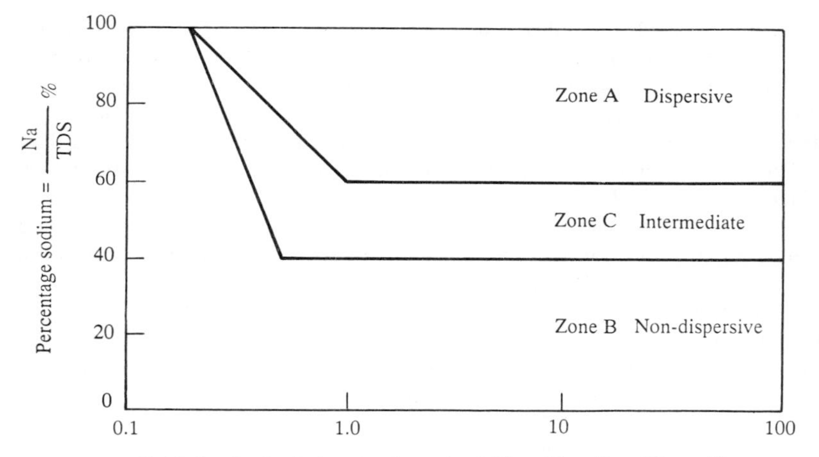

Figure 5.12 *Potential dispersiveness chart (after Sherard* et al., *1976b)*

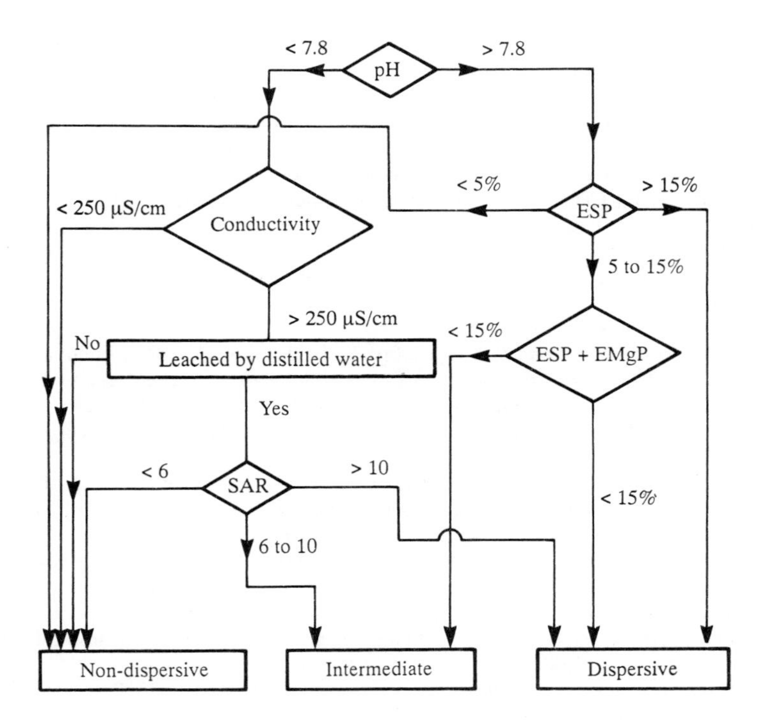

Figure 5.13 *Procedure for identification of dispersive soils (after Gerber and Harmse, 1987)*

are dispersive. They also noted that most soils with low TDS (less than 1.0 meq/l) were non-dispersive in the pinhole test even when the sodium content was well above 60%. None the less Gerber and Harmse (1987) noted that the determination of cation exchange capacity by the summation of extractable cations is not valid if soluble salts occur in soil solution or if the base saturation of the soil is less than 100%. They recommended another procedure for identifying dispersive soils which is outlined in Figure 5.13.

These various tests show that the boundary between deflocculated and flocculated states varies considerably among different clays so that the transition between dispersive and non-dispersive soils is wide. Because there is no clearly defined boundary between dispersive and non-dispersive soils it is usual to do all four tests mentioned above on at least a proportion of all samples, although the pinhole test supplemented by a few hydrometer and saturation extract tests usually can provide sufficient information to confirm the dispersivity of soils.

References

Allam, M. M. and Sridharan, A. (1981). 'Effect of wetting and drying on shear strength', *Proc. ASCE, Geot. Engg Div.*, **107**, GT4, 421–37.

Anon (1990). 'Tropical residual soils', Geological Society, Engineering Group Working Party Report. *Quarterly J. Engg Geol.*, **23**, 1–101.

Blight, G. E. (1990). 'Construction in tropical soils', *Proc. 2nd Int. Conf. Geomech. in Tropical Soils*, Singapore, A. A. Balkema, Rotterdam, **2**, 449–68.

Boyce, J. R. (1985). 'Some observations on the residual strength of tropical soils', *Proc. 1st Int. Conf. on Geomech. of Tropical Lateritic and Saprolitic Soils*, Brasilia, **1**, 235–51.

Braithwaite, C. J. R. (1983). 'Calcrete and other soils in Quaternary limestones: structures processes and applications', *J. Geol. Soc.*, **140**, 351–64.

Brand, E. W. (1985). 'Geotechnical engineering in tropical residual soils', *Proc. 1st Int. Conf. on Geomech. in Tropical Lateritic and Saprolitic Soils*, Brasilia, **3**, 23–100

Buchanan, F. (1807). *A Journey from Madras Through the Countries of Mysore, Canara and Malabar*, Vol. 3, Eastd India Company, London.

Charman, J. M. (1988). *Laterite in Road Pavements*. Special Publication 47, *CIRIA*, London.

Clare, K. E. (1957). 'Airfield construction on overseas soils. Part 1 – The formation, classification and characteristics of tropical soils', *Proc. Inst. Civ. Eng.*, **36**, 211–22.

Clare, K. E. (1957) 'Airfield construction on overseas soils. Part 2 – Tropical black clays, *Proc. Inst. Civ. Eng.*, **36**, 211–22.

De Graft-Johnson, J. W. S. and Bhatia, H. S. (1969). 'Engineering characteritics of lateritic soils', *Proc. Speciality Session on Engineering Properties of Lateritic Soils, 7th Int. Conf. on Soil Mechanics and Foundation Engineering*, Mexico City, **2**, 13–43.

De Graft-Johnson, J. W. S., Bhatia, H. S. and Yeboa, S. L. (1972). *Influence of Geology and Physical Properties on Strength Characteristics of Lateritic Gravels for Road Pavements.* Highway Research Record No 405, Highway Research Board, Washington D.C., 87–104.

Dixon, H. H. and Robertson, R. H. S. (1970). 'Some engineering experiences in tropical soils', *Q. J. Engg Geol.*, **3**, 137–50.

Duchaufour, P. (1982). *Pedology, Pedogenesis and Classification* (Translated by Paton, T. R.). George Allen and Unwin, London.

Dumbleton, M. J. (1967). 'Origin and mineralogy of African red clays and Keuper Marl', *Q. J. Engg Geol.*, **1**, 39–46.

Elges, H. F. W. K. (1985). 'Dispersive soils', *The Civil Engineer in South Africa,* **27**, 347–53.

Epps, R. J. (1980). 'Geotechnical practice and ground conditions in coastal areas of the United Arab Emirates', *Ground Engg,* **13**, No 3, 19–25.

Fookes, P. G. (1978). 'Engineering problems associated with ground conditions in the Middle East: inherent ground problems', *Q. J. Engg Geol.,* **11**, 33–50.

Foss, I. (1973). 'Red soil from Kenya as a foundation material', *Proc. 8th Int. Conf. Soil Mech. Found. Engg,* Moscow, **2**, 73–80.

Gerber, A. and Harmse, H. J. von M. (1987). 'Proposed procedure for identification of dispersive soils by chemical testing', *The Civil Engineer in South Africa,* **29**, 397–9.

Gidigasu, M. D. (1974). 'Degree of weathering and identification of laterite materials for engineering purposes – a review', *Engg Geol.,* **8**, 213–66.

Gidigasu, M. D. (1976). *Laterite Soil Engineering*. Elsevier, Amsterdam.

Hobbs, P. R. N., Culshaw, M. G., Northmore, K. J., Rachlan, A. and Entwhisle, D. C. (1990). 'Preliminary consolidation and triaxial strength test results for some undisturbed tropical red clay soils from West Java, Indonesia', *Proc. 2nd Int. Conf. on Geomechanics in Tropical Soils,* Singapore, A. A. Balkema, Rotterdam, **1**, 149–55.

Horn, A. (1982). 'Swell and creep properties of an African black clay', *Proc. ASCE Geot. Engg Div.,* Speciality Conference on Engineering and Construction in Tropical and Residual Soils, Honolulu, 199–215.

Horta, J. C. de S. O. (1980). 'Calcrete, gypcrete and soil classification', *Engg Geol.,* **15**, 15–52.

Lumb, P. (1965). 'The residual soils of Hong Kong', *Geotechnique,* **15**, 180–94.

Lupini, J. F., Skinner, A. E., and Vaughan, P. R. (1981). 'The drained residual strength of cohesive soils', *Geotechnique,* **31**, 181–213.

McDaniel, T. N. and Decker, R. S. (1979). 'Dispersive soil problem at Los Esteros Dam', *Proc. ASCE, J. Geot. Engg Div.,* **105**, GT9, 1017–30.

Madu, R. M. (1977). 'An investigation into the geotechnical properties of some laterites of eastern Nigeria', *Engg Geol.,* **11**, 101–25.

Malomo, S. (1989). 'Microstructural investigation on laterite soils', *Bull. Int. Assoc. Engg Geol.*, No 39, 105–9.

Mitchell, J. K. and Sitar, N. (1982). 'Engineering properties of tropical residual soils', *Proc. ASCE Geot. Engg Div.,* Speciality Conference on Engineering and Construction in Tropical and Residual Soils, Honolulu, 30–57.

Morin, W. J. (1982). 'Characteristics of tropical red soils', *Proc. ASCE Geot. Engg Div.,* Speciality Conference on Engineering and Construction in Tropical and Residual Soils, Honolulu, 172–98.

Morin, W. J. and Toder, P. C. (1975). 'Laterite and lateritic soils and other problem soils in the tropics', in *Engineering Evaluation and Highway Design Study*, US Agency for International Development, AID/csd 3682, Lyon Associates Inc, Baltimore, USA.

Netterberg, F. (1980). 'Geology of South African calcretes: 1 Terminology, description and classification', *Trans. Geol. Soc. South Africa,* **83**, 255–83.

Nixon, I. K. and Skipp, B. O. (1957). 'Airfield construction on overseas soils, Part 5 – Laterite', *Proc. Inst. Civ. Eng.,* **36**, 253–75.

Ola, S. A. (1978a). 'Geotechnical properties and behaviour of stabilized lateritic soils', *Q. J. Geol. Soc.,* **11**, 145–60.

Ola, S. A. (1978b). 'The geology and engineering properties of black cotton soils in north eastern Nigeria', *Engg Geol.,* **12**, 375–91.

Ola, S. A. (1980). 'Mineralogical properties of some Nigerian residual soils in relation with building problems', *Engg Geol.,* **15**, 1–13.

Sabba Rao, K. S., Sivapullaiah, P. V. and Padmanabha, J. R. (1990). 'Influence of climate on the properties of tropical soils', *Proc. 2nd Int. Conf. Geomech. in Tropical Soils*, Singapore, A. A. Balkema, Rotterdam, **1**, 49–54.

Sherard, J. L., Dunnigan, L. P., Decker, R. S. and Steele, E. F. (1976a). 'Pinhole test for identifying dispersive soils', *Proc. ASCE, J. Geot. Engg Div.*, **102**, GT1, 69–85.

Sherard, J. L., Dunnigan, L. P. and Decker R. S. (1976b). 'Identification and nature of dispersive soils', *Proc. ASCE, J. Geot. Engg Div.*, **102,** GT4, 287–301.

Sridharan, A. (1990). 'Engineering properties of tropical soils', *Proc. 2nd Int. Conf. on Geomechanics in Tropical Soils*, Singapore, A. A. Balkema, Rotterdam, **2,** 537–40.

Tadanier, R. and Ingles, O. G. (1985). 'Soil security test for water retaining structures', *Proc. ASCE, J. Geot. Engg Div.*, **111**, GT3, 289–301.

Townsend, F. C. (1985). 'Geotechnical characteristics of residual soils', *Proc. ASCE, J. Geot. Engg Div.*, **111**, 77–93.

Tuncer, R. E. and Lohnes, R. A. (1977). 'An engineering classification of certain basalt derived lateritic soils', *Engg Geol.*, **11**, 319–39.

Vargas, M. (1974). 'Engineering properties of residual soils from south-central region of Brazil', *Proc. 2nd Int. Conf. Int. Assoc. Eng. Geol.*, Sao Paulo, **1**.

Vargas, M. (1990). 'Collapsible and expansive soils in Brazil', *Proc. 2nd Int. Conf. on Geomechanics in Tropical Soils*, Singapore, A. A. Balkema, Rotterdam, **2,** 489–92.

Vaughan, P. R. (1990). 'Characterizing the mechanical propertis of in-situ residual soil', *Proc. 2nd Int. Conf. on Geomechanics in Tropical Soils*, Singapore, A. A. Balkema, Rotterdam, **2**, 469–87.

Vaughan, P. R., Maccarini, M. and Mokhtar, S. M. (1988). 'Indexing the properties of residual soil', *Q. J. Engg Geol.*, **21**, 69–84.

Weinert, H. H. (1973). 'A climatic index of weathering and its application to road engineering', *Geotechnique*, **23**, 471–94.

Chapter 6

Organic soils: peat

6.1 Introduction

Peat represents an accumulation of partially decomposed and disintegrated plant remains which have been preserved under conditions of incomplete aeration and high water content. It accumulates wherever the conditions are suitable, that is, in areas where there is an excess of rainfall and the ground is poorly drained, irrespective of latitude or altitude. Nonetheless peat deposits tend to be most common in those regions with a comparatively cold wet climate. Physico-chemical and biochemical processes cause this organic material to remain in a state of preservation over a long period of time. In other words waterlogged poorly drained conditions not only favour the growth of particular types of vegetation but they also help preserve the plant remains. The chemistry of the water associated with peatlands influences the type of vegetation which grows there; this especially applies to the pH value of the water.

Peat accumulates when the rate of addition of dry matter exceeds the rate of decay. Nonetheless the rate of peat accumulation is exceedingly slow and the surface of the peat rises at a slower rate than that of addition because of anaerobic decay in the catotelm and consolidation of the peat during drier periods. The acrotelm is the relatively thin upper layer of peat in which aerobic bacteria are active. Beneath this occurs the catotelm which extends downwards to the stratum on which the peat rests.

All present day surface deposits of peat in northern Europe, Asia and Canada have accumulated since the last ice age and therefore have formed during the last 20 000 years. On the other hand, some buried peats may have been developed during inter-glacial periods. Peats also have accumulated in post-glacial lakes and marshes where they are interbedded with silts and muds. Similarly they may be associated with salt marshes. Fen deposits are thought to have developed in relation to the eustatic changes in sea level which occurred after the retreat of the last ice sheets. In Britain the most notable fen deposits are found south of the Wash. Similar deposits are found in Suffolk and Somerset. These are areas where layers of peat interdigitate with wedges of estuarine silt and clay. However, the blanket bog is a more common type of peat deposit. These deposits are found on cool wet uplands.

Approximately 95% of all deposits of peat have been formed from plants growing under aerobic conditions. The high water-holding capacity of peat maintains a surplus of water, which ensures continued plant growth and consequent

Table 6.1 *Mire stages, morphology, flora and associated properties of some British peats (after Hobbs, 1986)*

Stage and mire type		*Morphology (simplified and diagrammatic)*	*Nutrient*	
			Source	*State*
THIRD STAGE Above perimeter ground surface	Blanket or hill bog	Bog peat	Rainfall only	Poor
	Raised bog or bog		Ombrogenous	Oligotrophic
SECOND STAGE Above perimeter g.w. level Below perimeter ground surface	Transition		Mixed	Mixed
	Basin bog	Basin bog peat poor fen peat transition peat	Rainfall and run-off	Mesotrophic
	Fen Carr	Fen peat		
FIRST STAGE Below ground water level	Fen	wl Mud	Groundwater run-off and rainfall	Rich Eutrophic Minerotrophic
	Swamp	wl	Flowing water run-off rainfall	ditto
	Lake	Mud	Rheotrophic	

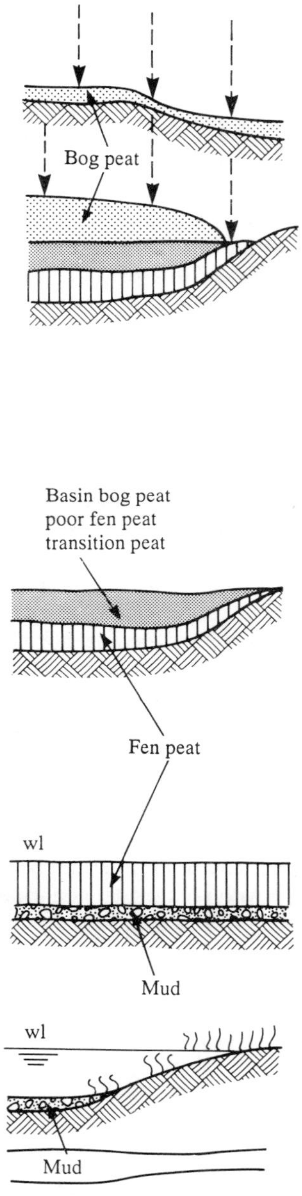

* More commonly known as cotton grass and deer grass, but are true sedges (Cyperaceae)
† The most prominent bog plant generally
‡ The common reed (*Phragmites australis*) is actually a member of the grass family (Poaceae)

Common plant communities		*CEA§*	*pH of peat***	*Specific gravity*	*Organic content*	*Water content (%)*	*Liquid limit (%)*	*Plastic limit test*	*Permeability*	*Remarks*
Cloudberry Ling Heather Purple moor grass Cotton sedge* Deer sedge* Sphagna† Bog asphodel	The Sphagnum cover plant community	High in bog plant communities	<4	1.4	>98%	2000 to 1000	1500 to 900	Not possible on pure bog peat		Bogs invaded by pine & birch under drier climatic conditions.
Willow Alder Sallow	Fen Carr invaded by Sphagna and other bog species	↑	Decreasing as conditions become more ombrogenous →	1.4 to 1.6	>80% ↑	1000 to 500	900 to 600		Permeability is higher in more fibrous less humified peats →	Transition peats can be very variable. Properties intermediate between fen & bog.
Fen mosses Spearwort Meadow rue Purple loosestrife Saw sedge Sedges		Low in fen and transition plant communities	>5	1.6	Increasing with height above the substrate ↑	500 to 200	600 to 200			
Common reed‡ Reedmaces Rushes								Generally possible on fen peats		Very rich in species particularly under calcareous conditions pH >6
Water lily Submerged plants				to 2.5	<10%	<200	<200			Acid fens also occur poorer in species pH <5

§ Cation exchange capacity
** Mire water tends to have somewhat higher values

peat accumulation. The rate of decomposition of plant detritus is relatively rapid under aerobic conditions but is slowed down several thousandfold under anaerobic conditions. Drying out, groundwater fluctuations and snow loading bring about compression in the upper layers of a peat deposit. Indeed these mechanisms are often more important as far as near surface compression is concerned than effective overburden pressure. This is because the unit weight of peak may be similar to that of water. As the water table in peat generally is near the surface, the effective overburden pressure is negligible.

Macroscopically peaty material has been divided by Radforth (1952) into three basic groups, namely, amorphous granular, coarse fibrous and fine fibrous peat. The amorphous granular peats have a high colloidal fraction, holding most of their water in an adsorbed rather than a free state, the absorption occurring around the grain structure. Landva and Pheeney (1980), however, maintained that the term *amorphous granular* should be reserved for non-fibrous, truly amorphous peats only. They contended that most such material referred to in engineering literature is in fact moss peat, actual amorphous granular peat being rare. They suggested, therefore, that the term should be used with caution. In the other two types the peat is composed of fibres, these usually being woody. In the coarse variety a mesh of second-order size exists within the interstices of the first-order network, whilst in fine fibrous peat the interstices are very small and contain colloidal matter.

6.2 Development of peatlands and types of peat deposit

Peatlands may pass through a number of morphological stages, each with its own particular plant communities which characterize the type of peat which develops (Table 6.1). These stages are very much affected by the existing hydrological conditions. The rheotrophic stage is associated with development in water in lakes, basins and valleys. Lakes and basins are gradually filled by sediments but surface run-off continues to bring both nutrients and sediments into the area allowing colonization by vegetation. The vegetation contributes plant remains in greater and greater abundance to the sedimentary deposit, the muds therefore contain an increasingly higher amount of peat. These highly organic muds (reed peat) eventually give way to sedge peat (i.e. the product of the plant community). The peat land eventually adopts a marsh-like landscape which has been referred to as fen.

Valley bogs form along the flatter parts of valley bottoms and generally occur as a result of water draining from relatively acidic rocks. The bog has a complex lateral zonation due to the differences in the vegetation developed, for example, it is richer along the borders of the bog and along streams flowing in the valley.

Peatlands may be formed by swamping, that is, paludification. These peatlands, unlike those which form as a result of terrestrialization, normally are not underlain by soft muds as the vegetation grows directly upon the ground beneath.

Climatic changes affect plant growth in peatlands. A period in which increased rainfall occurs enhances peat growth and development. By contrast a long drier period means that water levels fall. This gives rise to shrinkage and wastage of peat at the surface whilst the peat below is compacted.

Wastage of peat also occurs as a result of permanent drainage works. Drainage lowers the water table which normally reduces the growth of vegetation and means that oxygen begins to invade the anaerobic zone. This, together with the higher temperature, enhances aerobic decay and associated humification. Furthermore the effective pressure is raised when the water table declines causing compression of the peat. For instance, a fall of 1 m imposes an extra load of 10 kN/m^2 which can lead to approximately 1.5 m of settlement in a layer of peat 10 m in thickness if the water level is maintained at 1 m below the subsided surface for a year. Capillary suction and desiccation of peat above the water table lead to its shrinkage. As there is no loss of material these processes of reduction in thickness of peat are not included within wastage.

As can be inferred from above, fen peat frequently is underlain by very soft organic muds. Next follows a transition stage when, because of upward growth, the water to the peatlands is supplied more and more by direct precipitation. The peat generally is mixed and woody. Lastly, the ombrotrophic stage is reached when the peatland grows beyond the maximum physical limits of its groundwater supply and therefore relies entirely for its water supply on direct precipitation. The peat itself acts as a reservoir holding water above groundwater level. The water associated with such peatlands is typically acidic and they are termed raised bogs.

The differences between fen and bog peat are attributable to the types of plant remains which occur in the peat and their modes of origin. The differences involve the degree of humification, structure, fabric and proportion of mineral material contained in the peat and these, in turn, affect its engineering behaviour.

Some fen peats in Britain, because they occur in areas where carbonate rocks such as chalk and limestone are present, are not associated with acidic water, the acidity being reduced by the lime-rich water contributed to the fen by run-off. Acid fens have poorer plant communities and are less humified because of their acidity than those fen peats where the water is slightly alkaline. In fact where there is a rich supply of nutrients brought into the peatland by lime-rich surface run-off the plant communities are much more diverse and give rise to what is sometimes called rich fen. Rich fen peat develops a much higher degree of humification than acid peat. Because the strength and permeability of peat declines significantly as humification increases, rich fen peat presents more problems to the engineer than acid fen peat.

Blanket bogs are associated with wet upland areas. The peat develops where slopes are not excessive and drainage is impeded. The process is often one of paludification and it may start in shallow waterlogged depressions. Bogs extend downslope if poorly drained surface water gives rise to waterlogging. In fact high rainfall in such cool upland areas gives rise to leaching which results in the accumulation of impervious humus colloids and iron pan at small distances below the surface, usually between 0.3 and 1.0 m. Such an impermeable layer gives rise to waterlogging which represents ideal conditions for the development of ombrotrophic peat. Blanket bogs generally are not underlain by soft sands and clays. Generally these peats are thinner at higher altitudes and on steeper slopes. Significant thicknesses are attained only in large deep depressions in the surface topography.

Water flows very slowly along well defined water courses on the surface of large bogs. These water courses give rise to zones of richer vegetation. Hence the character of peat formed along these zones is somewhat different and so the peat beneath a blanket bog is heterogeneous.

6.3 Humification of peat

Humification involves the loss of organic matter either as gas or in solution, the disappearance of physical structure and change in chemical state. Breakdown of the plant remains is brought about by soil microflora, bacteria and fungi which are responsible for aerobic decay. Hence the end products of humification are carbon dioxide and water, the process being essentially one of biochemical oxidation. Immersion in water reduces the oxygen supply enormously which, in turn, reduces aerobic microbial activity and encourages anaerobic decay which is much less rapid. This results in the accumulation of partially decayed plant material as peat.

The acrotelm is the uppermost layer of peat, usually between 100 and 600 mm in thickness, which is composed of undecayed fibrous plant matter. The catotelm occurs beneath the acrotelm. Because anaerobic conditions exist in the catotelm the gaseous products of decomposition are methane (CH_4), ammonia (NH_3), sulphuretted hydrogen (H_2S); and other sulphides and compounds devoid of oxygen also are produced.

Metabolic activity, apart from the supply of oxygen, is very much influenced by temperature, acidity and availability of nitrogen. Normally the higher the temperature and pH value the faster decomposition occurs, hence the slower the accumulation of peat in relation to plant production. The optimum temperature for the decay of plant debris seems to fall within the range 35–40°C. Turning to pH value, decomposition generally tends to be most active in neutral to weakly alkaline conditions (pH values 7–7.5). The more acid the peat, the better are the plant remains preserved. The acidity of a peatland depends upon the rock types in the area draining into the peatland, the types of plants growing there, the supply of oxygen and the concentration of humic acids. In blanket and raised bogs the rate of decomposition is slow since the pH values frequently are in the range 3.3–4.3. As mentioned above, the pH value in fen peats may be neutral or slightly alkaline. Normally decomposition takes place more rapidly as the amount of available nitrogen increases. Plants associated with ombrogenous peats have a low nitrogen content in relation to carbon, this being another reason for their slow decay.

The loss of organic matter and change of chemical state accompanies the breakdown of cellulose within plant tissues so that detritus gradually becomes increasingly finer until all trace of fibrous structure disappears. The peat then has an amorphous granular appearance, the material consisting principally of gelatinous organic acids which have a sponge-like fabric. The degree of humification varies throughout peats since some plants are more resistant than others and certain parts of plants are more resistant than others.

Decay is most intense at the surface where aerobic conditions exist, it occurs at a much slower rate throughout the mass of the peat. Because different types of plants decay at different rates peat may be dominated by a small number of species.

Table 6.2 *Humification of peat (after Von Post, 1922)*

Degree of humification	*Decomposition*	*Plant structure*	*Content of amorphous material*	*Material extruded on squeezing (passing between fingers)*	*Nature of residue*
H_1	None	Easily identified	None	Clear, colourless water	
H_2	Insignificant	Easily identified	None	Yellowish water	
H_3	Very slight	Still identifiable	Slight	Brown, muddy water; no peat	Not pasty
H_4	Slight	Not easily identified	Some	Dark brown, muddy water; no peat	Somewhat pasty
H_5	Moderate	Recognizable, but vague	Considerable	Muddy water and some peat	Strongly pasty
H_6	Moderately strong	Indistinct (more distinct after squeezing)	Considerable	About one-third of peat squeezed out; water dark brown	
H_7	Strong	Faintly recognizable	High	About half of peat squeezed out; any water very dark brown	Fibres and roots more resistant to decomposition (applies to H_6–H_8)
H_8	Very strong	Very indistinct	High	About two-thirds of peat squeezed out; also some pasty water	
H_9	Nearly complete	Almost unrecognizable		Nearly all the peat squeezed out as a fairly uniform paste	
H_{10}	Complete	Not discernible		All the peat passes between the fingers; no free water visible	

The change undergone in peat as a result of increasing humification is not uniform since the fibres are reduced in size and strength in an irregular manner as the quantity of totally humified peat increases. Generally the fresher the peat, the more fibrous material it contains and, as far as engineering is concerned, the more fibrous the peat, the higher the tensile and shear strength, void ratio and water content.

6.4 Description and classification of peat

As far as the description of peat is concerned Hobbs (1986) suggested that 10 characteristics should be included in a full description of peat. These were colour, which can provide a guide to the state of the peat; degree of humification on a scale of 1 to 10 (Table 6.2); moisture content (the water content usually increases in amount from fen peat, through transition peat to bog peat, that is, from wet to extremely wet); principal plant components, namely coarse fibre, fine fibre, amorphous granular material and woody material; amount of mineral constituents; smell, notably the distribution of H_2S (methane has no smell); pH value; tensile strength; and lastly any special characteristics. Hobbs (1986) suggested the following grades of ignition loss determination of organic content: N_1 = 20–40%; N_2 = 40–60%, N_3 = 60–80%, N_4 = 80–95%. In addition, he recommended the following scale for tensile strength: T_0 = zero strength; T_1 = low, less than 2 kN/m^2; T_2 = moderate, 2 to 10 kN/m^2; T_3 = high, greater than 10 kN/m^2. The tensile strength may differ in the vertical and horizontal directions. Lastly the pH value can simply be described as acid, neutral or alkaline.

Previously Landva and Pheeney (1980) had suggested that the proper identification of peat should include a description of the constituents since the fibres of some plant remains are much stronger than others and are consequently of interest to the engineer. The degree of humification and water content should also be taken into account. They proposed a modified form of the Von Post (1922) classification of peat (Table 6.3).

6.5 Basic properties of peat

The void ratio of peat ranges between 9, for dense amorphous granular peat, up to 25, for fibrous types with high contents of sphagnum. It usually tends to decrease with depth within a peat deposit. Such high void ratios give rise to phenomenally high water content (Figure 6.1). The latter is the most distinctive characteristic of peat. Indeed most of the differences in the physical characteristics of peat are attributable to the amount of moisture present.

The reason for the phenomenally high water content is because the walls of cell tissue are microscopically thin, which means that very little solid material can be disseminated throughout an essentially liquid mass. Generally most water, especially in fibrous peats, occurs as free water in the large pores; it also occurs as capillary water in the small pores; and as adsorbed, chemically bound, colloidal or

Table 6.3 *Classification of peat (from Landva and Pheeny, 1980)*

(1) *Genera*		(2) *Designation*	(3) *The degree of humidification (H)*		
			H	*Decomposition*	*Plant structure*
Bryales (moss)	= B	With few exceptions peats consist of a mixture of two or more genera. These are listed in decreased order of content, i.e. the principal component first, e.g. ErCS.	H_1	None	Easily identified
Carex (sedge)	= C		H_2	Insignificant	Easily identified
Equisetum (horsetails)	= Eq		H_3	Very slight	Still identifiable
Eriophorum (cotton grass)	= Er		H_4	Slight	Not easily identified
Hypnum (moss)	= H		H_5	Moderate	Recognizable but vague
Lignidi (wood)	= W		H_6	Moderately strong	Indistinct
Nanolignidi (shrubs)	= N		H_7	Strong	Faintly recognizable
Phragmites	= Ph		H_8	Very strong	Very indistinct
Scheuchzeria (aquatic herbs)	= Sch		H_9	Nearly complete	Almost unrecognizable
Sphagnum (moss)	= S		H_{10}	Complete	Not discernible

(4) *Water content (B)*	(5) *Fine fibres (F)*	(6) *Coarse fibres (R)*	(7) *Wood (W) and shrub remnants*
Estimated from a scale of 1 (dry) to 5 (very high) and designated B_1, B_2, etc. Landva and Pheeney suggested the following ranges B_2 less than 500% B_3 500–1000% B_4 1000–2000% B_5 Over 2000%	These are fibres and stems less than 1 mm in diameter, F_0 = nil; F_1 = low content; F_2 = moderate content; F_3 = high content	These fibres have a diameter exceeding 1 mm; R_0 = nil; R_1 = low content; R_2 = moderate content; R_3 = high content.	Wood and shrub is similarly graded: W_0 = nil; W_1 = low content; W_2 = moderate content; W_3 = high content.

The classification is effected by combining the letters from the groups, in the order 1, 2, . . . 7, as appropriate.

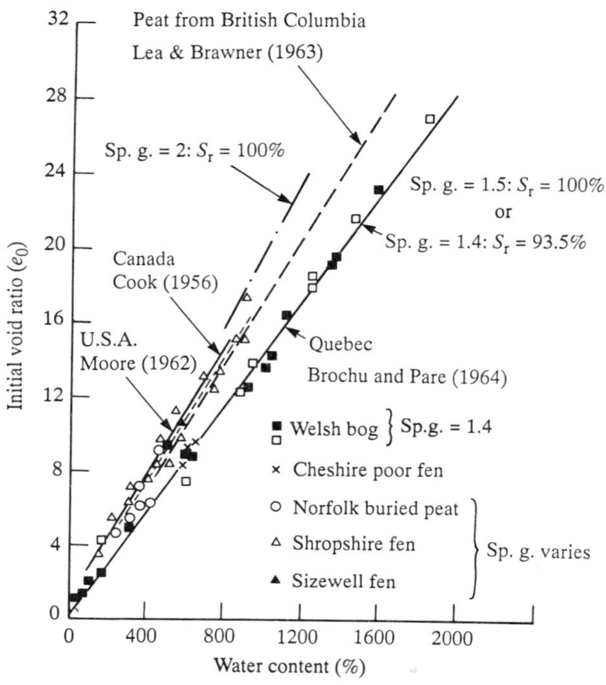

Figure 6.1 *Variation of water content in peat with void ratio (after Hobbs, 1986)*

osmotic water. The free water or intracellular water is under a suction pressure of less than 10 kN/m^2. On the other hand capillary forces hold interparticle water at a suction pressure greater than 10 kN/m^2. The suction pressure does not exceed 20 kN/m^2 in the case of adsorbed water. Only intracellular and interparticle water can be expelled by consolidation. The proportion of water held in each of these two states, as well as total amount, depends primarily on the morphology and structure of the material present and on the degree of humification. Fen peats have lower interparticle and total water contents than bog peats, mineral constituents reducing the quantity of interparticle water in fen peats. The degree of humification also reduces the proportion of water held in peat.

According to Wilson (1978) the water content of peat is held in the cells of plant remains, as well as in the voids. Water is also absorbed by the cell walls of the plant detritus. Indeed Landva and Pheeney (1980) estimated that about one-third of the water content of sphagnum peat is located in the voids, the remainder being in the cellular plant material. They further estimated that about half the water in sedge peat was contained in the plant remains. These three types of held water have different drainage characteristics. Water is forced out of the voids when peat undergoes stress. With continuing stress the particles are brought into contact and the cell structure begins to be distorted. Hence the water in the plant cells is pressurized. Some of this water moves through openings in the cell walls, but with increasing stress these begin to rupture. Water is thereby expelled, giving rise to increasing pore pressure in the voids. Wilson indicated that at this point the peat behaves as a material which has become rapidly softened. Further straining and rupture of the cell walls means that shear failure is imminent.

The rigidity and thickness of the zone of adsorbed water is governed by the cation exchange capacity of the tissue and the chemistry of the water in that the higher the cation exchange capacity, the stronger the adsorption complex and the greater the interparticle adherence. The cation exchange capacity has an inverse relationship with the mineral concentration in the water supply. As the supply of nutrients declines a change occurs in the type of plants, the cation exchange capacity increases, as does the water content, and the adsorption complex strengthens.

Most of the cation exchange ability in the less organic soils is saturated by metallic cations (Ca, Mg, Na, K) from mineral matter in the soil. As the organic content rises the quantity of exchangeable hydrogen ions slowly increases. In peats most of the ions are strongly adsorbed into the exchange complex. The cation exchange capacity of peats present in ombrotrophic bogs is very similar to that of Na montmorillonite. Fen peats have similar cation exchange capacities to illite. Because the specific gravity of the cell walls of plants is half that of clay minerals the adsorption complex in peat is approximately twice as effective as that in clay. This explains why peats possess very high liquid limits as compared with clays of similar cation exchange capacity. The liquid limit declines as the degree of humification increases, in other words as the adsorption complex is weakened due to the destruction of plant material. Hence fibrous peats have higher liquid limits than amorphous peats.

The water content of peats varies from a few hundreds per cent dry weight (e.g. 500% in some amorphous granular peats) to over 3000% in some coarse fibrous varieties. Put another way the water content may range from 75 to 98% by volume of peat. Moreover changes in the amount of water content can occur over very small distances. This is due to the fact that plant communities differ over the surface of a peat and differences in the character of the peat occur as a result. In addition the decay of peat is by no means uniform and the water content is reduced with increasing humification. The water content also declines with increasing mineral content. Hence fen peats have lower and less variable water contents than bog peats.

A strong relationship exists between the type of peat and the chemistry of the associated water. In the United Kingdom the pH value of fen peat frequently is in excess of 5, whilst that of bog peat is usually less than 4.5 and may be less than 3.

Gas is formed in peat as plant material decays and this tends to take place from the centre of stems, so gas is held within stems. The volume of gas in peat varies and figures of around 5–7.5% have been quoted (Hanrahan, 1954). At this degree of saturation most of the gas is free and so has a significant influence on initial consolidation, rate of consolidation, pore pressure under load and permeability.

The mineral content of organic soils varies appreciably from some peat deposits which are more or less completely free of mineral matter (dry ash contents as low as

2%) to organic muds which may contain some 10% of organic detritus. Bell (1978) mentioned ash contents as high as 50% in some peats found on moors in Yorkshire. The mineral material is usually quartz sand and silt. In many peats the mineral content increases with depth. The specific gravity of peats varies according to the amount of mineral matter contained. Turning to the organic content, this provides some indication of how the peat was formed. As far as engineering is concerned the organic content is important in that it influences the water-holding capacity of organic soils. In the United Kingdom fen peats may be distinguished from bog peats by their specific gravity, as well as their water content (Figure 6.2a).

The bulk density of peat is both low and variable, being related to the organic content, mineral content, water content and degree of saturation. The relationship between bulk density and water content for bog peat (Sp.g. = 1.5) and fen peat with a high mineral content (Sp.g. = 2) for varying degrees of saturation are illustrated in Figure 6.2(b). This shows that above 600% both the specific gravity and water content do not greatly influence bulk density. The primary influence is the degree of saturation or gas content. Peats frequently are not saturated and may be buoyant under water due to the presence of gas. Except at low water contents (less than 500%) with high mineral contents, the average bulk density of peats is slightly lower than that of water. Amorphous granular peat has a higher bulk density than the fibrous types. For instance, in the former it can range up to 1.2 Mg/m^3, whilst in woody fibrous peats it may be half this figure.

However, the dry density is a more important engineering property of peat, influencing its behaviour under load. The dry density itself is influenced by the effective load to which a deposit of peat has been subjected. Hanrahan (1954) recorded dry densities of drained peat within the range 65–120 kg/m^3. The dry density is influenced by the mineral content and higher values than that quoted can be obtained when peats possess high mineral residues. The specific gravity of peat has been found to range from as low as 1.1 up to about 1.8, again being influenced by the content of mineral matter. MacFarlane (1969) reported that fibrous peats in Canada in which the water content was greater than 500% and organic contents in excess of 80%, had specific gravities in the range 1.4–1.7.

Plastic limits can only be obtained from peats which contain a given amount of clay, the clay content required decreasing as the degree of humification increases. According to Hobbs (1986) it is not possible to carry out plastic limit tests on pure bog peats on the one hand even if they are highly humified, or peat whose liquid limit is greater than 1000% on the other. In fact he concluded that there was little point in performing plastic limit testing on peat soils since the plasticity gives little indication of their character.

Liquid limit tests have been performed on peat soils, the liquid limit depending on the type of plant detritus contained (this determines the initial cation exchange capacity), on the degree of humification, and on the proportion of clay soil present. Generally the liquid limit of fen peat, according to Hobbs (1986), ranges from 200 to 600% and bog peat from 800 to 1500%, with transition peats between. The liquid limit, in other words, is reduced by increasing degree of humification. In addition, as the organic content declines so lower values of liquid limit are obtained. Usually fen peats have water contents at or somewhat below their liquid limits. This is

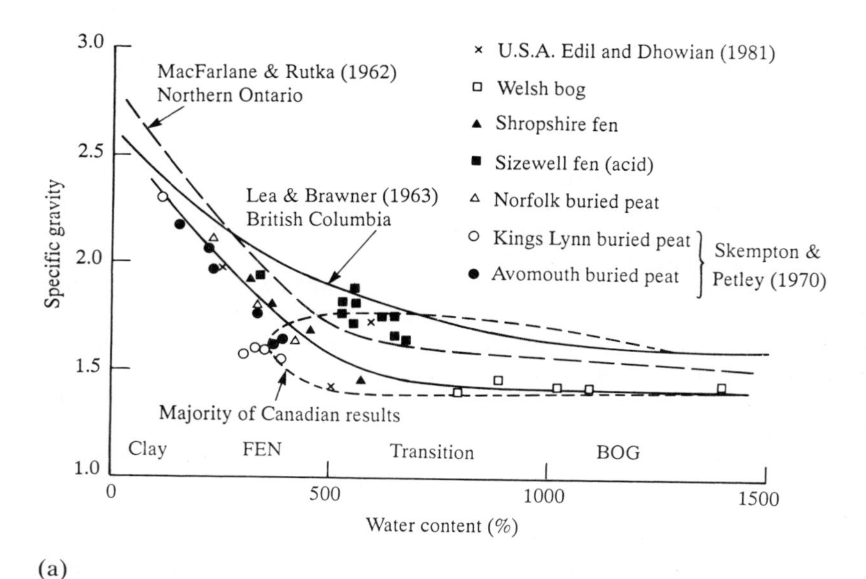

(a)

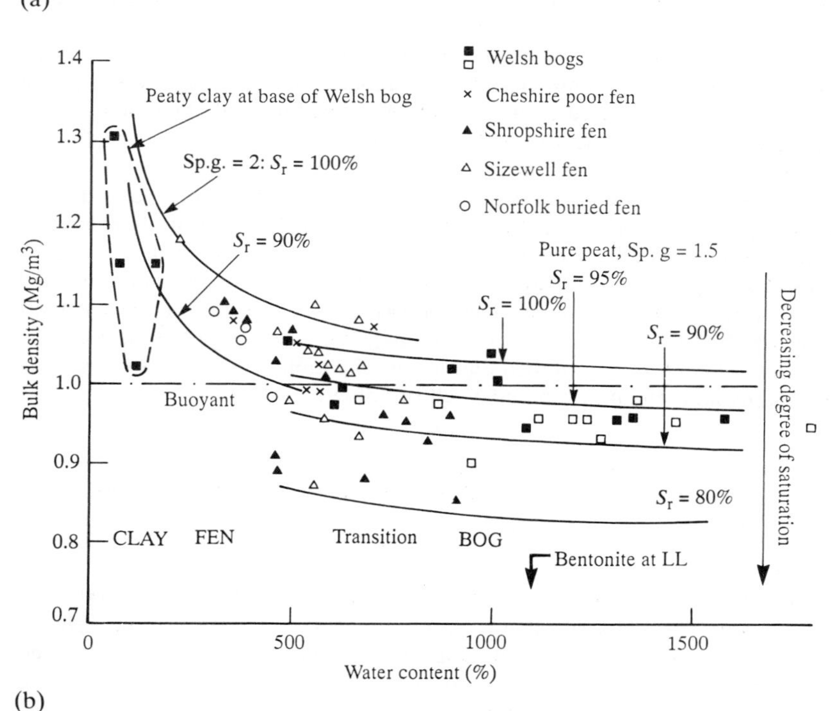

(b)

Figure 6.2 *Variation of water content with* (a) *specific gravity;* (b) *bulk density*

because partially decomposed plant material has a higher cation exchange capacity than any clay which occupies the pores. Bog peats contain less mineral matter and so their water contents exceed their liquid limits.

Since peat has such a high void ratio and water content it undergoes significant shrinkage on drying out. None the less the volumetric shrinkage of peat increases up to a maximum and then remains constant, the volume being reduced almost to the point of complete dehydration. The amount of shrinkage which can occur ranges between 10 and 75% of the original volume of the peat and it can involve reductions in void ratio from over 12 down to about 2. The change in peat is permanent in that it cannot recover all the water lost when wet conditions return. Hobbs (1986) noted that the more highly humified peats, even though they have lower water contents, tend to shrink more than the less humified fibrous peat. He quoted values of linear shrinkage on oven drying between 35 and 45%.

The acrotelm, depending on the nature of the plant remains, has appreciable tensile strength and frequently possesses a high permeability. Permeability tends to decline with depth, from, for example, 10^{-1} m/s to 3×10^{-5} m/s in slightly humified bog peat; in fen peat values as low as 6×10^{-7} m/s have been recorded at the base of the acrotelm.

The permeability of the catotelm is influenced by several factors such as the type of plant detritus and its degree of humification, the fibre content which in turn affects the porosity, the bulk density and the amount of surface loading. The macro- and micro-structure of peat are of obvious importance as far as its permeability is concerned, although such differences are of less significance as the degree of humification increases. Permeability values for highly humified peat in the catotelm may range down to 6×10^{-10} m/s, most values for this layer falling in the range 1×10^{-5} to 5×10^{-8} m/s.

One of the notable characteristics of peat is its appreciable reduction in permeability as its porosity is reduced, for instance, if the porosity is halved its permeability is lowered by some three orders of magnitude. When peat is loaded the particles of plant detritus are compressed closer together and the adsorbed layers of water tend to coalesce. This increases the tortuosity of the channels through which water flows. Hanrahan (1954), for example, showed that the permeability of peat, as determined during consolidation testing, varied according to the loading and length of time involved as follows:

(1) Before test: void ratio = 12; permeability = 4×10^{-6} m/s,
(2) After 7 months of loading at 55 kN/m^2: void ratio = 4.5; permeability = 8×10^{-11} m/s.

Thus after seven months of loading, the permeability of the peat was 50 000 times less than it was originally. Miyahawa (1960) and Adams (1965) also have shown that there is a marked change in the permeability of peat as its volume is reduced under compression. The magnitude of construction pore water pressure is particularly significant in determining the stability of peat. Adams showed that the development of pore pressures in peat beneath embankments was appreciable; in one instance it approached the vertical unit weight of the embankment.

Peat is hydraulically anisotropic in that the horizontal permeability tends to be greater than the vertical. The degree of humification reduces this anisotropy because as humification increases the difference between horizontal and vertical permeability decreases so that in highly humified peats it approaches uniformity. Anisotropic ratios can vary up to 7.5 and tend to increase with loading.

Because of the variability of peat in the field the value of permeability as tested in the laboratory can be misleading.

When loaded, peat deposits undergo high deformations. Nevertheless their values of Young's modulus tend to increase with increasing load. If peat is very fibrous it appears to suffer indefinite deformation without planes of failure developing. On the other hand failure planes nearly always form in dense amorphous granular peats. Hanrahan and Walsh (1965) found that the strain characteristics of peat were independent of the rate of strain and that flow deformation, in their tests, was negligible. Strain often takes place in an erratic fashion in a fibrous peat. This may be due to the different fibres reaching their ultimate strengths at different strain values, the more brittle, woody fibres failing at low strain whilst the non-woody types maintain the overall cohesion of the mass up to much higher strains. The viscous behaviour of peat is generally recognized as being non-Newtonian and the relationship between stress and strain is a function of the void ratio. As the void ratio increases so the effective viscosity increases and hence a certain value of stress produces a correspondingly smaller value of strain rate.

Apart from its moisture content and dry density, the shear strength of a peat deposit appears to be influenced, firstly, by its degree of humification and, secondly, by its mineral content. As both these factors increase so does the shear strength. Conversely the higher the moisture content of peat the lower is its shear strength. As the effective weight of 1 m^3 of undrained peat is approximately 45 times that of 1 m^3 of drained peat the reason for the negligible strength of the latter becomes apparent. Due to its extremely low submerged density, which may be between 15 and 35 kN/m^3, peat is especially prone to rotational failure or failure by spreading, particularly under the action of horizontal seepage forces.

In an undrained bog the unconfined compressive strength is negligible, the peat possessing a consistency approximating to that of a liquid. The strength is increased by drainage to values between 20 and 30 kN/m^2 and the modulus of elasticity to between 100 and 140 kN/m^2. According to Hanrahan (1964) unconfined compressive strengths of up to 70 kN/m^2 are not uncommon in peats consolidated under pavements, a typical modulus of elasticity in such a situation being 700 kN/m^2 (see also Wilson, 1978).

Hobbs (1986) maintained that peat behaves in a similar way to normally consolidated clay, notwithstanding its extremely high water content. He further stated that fen peats, in some respects, appear to be similar to remoulded clays of normal sensitivity whereas bog peats are similar to undisturbed extra-sensitive or quick clays.

6.6 Consolidation and settlement of peat

Peat is a highly compressible material with time-dependent features which, in part, are due to the way in which the pore water is held and then released. As mentioned above, water is held in macropores (intercellular water), in micropores (interparticle water) and as adsorbed water, the amounts of each being governed by the structure of the peat and its degree of humification. Hence the consolidation and rheological behaviour of peat are affected by the distribution of water. As peat decomposes the void ratio declines, so the peat becomes stiffer and its structure is steadily deformed under secondary strain.

If the organic content of a soil exceeds 20% by weight, consolidation becomes increasingly dominated by the behaviour of the organic material. For example, on loading, peat undergoes a decrease in permeability of several orders of magnitude which invalidates the Terzaghi theory of consolidation. The decreasing coefficient of compressibility on loading and thixotropy, as well as the surface activity of organic material, also militate against a precise settlement analysis based on the Terzaghi theory. Moreover residual pore water pressure affects primary consolidation and considerable secondary consolidation further complicates settlement prediction. It must be remembered that primary and secondary consolidation are empirical divisions of a continuous compression process both of which occur simultaneously during part of that process. In fact accurate prediction of the amount and rate of settlement of peat cannot be derived directly from laboratory tests. Hence large-scale field trials seem to be essential for important projects. However, a modified method of settlement prediction for organic soils has been developed by Andersland and Al-Khafaji (1980) which is related to the amount of organic matter in soil. The latter has to be accurately determined in order to use their method.

Consolidation of peat takes place when water is expelled from the pores and the particles undergo some structural rearrangement. Initially the two processes occur at the same time but as the pore water pressure is reduced to a low value, the expulsion of water and structural rearrangement occur as a creep-like process. In other words the initial stage of drainage can be regarded as primary consolidation whereas the stage of continuing creep represents secondary compression.

The quantity of water removed from peat in the later stages of consolidation results in an increase in strength considerably greater than that following the removal of the same quantity during the early stages. What happens to a peat is therefore very largely a function of the structure of the material since this affects the retention and expulsion of water and affords it its strength.

Berry and Vickers (1975) undertook a series of single-increment consolidation tests and found that the higher the load increment ratio, the faster the pore pressure was dissipated. This suggests that embankments should be constructed rapidly over expanses of peat. Previously Barden (1969) had demonstrated from multi-increment tests (such tests indicate the influence of the previous loading cycle on the present cycle) that as the structural viscosity increases with decreasing void ratio the pore water pressure falls very rapidly on subsequent loading. Stability, it has been suggested (Lee and Brawner, 1963), is not a problem when constructing embankments across fibrous peat if its initial permeability is high and it is not underlain by soft clay. Conversely stability problems are likely to be encountered when construction occurs on fen peats with a high degree of humification and lower permeability.

Adams (1963) maintained that the macro- and micro-pores of fibrous peat influence its consolidation. He considered that primary consolidation of such peats took place due to a drainage of water from the macro-pores whilst secondary consolidation was due to the extremely slow drainage of water from the micro-pores into the macro-pores. Because of its higher permeability the rate of primary consolidation of a fine fibrous peat is higher than that of an amorphous granular peat.

Due to the highly viscous water adsorbed around soil particles, amorphous granular peat exhibits a plastic structural resistance to compression and hence has a similar rheological behaviour to that of clay. In this case secondary consolidation is believed to occur as a result of the gradual readjustment of the soil structure to a more stable configuration following the breakdown which occurs during the primary phase due to dissipation of pore water pressure. The rate at which this process takes place is controlled by the highly viscous adsorbed water surrounding each soil particle, the colloidal material which the former contains tending to plug the interstices and thereby reduces permeability.

Wilson *et al.* (1965) suggested that amorphous granular peats exhibit considerable secondary consolidation and therefore settlement. Because of the highly complex structure of such peat they further suggested that it may also exhibit phases of tertiary and quaternary consolidation.

One-dimensional consolidation theories have been developed by Berry and Postkitt (1972) for both amorphous granular and fibrous peat. These consider finite strain, decreasing permeability, compressibility and the influence of secondary compression with time. The different mechanisms involved in secondary compression of these two types of peat were found to give similar non-linear rheological models but their relative creep equations were fundamentally different. That for amorphous granular peat predicts an exponential increase in strain with incremental loading whilst that for fibrous peat predicts a linear increase.

With few exceptions improved drainage has no beneficial effect on the rate of consolidation. This is because efficient drainage only accelerates the completion of primary consolidation which anyhow is completed rapidly.

Peat has a high coefficient of secondary compression, the latter being the dominant process in terms of settlement of peat, and in terms of strain, this is virtually independent of water content and degree of saturation. The short phase of primary consolidation is responsible for little distortion. Bog peats appear to possess lower values of secondary compression than fen peats. This probably is because of their non-plastic, highly frictional character.

The relationships between void ratio and compression index, compression index and liquid limit, and water content and compression index are illustrated in Figures 6.3(a)–(c). In all three cases fen peats can be distinguished from bog peats. There is no continuous trend from fen to bog peats and these two types belong to quite separate domains.

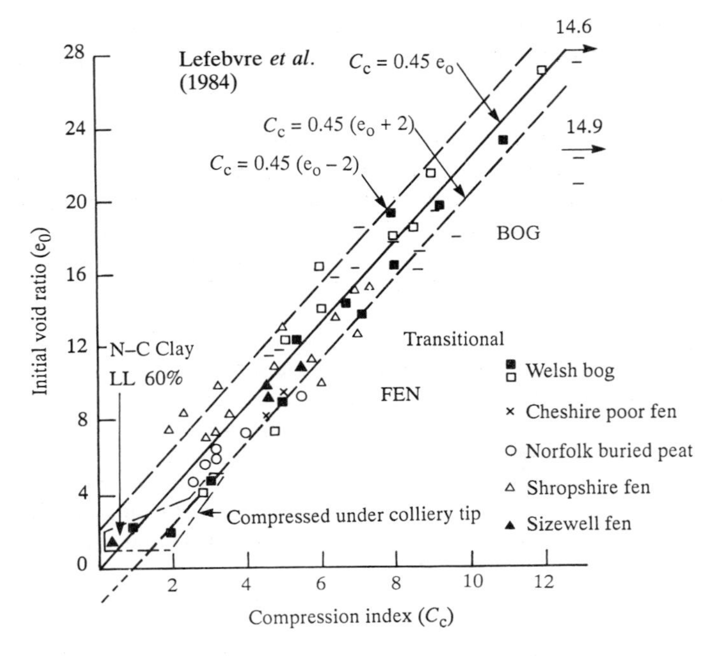

Lefebvre et al. (1984)
Cc = 0.45 eo
Cc = 0.45 (eo + 2)
Cc = 0.45 (eo – 2)
14.6
14.9
BOG
Transitional
FEN
N–C Clay
LL 60%
Compressed under colliery tip
Welsh bog
Cheshire poor fen
Norfolk buried peat
Shropshire fen
Sizewell fen
Initial void ratio (e0)
Compression index (Cc)
0
4
8
12
16
20
24
28
2
6
10

(a)

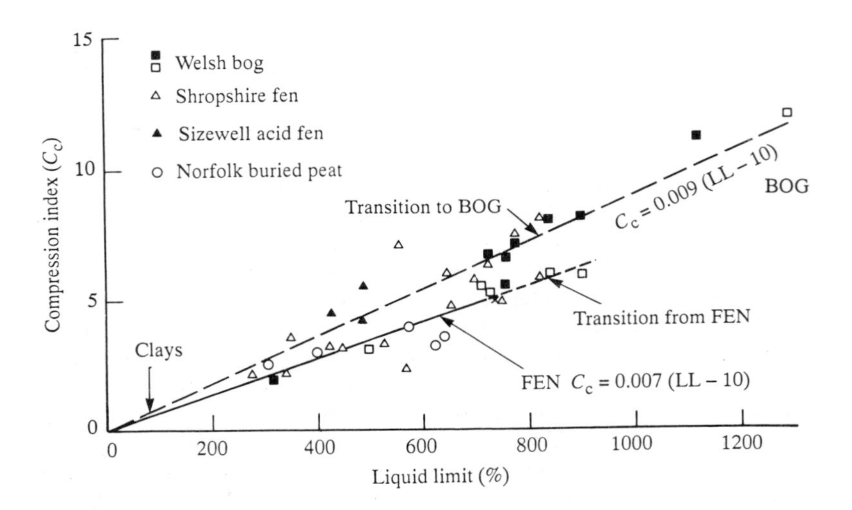

Welsh bog
Shropshire fen
Sizewell acid fen
Norfolk buried peat
Transition to BOG
Cc = 0.009 (LL – 10)
BOG
Transition from FEN
FEN Cc = 0.007 (LL – 10)
Clays
Compression index (Cc)
Liquid limit (%)
0
5
10
15
200
400
600
800
1000
1200

(b)

Differential and excessive settlement is the principal problem confronting the engineer working on a peaty soil. When a load is applied to peat, settlement occurs because of the low lateral resistance offered by the adjacent unloaded peat. Serious shearing stresses are induced even by moderate loads. Worse still, should the loads exceed a given minimum, then settlement may be accompanied by creep, lateral spread, or in extreme cases by rotational slip and upheaval of adjacent ground. At any given time the total settlement in peat due to loading involves settlement with and without volume change. Settlement without volume change is the more serious for it can give rise to the types of failure mentioned. What is more it does not enhance the strength of peat.

Creep does not take place in peat at a constant rate. This is probably due to the increase in density consequent upon consolidation. A good example of the long-term behaviour of peat was given by Buisman (1936) who cited examples of embankments on peat in the Netherlands in which continuous settlement, linear with the logarithm of time, was recorded for more than 80 years.

When peat is compressed the free pore water is expelled under excess hydrostatic pressure. Since the peat is initially quite pervious and the percentage of pore water is high, the magnitude of settlement is large and this period of initial settlement, as remarked above, is short (a matter of days in the field). Adams (1965) showed that the magnitude of initial settlement was directly related to peat thickness and applied load. The original void ratio of a peat soil also influences the rate of initial settlement. Excess pore pressure is almost entirely dissipated during this period. Settlement subsequently continues at a much slower rate which is approximately linear with the logarithm of time. This is because the permeability of the peat is

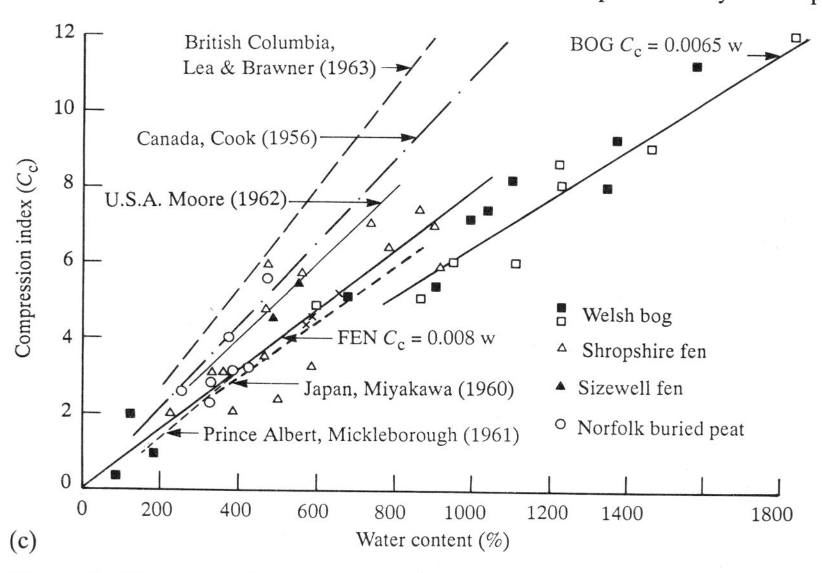

Figure 6.3 *Variation of compression index in peat with* (a) *initial void ratio;* (b) *liquid limit;* (c) *water content (after Hobbs, 1986)*

significantly reduced due to the large decrease in volume. During this period the effective consolidating pressure is transferred from the pore water to the solid peat fabric. The latter is compressible and will only sustain a certain proportion of the total effective stress, depending on the thickness of the peat mass.

Berry *et al.* (1985) gave an account of extensive settlements which have occurred at two housing estates at St Annes, Lancashire, which have resulted from the compression of a layer of peat, 1.3–2.6 m thick, which occurs just below the surface. The two-storey blocks of flats were built between 1957 and 1971. Total settlements ranged from 60 to 260 mm and were estimated from single-increment consolidation tests to continue at 2 mm per year over the following 10 years. The present settlement should allow the blocks to remain serviceable over 10 to 15 years without having to be underpinned.

The use of precompression involving surcharge loading in the construction of embankments across peatlands involves the removal of the surcharge after a certain period of time. This gives rise to some swelling in the compressed peat. Uplift or rebound can be quite significant depending on the actual settlement and surcharge ratio (i.e. the mass of the surcharge in relation to the weight of the fill once the surcharge has been removed). Rebound also is influenced by the amount of secondary compression induced prior to unloading. The swelling index, C_s, is related to the compression index, C_c, in that an average C_s is around 10% of the C_c, within a range of 5–20%. Rebound undergoes a marked increase when surcharge ratios are greater than about 3. Normally rebound in the field is between 2 and 4% of the thickness of the compressed layer of peat before surcharge has been removed. Put another way, Lee and Brawner (1963) found that rebound in the field was approximately 5% of settlement. Subsequently Samson and La Rochelle (1972) showed that rebound during the first one or two days after surcharge had been removed accounted for around one third of the total rebound and that the period over which swelling took place was more or less the same as the length of time the surcharge was imposed.

It follows from above that there is only a small increase in the void ratio after the reduction of the load on a peat deposit, in other words the voids are not restored to their original value, and the compressibility of preconsolidated peat is greatly reduced (Bell, 1978). This can be illustrated from the following figures:

Coefficient of volume compressibility of peat for a range of loading from 13.4 to 26.8 kN/m^2:

(1) Normally loaded m_v = 12.214 m^2/MN
(2) Pre-consolidated m_v = 0.599 m^2/MN

References

Adams, J. I. (1963). 'A comparison of field and laboratory measurement of peat', *Proc. 9th Muskeg Research Conf. NRC-ACSSM*, Tech Memo. 81, 117–35.

Adams, J. I. (1965). 'The engineering behaviour of Canadian muskeg', *Proc. 6th Int. Conf. Soil Mech. Found. Engg*, Montreal, **1**, 3–7.

Al-Khafagi, A. W. N. and Andersland, O. B. (1981). 'Compressibility and strength of decomposing fibre', *Geotechnique,* **31**, 497–508.

Barden, L. (1969). 'Time dependent deformation of normally consolidated clays and peats', *Proc. ASCE, J. Soil Mech. Found. Div.,* **95**, SM1, 1–31.

Bell, F. G. (1978) 'Peat: a note on its geotechnical properties', *Civil Engg*, January, 45–9; February, 49–53.

Berry, P. L. and Poskitt, T. J. (1972). 'The consolidation of peat', *Geotechnique,* **22**, 27–52.

Berry, P. L. and Vickers, B. (1975). 'The consolidation of fibrous peat', *Proc. ASCE, J. Geot. Engg Div.,* **101**, GT8, 741–53.

Berry, P. L., Illsley, D. and McKay, I. R. (1985). 'Settlement of two housing estates at St Annes due to consolidation of a near surface peat stratum', *Proc. Inst. Civ. Engrs.*, Part I, **77**, 111–36.

Brochu, P. A. and Pare, J. J. (1964). 'Construction des routes sur tourbienes deus la Province de Quebec', *Proc. 9th Muskeg Res. Conf.*, NRC-ACSSM, Tech. Memo 81.

Buisman, A. S. K. (1936). 'Results of long duration settlement tests', *Proc. 1st Int. Conf. Soil Mech. Found. Engg*, Cambridge, Mass, **1**, 103–5.

Cook, P. M. (1956) 'Consolidation characteristics of organic soils', *Proc. 9th Canad. Soil Mech. Conf.*, NRC-ACSSM, Tech. Memo. 41.

Edil, T. B. and Dhowian, A. W. (1981). 'At-rest lateral pressure in peat soils', *Proc. ACSE, J. Geot. Engg, Div.*, **102**, GT2.

Hanrahan, E. T. (1954). 'An investigation of some physical properties of peat', *Geotechnique,* **4**, 108–23.

Hanrahan, E. T. (1952). 'The mechanical properties of peat with special reference to road construction', *Trans. Inst. Civ. Engrs*, Ireland, **78**, 179–215.

Hanrahan, E. T. (1964). 'A road failure on peat', *Geotechnique,* **14**, 185–203.

Hanrahan, E. T. and Welsh, J. A. (1965). 'Investigations of the behaviour of peat under varying conditions of stress and strain', *Proc. 6th Int. Conf. Soil Mech. Found. Engg*, Montreal, **1**, 226–30.

Hobbs, N. B. (1986). 'Mire morphology and the properties and behaviour of some British and foreign peats', *Q. J. Engg Geol.*, **19**, 7–80.

Landva, A. O. and Pheeney, P. E. (1980). 'Peat fabric and structure', *Can. Geot. J.,* **17**, 416–35.

Lea, N. D. and Brawner, C. O. (1963). *Highway Design and Construction over Peat Deposits in Lower British Columbia*, Highway Research Record No. 7., Highway Research Board, Washington, D.C.

Lefebvre, G., Langlois, P., Lupien, C. and Lavelle, J. (1984). 'Laboratory testing on *in situ* behaviour of peat as embankment foundation', *Can. Geot. J.,* **21,** 322–37.

MacFarlane, I. C. (1969). *Muskeg Engineering Handbook,* University of Toronto Press, Toronto.

MacFarlane, I. C. and Rutka, A. (1962). *An Evaluation of Pavement Performance Over Muskeg in Northern Ontatio*, Highway Research Board, Bulletin 316, Washington, D.C.

Mickleborough, W. B. (1961). 'Embankment construction in Muskeg at Prince Albert', *Proc. 7th Muskeg Res. Conf.*, NRC-ACSSM, Tech. Memo. 71.

Miyahawa, I. (1960). *Some Aspects of Road Construction Over Peaty and Marshy Areas in Hokkaido, with Particular Reference to Filling Methods*, Civil Engineering Research Institute, Hokkaido Development Bureau, Sapporo, Japan.

Moore, L. H. (1962). *A Correlation of Engineering Characteristics of Organic Soils in New York State*. New York State Dept. of Public Works Technical Report, New York.

Radforth, N. W. (1952). 'Suggested classifications of muskeg for the engineer', *Eng J.* (Canada), **35**, 1194–210.

Samson, L. and La Rochelle, P. (1972). 'Design and performance of an expressway constructed over peat by preloading', *Can. Geot. J.,* **9**, 447–6.

Skempton, A. W. and Petley, D. J. (1970). 'Ignition loss and other properties of peats and clays from Avonmouth, King's Lynn and Cranberry Moss', *Geotechnique,* **20**, 343–56.

Von Post, L. (1922). 'Sveriges Geologiska Undersøkings torvinventering och nogra av dess hittils vunna resultat (SGU peat inventory and some preliminary results)', *Svenska Morskulturføreningens Tidskift*, Jønkøping, Sweden, **36**, 1–37.

Wilson, N. E. (1978). 'The contribution of fibrous interlock to the strength of peat', *Proc. 17th Muskeg Res. Conf.*, NRC-ACGR, Tech Memo **122**, 5–10.

Wilson, N. E., Radforth, N. W., MacFarlane, I. C. and Lo, M. B. (1965). 'The rates of consolidation of peat', *Proc. 6th Int. Conf. Soil Mech. Found. Engg*, Montreal, **1**, 407–12.

Chapter 7

Engineering behaviour of rocks and rock masses

7.1 Factors controlling the mechanical behaviour of rocks

The factors which influence the deformation characteristics and failure of rock can be divided into internal and external categories. The internal factors include the inherent properties of the rock itself, whilst the external factors are those of its environment at a particular point in time. As far as the internal factors are concerned the mineralogical composition and texture are obviously important, but planes of weakness within a rock and the degree of mineral alteration are frequently more important. The temperature–pressure conditions of a rock's environment significantly affect its mechanical behaviour, as does its pore water content. In this respect the length of time which a rock suffers a changing stress and the rate at which this is imposed affects its deformation characteristics.

7.1.1 Composition and texture

The composition and texture of a rock are governed by its origin. For instance, the olivines, pyroxenes, amphiboles, micas, feldspars and silica minerals are the principal components in igneous rocks. These rocks have solidified from magmas. Solidification involves a varying degree of crystallization, the greater the length of time involved, the greater the development of crystallization. Hence glassy, microcrystalline, fine, medium and coarse-grained types of igneous rocks can be distinguished. In metamorphic rocks either partial or complete recrystallization has been brought about by changing temperature–pressure conditions. Not only are new minerals formed in the solid state but the rocks may develop certain lineation structures. A varying amount of crystallization is found within the sedimentary rocks, from almost complete, as in the case of certain chemical precipitates, to slight, as far as diagenetic crystallization in the pores of, for example, certain sandstones.

Few rocks are composed of only one mineral species and even when they are the properties of that species vary slightly from mineral to mineral. Such variations within minerals may be due to cleavage, twinning, inclusions, cracking and alteration, as well as to slight differences in composition. This in turn is reflected in their physical behaviour. As a consequence few rocks can be regarded as homogeneous, isotropic substances. The size and shape relationships of the component minerals are also significant in this respect; generally the smaller the grain size, the stronger the rock.

One of the most important features of texture as far as physical behaviour, particularly strength, is concerned, is the degree of interlocking of the component grains. Fracture is more likely to take place along grain boundaries (intergranular fracture) than through grains (transgranular fracture) and therefore irregular boundaries make fracture more difficult. The bond between grains in many sedimentary rocks is provided by the cement and/or matrix, rather than by grains interlocking. The amount and, to a lesser extent, the type of cement/matrix is important, not only influencing strength and elasticity, but also density, porosity and permeability.

Rocks are not uniformly coherent materials, but contain defects which occur as visible or microscopic linear or planar discontinuities associated with certain minerals. Defects include microfractures, grain boundaries, mineral cleavages, twinning planes, inclusion trains and elongated shell fragments. As is to be expected, defects influence the ultimate strength of a rock and may act as surfaces of weakness which control the direction in which failure occurs.

Grain orientation in a particular direction facilitates breakage along that direction. This applies to all fissile rocks whether they are cleaved, schistose, foliated, laminated or thinly bedded. For instance, tests carried out by Brown *et al.* (1977) showed that the compressive strength of the Delabole Slate is highly directional, varying continuously with the angle made by the cleavage planes and the direction of loading.

7.1.2 Temperature–pressure conditions

Although all rock types show decrease in strength with increasing temperature and an increase in strength as the confining pressure is increased, the combined effect of these factors, as with increasing depth of burial, is notably different for different rock types. Experimental investigation has shown that the effects of temperature changes on sedimentary rocks are of less consequence than those of pressure down to depths of 10 000 m. Griggs (1936) found that the ultimate strength of the Solenhofen Limestone was increased by 360% under 10 000 atmospheres (100 MN/m^2). With increasing temperatures there is a reduction in yield stress and strain hardening decreases. Heating particularly enhances the ductility of calcareous and evaporitic rocks, and their ability to deform permanently without loss of cohesion.

With high confining pressures rocks become effectively stronger and so more difficult to fracture. This is particularly the case with calcareous rocks. At high pressures incipient fractures are closed and indeed the total flow of material without rupture may be indefinitely increased with increasing confining pressure.

The transition from brittle to ductile deformation in porous rocks is characterized by an abrupt change from dilational behaviour at low pressures to compaction during inelastic axial strain at high pressures. This type of behaviour differs from that of rocks with low porosity. In the latter, dilatancy persists well into the ductile field. The compaction which occurs during ductile deformation in porous rocks at high confining pressure is due to the collapse of pore space and the rearrangement of grains to give denser packing. At lower pressures, the dilation is attributed to fracture along grain boundaries as well as to fracturing of grains, and to the rearrangement of grains.

7.1.3 Pore solutions

The presence of moisture in rocks adversely affects their engineering behaviour. For instance, moisture content increases the strain velocity and lowers their fundamental strength. Over 50 years ago Griggs (1940) demonstrated in experiments with alabaster, subjected to a load of $20\,MN/m^2$, that a dry specimen soon reached its maximum strain of approximately 0.03%, whereas when a specimen had access to water the strain attained 1.75% in 36 days.

Perhaps the most frequently quoted work in this context is that carried out by Colback and Wiid (1965) who undertook a number of uniaxial and triaxial compression tests at eight different moisture contents, on quartzitic shale and quartzitic sandstone with porosities of 0.28 and 15% respectively. The moisture contents of the rock samples were controlled by keeping them in desiccators over saturated solutions of $CaCl_2$ at a constant temperature. The tests indicated that the compressive strengths of both rocks under saturated conditions were approximately half what they were under dry conditions. From Figure 7.1 it will be noted that the

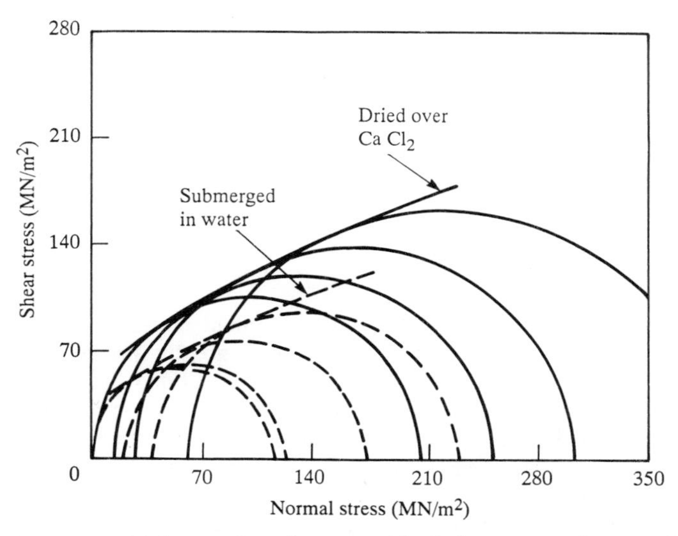

Figure 7.1 *Mohr envelope for quartzitic shale at two moisture contents (after Colback and Wiid, 1965) (© Crown Copyright reserved. Queen's Printer, Canada, 1965)*

slopes of the Mohr envelopes are not sensibly different, indicating that the coefficient of internal friction is not significantly affected by changes in moisture content. Colback and Wiid therefore tentatively concluded that the reduction in strength witnessed with increasing water content was primarily due to a lowering of the tensile strength, which is a function of the molecular cohesive strength of the material. Tests on specimens of quartzitic sandstone showed that their uniaxial compressive strength was inversely proportional to the surface tension of the different liquids into which they were placed. As the surface free energy of a solid submerged in a liquid is a function of the surface tension of the liquid, and since the uniaxial compressive strength is directly related to the uniaxial tensile strength, and this to the molecular cohesive strength, it was postulated that the influence of the immersion liquid was to reduce the surface free energy of the rock and hence its strength. The authors therefore concluded that the reduction in strength from the dry to the saturated condition of predominantly quartzitic rocks was a constant which was governed by the reduction of the surface free energy of the quartz due to the presence of any given liquid.

Vutukuri (1974) investigated the effects of liquids on the tensile strength of limestone. He reached a similar conclusion to that of Colback and Wiid (1965) regarding the reduction of strength on saturation. In other words he recognized that liquids influence the surface free energy of the rocks and because new surfaces are developed on fracturing, that the strength will depend upon the decrease or increase in surface energy due to the liquid present. For example, he found that as the dialectic constant and surface tension of the liquid increased, the tensile strength of the limestone decreased.

More recently it has been recognized that small changes in moisture content can bring about large changes in strength and deformability. For instance, Dobereiner and de Freitas (1986) found that in the case of weak sandstones a 10% change in moisture content may cause a change of approximately 20 MN/m^2 in strength and 3000 MN/m^2 in deformability. More dramatically Priest and Selvakumar (1982) reported a reduction in the strength of the Bunter Sandstone from 57 MN/m^2 to 38 MN/m^2 for an increase of only 1% in moisture content above the totally dry state.

7.1.4 Time-dependent behaviour

Most strong rocks, like granite, exhibit little time-dependent strain or creep: however, creep in evaporitic rocks, notably salt, may greatly exceed the instantaneous elastic deformation. The time–strain pattern exhibited by a wide range of materials subjected to a constant uniaxial stress can be represented diagramatically as shown in Figure 7.2. The instantaneous elastic strain, which takes place when a load is applied, is represented by OA. There follows a period of primary or transient creep (AB) in which the rate of deformation decreases with time. Primary creep is the elastic effect attributable to intragranular atomic and lattice displacements. If the stress is removed the specimen recovers. At first this is instantaneous (BC), but this is followed by a time elastic recovery, illustrated by curve CD. On the other hand, if the loading continues the sample begins to exhibit

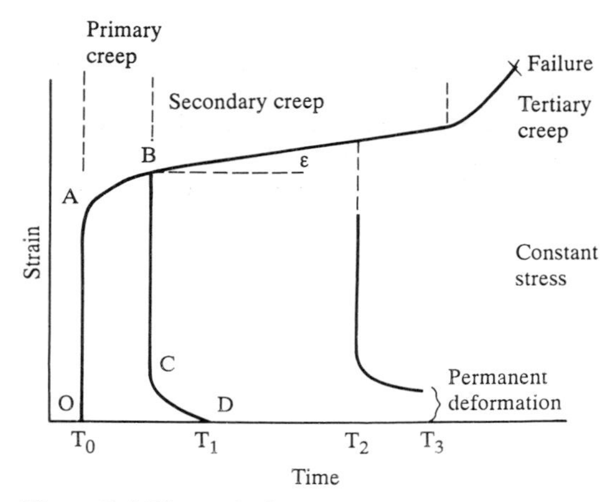

Figure 7.2 *Theoretical strain curve at constant stress (after Price, 1966)*

secondary or pseudo-viscous creep. This type of creep represents a phase of deformation in which the rate of strain is constant and is due principally to movements which occur on grain boundaries. The deformation is permanent and is proportional to the length of time over which the stress is applied. If the loading is further continued, then the specimen suffers tertiary creep in which the strain rate accelerates with time and ultimately leads to failure. Creep deformation is limited at low temperatures and pressures but it may greatly exceed normal plastic flow when the pressures approach the limit of rupture. High temperatures also favour an increase in the rate and extent of creep.

In experiments in which he applied stress to the Solenhofen Limestone Griggs (1936) introduced pauses in the rise of stress. He noted that during these pauses small non-elastic deformation occurred once the differential stress had reached a high enough threshold value (Figure 7.3). It also was observed that for the lowest stress application the strain did not increase with time as the threshold value was

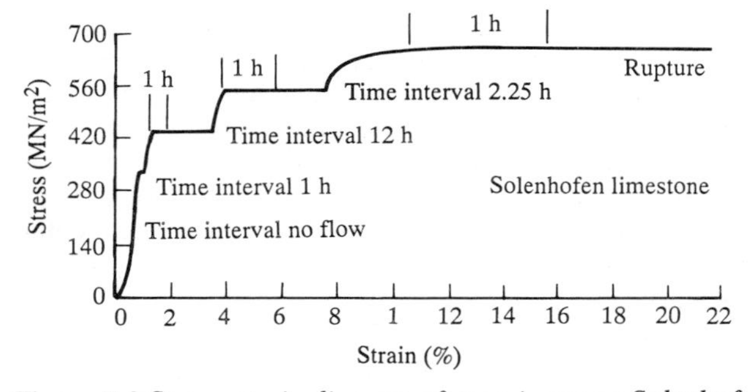

Figure 7.3 *Stress–strain diagram of experiment on Solenhofen Limestone with pauses in increase of stress (after Griggs, 1936)*

not attained. For each successive pause at a higher stress level the velocity of strain increased. It appeared that flowage in many experiments was masked by the rapid application of increased stress. Griggs also performed experiments which tested the change in ultimate strength with time. He found that time reduced the ultimate strength up to a certain point, beyond which there was no change. Moreover, he showed that the amount of plastic deformation before rupture decreased with the duration of time.

7.2 Deformation and failure of rocks

7.2.1 Stages of deformation

Four stages of deformation have been recognized, elastic, elastico-viscous, plastic and rupture. The stages are dependent upon the elasticity, viscosity and rigidity of a rock, as well as on stress history, temperature, time, pore water and anisotropy. An elastic deformation is defined as one which disappears when the stress responsible for it ceases. Ideal elasticity would exist if the deformation on loading and its disappearance on unloading were both instantaneous. This is never the case since there is always some retardation, known as hysteresis, in the unloading process. With purely elastic deformation the strain is a linear function of stress, that is, the material obeys Hooke's law. Therefore the relationship between stress and strain is constant and is referred to as Young's modulus (E). Rocks, however, only approximate to the ideal Hookean solid, the stress–strain relationships generally are not linear. Consequently Young's modulus is not a simple constant but is related to the level of applied stress.

The change at the elastic limit from elastic to plastic deformation is referred to as the yield point or yield strength. If the stress on a material exceeds its elastic limit, then it is permanently strained, the latter being brought about by plastic flow. Within the field of plastic flow there is a region where elastic stress is still important and this is referred to as the field of elastico-viscous flow. This term has been used to describe creep or continuous deformation which occurs in rocks when they are subjected to constant stress. Plasticity may be regarded as time independent, non-elastic, non-recoverable, stress-dependent deformation under uniform sustained load. Solids are classified as brittle or ductile according to the amount of plastic deformation they exhibit. In brittle materials the amount of plastic deformation is zero or very little whilst it is large in ductile substances.

Rupture, or ultimate strength, occurs when the stress exceeds the strength of the material involved. It represents the maximum stress difference a body is able to withstand prior to loss of cohesion by fracturing for constant experimental conditions, fracturing being conceived as the breaking process leading to rupturing. The initiation of rupture is marked by an increasing strain velocity.

Young's modulus is the most important of the elastic constants and can be derived from the slope of the strain–stress curve obtained when a rock specimen is subjected to unconfined compression, it being the ratio of stress to strain. Most crystalline rocks have S-shaped stress–strain curves (Figure 7.4). At low stresses

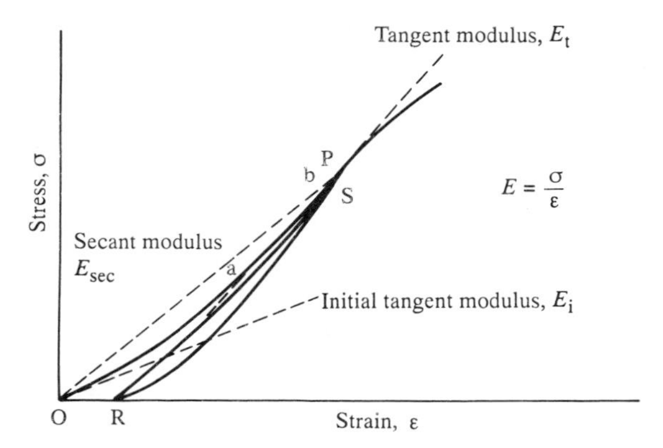

Figure 7.4 *Representative stress–strain curve for rock in uniaxial compression*

the curve is non-linear and concave upwards, that is, Young's modulus increases as the stress increases. The initial tangent modulus is given by the slope of the stress–strain curve at the origin. Gradually a level of stress is reached where the slope of the curve becomes approximately linear. In this region Young's modulus is defined as the tangent modulus, or secant modulus. At this stress level the secant has a lower value than the tangent modulus because it includes the initial 'plastic' history of the curve.

Deere and Miller (1966) classified the uniaxial stress–strain curves into six types (Figure 7.5). Types III, IV and V, however, are modifications of the representative S-curve. Type I represents the classical straight-line behaviour of brittle materials which is typical of the more explosive failures of basalts, dolerites, quartzites, and strong dolostones and limestones. Softer limestones, siltstones and tuffs exhibit a more concave downwards curve as illustrated in Type II. These are usually somewhat more linear in the earlier and central portions, yielding 'plastically' as failure approaches. Type III is typical of sandstone, granite, some dolostones and dolerites, and schist cut parallel to the schistosity. Metamorphic rocks like marbles and gneiss are represented by Type IV. Schist cored along the schistosity has the long, sweeping S-shaped curve of Type V. Types III, IV and V are characterized by initial 'plastic' crack closure, followed by a steeper linear section. The upper part of such curves exhibit varying degrees of plastic yield as failure is approached. Type III rocks do not yield significantly, being more explosive with brittle-type fractures (similar to Type I) than Types IV and V. The Type VI curve for rock salt has an initial small elastic straight-line portion followed by plastic deformation and continuous creep.

In addition to their non-elastic behaviour most rocks exhibit hysteresis. Under uniaxial stress the slope of the stress–strain curve during unloading is initially greater than during loading for all stress values (Figure 7.4). As stress is decreased to zero a residual strain, OR, is often exhibited. On reloading the curve RS is produced, which in turn is somewhat steeper than OP. Further cycles of unloading and reloading to the same maximum stress give rise to hysteresis loops, which are

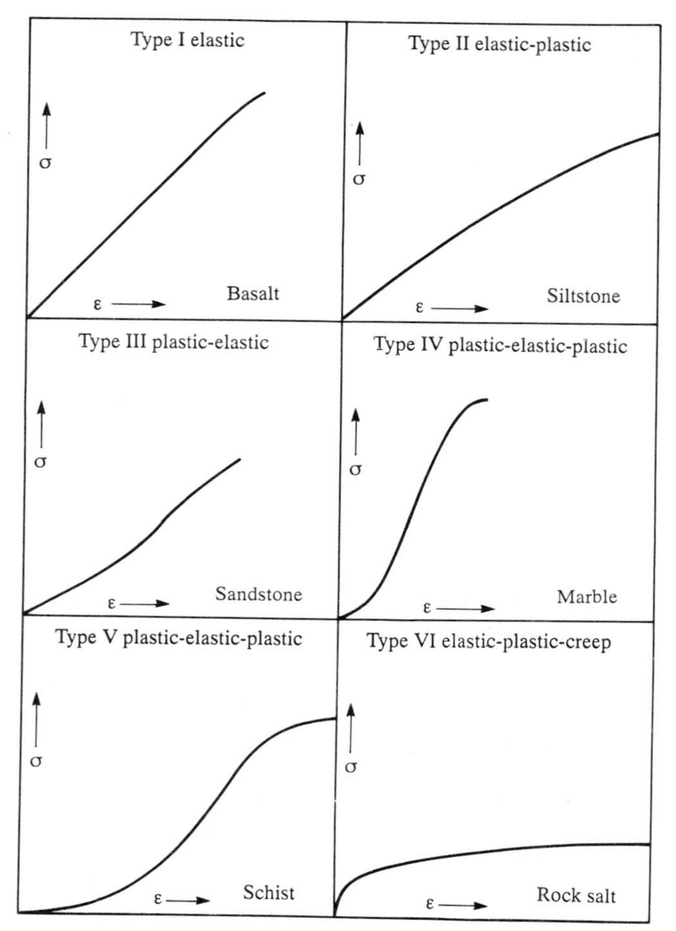

Figure 7.5 *Typical stress–strain curves for rock in uniaxial compression loaded to failure (after Deere and Miller, 1966)*

shifted slightly to the right. These effects are associated with transient creep. The non-linear elastic behaviour and elastic hysteresis of brittle rocks under uniaxial compression has been explained as due to the presence of flaws or minute cracks in the rocks (Walsh, 1965). At low stresses these cracks are open but they close as the stress is increased and the rock becomes elastically stiffer, that is, E increases with stress. Once the cracks are closed the stress–strain curve becomes linear. Nevertheless E is still lower in this portion of the curve than it would be for an uncracked solid and this has been attributed to sliding along crack surfaces. Since these cracks do not immediately slide in the opposite sense as the load is reduced, hysteresis loops are produced.

When hysteresis is large it is difficult to distinguish between elastic and plastic deformation; however, an elastic strain is related only to stress whereas a permanent strain also is related to the period over which the stress is applied. As a consequence, in a strain experiment, with constant load, the elastic deformation is characterized by a gradual decrease of the strain velocity which ultimately leads to a halt in the process. On the other hand a permanent or plastic deformation continues indefinitely with a constant strain velocity. Under high temperature–pressure conditions permanent deformations also may take place by creep.

When a specimen undergoes compression it is shortened and this generally is accompanied by an increase in its cross-sectional area. The ratio of lateral unit deformation to linear unit deformation, within the elastic limit, is known as Poisson's ratio. An idealized value for Poisson's ratio can be obtained by considering an idealized crystal structure, where contraction in one direction automatically leads to extension of the lattice in a perpendicular direction. In such a case, by considering the geometry of the structure, it can be shown that Poisson's ratio is 0.333. The work carried out by Deere and Miller (1966) showed that values of Poisson's ratio for rock must be regarded with some suspicion. These two authors gave the average initial tangent value of Poisson's ratio for all the rocks studied as 0.125 and at a stress level of 50% ultimate stress as 0.341.

Rocks subjected to uniaxial compression tend to exhibit a common behaviour in that both Young's modulus and Poisson's ratio increase to more or less constant values as the stress is increased. As the compressive stress approaches the failure limit, Young's modulus falls, eventually reaching zero, while Poisson's ratio increases to a value nearing or exceeding the theoretical maximum of 0.5 for an incompressible solid body. The opposite trend is observed when rocks are placed under uniaxial tension, namely, both Young's modulus and Poisson's ratio are initially high and they fall continuously as stress increases to the failure point. Hawkes *et al.* (1973) found that the initial tangent moduli for the rocks they tested were similar in compression and tension. The value of Young's modulus in compression at half the load failure (E_{t50}) is usually greater than the value in tension but there is considerable variation from one rock type to another. At very low stresses Poisson's ratio of a rock in tension can be greater than 0.5, indicating an initial decrease in volume, that is, a decrease in porosity. However, as stress is increased the ratio falls to comparatively low values (0.1).

7.2.2 Theories of brittle failure

Brittle failure is regarded as the sudden loss of cohesion across a plane that is not preceded by any appreciable permanent deformation. It may occur in rock on both microscopic and macroscopic scales.

One of the most popular theories which was proposed to explain shear fractures was advanced by Coulomb (1773). The Coulomb criterion of brittle failure is based upon the idea that shear failure occurs along a surface if the shear stress acting in that plane is high enough to overcome the cohesive strength of the material and the resistance to movement. The latter is equal to the stress normal to the shear surface multiplied by the coefficient of internal friction of the material, whilst the cohesive strength is its inherent shear strength when the stress normal to the shear surface is zero. The relation between the failure criterion, the friction and the cohesion is then expressed by Coulomb's Law

$$\tau = c + \sigma_n \tan \phi \tag{7.1}$$

where τ is the shearing stress, c is the apparent cohesion, σ_n is the normal stress and ϕ is the angle of internal friction or shearing resistance. It can be shown that under triaxial conditions (Figure 7.6)

$$\sigma_n = \tfrac{1}{2}(\sigma_1 + \sigma_3) + \tfrac{1}{2}(\sigma_1 - \sigma_3)\cos 2\beta \tag{7.2}$$

and that

$$\tau = \tfrac{1}{2}(\sigma_1 - \sigma_3)\sin 2\beta \tag{7.3}$$

where σ_1 and σ_3 are the stress at failure and the confining pressure respectively.

The Coulomb criterion has been shown to agree with experimental data for rocks in which the relationship between the principal stresses at rupture is, to all intents, linear. However, experimental evidence indicates that peak strength envelopes generally are non-linear. This may be due to the area of the grains in frictional contact increasing as the normal pressure increases. What is more the criterion implies that a major shear fracture exists at peak strength and this is not always the case (Wawersik and Fairhurst, 1970). It also implies a direction of shear failure which does not always agree with experimental observations.

Coulomb's concept was subsequently modified by Mohr (1882). Mohr's hypothesis states that when a rock is subjected to compressive stress shear fracturing occurs parallel to those two equivalent planes for which shearing stress is as large as possible whilst the normal pressure is as small as possible. This statement assumes that a triaxial state of external stress is applied to a substance and that the maximum external stress is resolved into shear and normal components for any inclined potential shear planes existing in the stressed material.

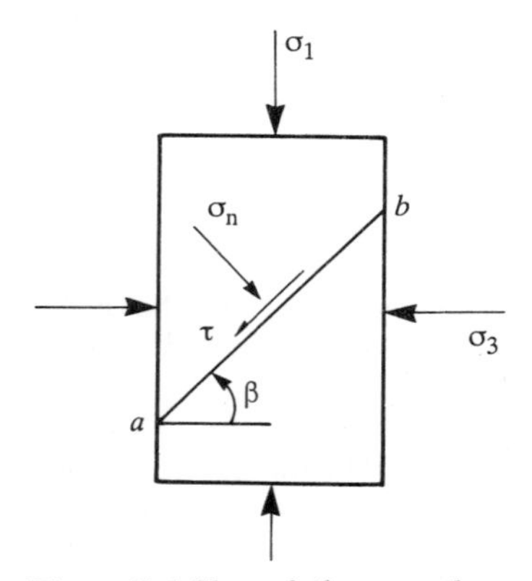

Figure 7.6 *Shear failure on plane a, b*

Griffith (1920) claimed that because of the presence of minute cracks or flaws, particularly in surface layers, the measured tensile strengths of most brittle materials are much less than those which would be inferred from the values of their molecular cohesive forces. Although the mean stress throughout a body may be relatively low, local stresses developed in the vicinity of the flaws were assumed to attain values equal to the theoretical strength. Under tensile stress, the stress magnification around a flaw is concentrated where the radius of curvature is smallest, that is, at its ends. Hence the tensile stresses which develop around a flaw have most influence when the tensile stress zone coincides with the zone of minimum radius of curvature. The concentration of stress at the ends of flaws causes them to be enlarged and presumably with time they develop into fractures.

Although there is an encouraging agreement between experimental and theoretical results, the Griffith theory does not provide a complete description of the mechanism of rock failure. For instance, Hoek (1968) was able to demonstrate that the original and modified Griffith theories, although adequate for the prediction of fracture initiation in rocks, were unable to describe its propagation and subsequent failure of rocks.

More recently a number of empirical strength criteria have been advanced because the classical theories do not apply to rock over a wide range of applied compressive stress conditions. These criteria usually take the form of a power law in recognition of the fact that peak σ_1 vs σ_3 and τ vs σ_n envelopes for rocks are non-linear, that is, they are generally concave downwards. In order to ensure that the parameters used in the power laws are dimensionless, these criteria are best written in normalized form with all stress components being divided by the uniaxial compressive strength of the rock (Brady and Brown, 1985).

One of the most recent empirical laws is that developed by Hoek and Brown (1980) who proposed that the peak triaxial compressive strengths of a wide range of isotropic rock materials could be described by the expression

$$\sigma_1 = \sigma_3 + (m\,\sigma_c\,\sigma_3 + s\sigma_c^2)^{1/2} \qquad (7.4)$$

where σ_1 is the major principal stress at failure, σ_3 is the minor principal stress (or, in the case of the triaxial test, the confining pressure), σ_c is the uniaxial compressive strength of the intact rock, and m and s are dimensionless constants which are approximately analogous to the angle of friction and cohesive strength of the conventional Mohr-Coulomb failure criterion. The constant m varies with rock type, ranging from about 0.001 for highly disturbed rock masses to about 25 for hard intact rock (Table 7.1). Large values of m (that is, 15–25) give steeply inclined Mohr envelopes and high instantaneous friction angles at low effective normal stress levels and are associated with brittle igneous and metamorphic rocks. Lower values of m, around 7, yield lower instantaneous friction angles and tend to be associated with carbonate rocks. For intact rock $s = 1$. Equation (7.4) when normalized becomes

$$\frac{\sigma_1}{\sigma_c} = \frac{\sigma_3}{\sigma_c} + \left(m\frac{\sigma_3}{\sigma_c} + s\right)^{1/2} \qquad (7.5)$$

Table 7.1 *Approximate relationship between rock mass quality and material constants (after Hoek, 1983)*

Empirical failure criterion $\sigma_1' = \sigma_3' + (m\sigma_c\sigma_3' + s\sigma_c^2)^{1/2}$ σ_2' = *major principal stress* σ_3' = *minor principal stress* σ_c' = *uniaxial compressive strength of intact rock* $m.s$ = *empirical constants*		*Carbonate rocks with well developed crystal cleavage, e.g. dolostone, limestone and marble*	*Lithified argillaceous rocks, e.g. mudstone, siltstone, shale and slate (tested normal to cleavage)*	*Arenaceous rocks with strong crystals and poorly developed crystal cleavage, e.g. sandstone and quartzite*	*Fine grained polymineralic igneous crystalline rocks, e.g. andesite, dolerite, diabase and rhyolite*	*Coarse grained polymineralic igneous and metamorphic crystalline rocks, e.g. amphibolite, gabbro, gneiss, granite, norite and quartz diorite*
Intact rock samples Labratory size samples free from pre-existing fractures		$m = 7$ $s = 1$	$m = 10$ $s = 1$	$m = 15$ $s = 1$	$m = 17$ $s = 1$	$m = 25$ $s = 1$
Geomechanics system (CSIR)* rating	100					
Q System (NGI)† rating	500					
Very good quality rock mass Tightly interlocking undisturbed rock with unweathered joints spaced at 1 to 3 m		$m = 3.5$ $s = 0.1$	$m = 5$ $s = 0.1$	$m = 7.5$ $s = 0.1$	$m = 8.5$ $s = 0.1$	$m = 12.5$ $s = 0.1$
Geomechanics system (CSIR) rating	85					
Q System (NGI) rating	100					
Good quality rock mass Fresh to slightly weathered rock, slightly disturbed with joints spaced at 1 to 3 m		$m = 0.7$ $s = 0.004$	$m = 1$ $s = 0.004$	$m = 1.5$ $s = 0.004$	$m = 1.7$ $s = 0.004$	$m = 2.5$ $s = 0.004$
Geomechanics system (CSIR) rating	65					
Q System (NGI) rating	10					
Fair quality rock mass Several sets of moderately weathered joints spaced at 0.3–1 m disturbed		$m = 0.14$ $s = 0.0001$	$m = 0.20$ $s = 0.0001$	$m = 0.30$ $s = 0.0001$	$m = 0.34$ $s = 0.0001$	$m = 0.50$ $s = 0.0001$
Geomechanics system (CSIR) rating	44					
Q System (NGI) rating	1					
Poor quality rock mass Numerous weathered joints at 30 to 500 mm with some gouge. Clean, compacted rockfill		$m = 0.04$ $s = 0.00001$	$m = 0.05$ $s = 0.00001$	$m = 0.08$ $s = 0.00001$	$m = 0.09$ $s = 0.00001$	$m = 0.13$ $s = 0.00001$
Geomechanics system (CSIR) rating	23					
Q System (NGI) rating	0.1					
Very poor quality rock mass Numerous heavily weathered joints spaced at 50 mm with gouge. Waste rock		$m = 0.007$ $s = 0$	$m = 0.010$ $s = 0$	$m = 0.015$ $s = 0$	$m = 0.017$ $s = 0$	$m = 0.025$ $s = 0$
Geomechanics system (CSIR) rating	3					
Q System (NGI) rating	0.01					

This expression is useful when comparing the shape of Mohr failure envelopes for different rocks.

Rock strength and fracture are influenced by mineral composition; grain size, shape and packing; amount and type of cement/matrix; degree of grain interlock, etc. If these factors are relatively uniform within a given rock type, then a single curve probably will give a good fit to the normalized strength data (e.g. granites; see Hoek, 1983). On the other hand if these factors are quite variable, as in limestones or sandstones, then a single curve will give a poorer fit. Nonetheless the empirical criterion formulated by Hoek and Brown (1980) allows preliminary design calculations to be made without testing by using an approximate value of m for a particular rock and by determining a value of uniaxial compressive strength.

According to Hoek (1983) under triaxial conditions a transition from brittle to ductile behaviour usually occurs somewhere between a principal stress ratio (σ_1'/σ_3') of 3 and 5 (Figure 7.7). He suggested a rough rule of thumb, that is, that the confining pressure should not exceed the unconfined compressive strength of the rock for behaviour to be regarded as brittle. However, for those rocks with very low values of m the principal stress ratio may fall beyond the brittle–ductile transition.

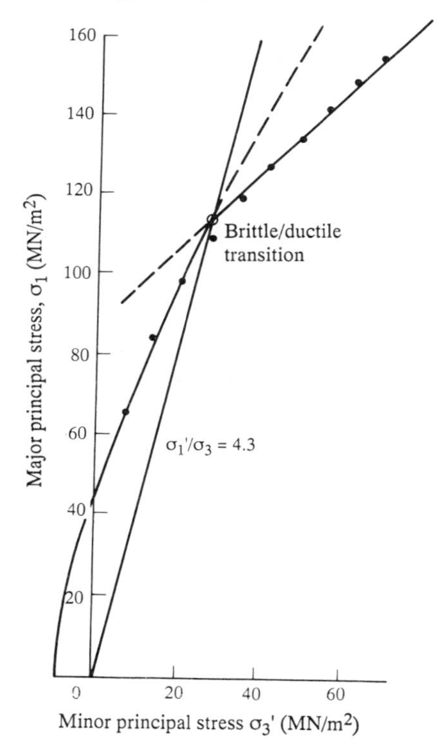

Figure 7.7 *Results of triaxial tests on Indiana Limestone illustrating brittle ductile transition (after Schwartz, 1964)*

7.3 Strength of jointed rock masses and its assessment

Joints in a rock mass reduce its effective shear strength at least in a direction parallel with the discontinuities. Hence the strength of jointed rocks is highly anisotropic. Joints offer no resistance to tension whereas they offer high resistance to compression. Nevertheless they may deform under compression if there are crushable asperities, compressible filling or apertures along the joint or if the wall rock is altered.

John (1965) considered that when a jointed rock mass failed by sliding along one joint or a set of joints, the limiting stress ratios could be determined based on the parameters of the joints and the confining pressure. In other words, where a load is applied in a direction parallel or sub-parallel to the joint direction the shear strength depends on the shearing resistance along the joint surfaces. At low normal pressure shearing stresses along a joint with relatively smooth asperities produce a tendency for one block to ride up onto and over the asperities of the other, whereas at high normal pressures shearing takes place through the asperities. When a jointed rock mass undergoes shearing this may be accompanied by dilation especially at low pressures and small shear displacements probably occur as shear stress builds up.

It has been suggested (Hoek, 1983) that the shear strength, τ, along a surface of failure can be obtained from

$$\tau = (\cot \phi_i' - \cos \phi_i') \frac{m\sigma_c}{8} \tag{7.6}$$

where ϕ_i' is the instantaneous angle of friction at given values of τ and σ' (i.e. the inclination of the tangent to the Mohr failure envelope at the point (σ', τ) shown in Figure 7.8. Figure 7.8 also includes the equations by which ϕ_i', c_i' (instantaneous cohesion) and β (inclination of failure plane) are derived.

Under triaxial conditions the peak strengths developed by anisotropic rocks (e.g. those characterized by lamination such as shales, or cleavage such as slates) depend on the orientation of these planes of relative weakness to the principal stress directions. Figure 7.9 shows variations in peak stress in relation to the angle of inclination of the major principal stress to the plane of weakness. Each plane of weakness possesses a limiting value of shear strength in accordance with Coulomb's equation (7.1) and Equations (7.2) and (7.3) allow the normal and shear stresses on the plane to be determined. Substituting for normal stress (σ_n, Equation (7.2)) and shear strength (τ, Equation (7.3)) in the Coulomb equation and rearranging provides the axial strength,σ', of a triaxial specimen from the following equation:

$$\sigma_1' = \sigma_3' - \frac{2(c_1' + \sigma_3' \tan \phi_i')}{(1 - \tan \phi_i' \tan \beta) \sin 2\beta} \tag{7.7}$$

Equation (7.7), however, can only be solved for values of β which are within about 25° of the friction angle, ϕ'. Very high values of σ_1' are obtained from very small

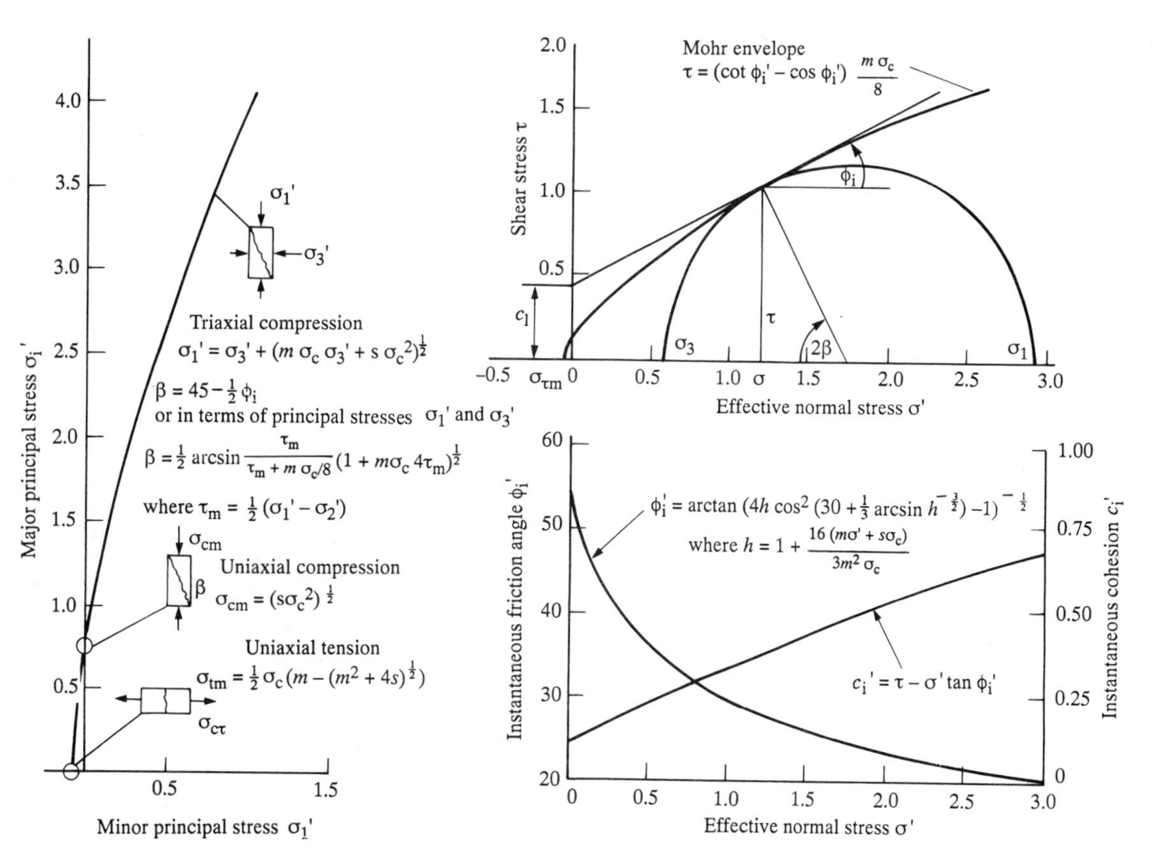

Figure 7.8 *Summary of equations associated with the non-linear failure criterion proposed by Hoek and Brown (1980)*

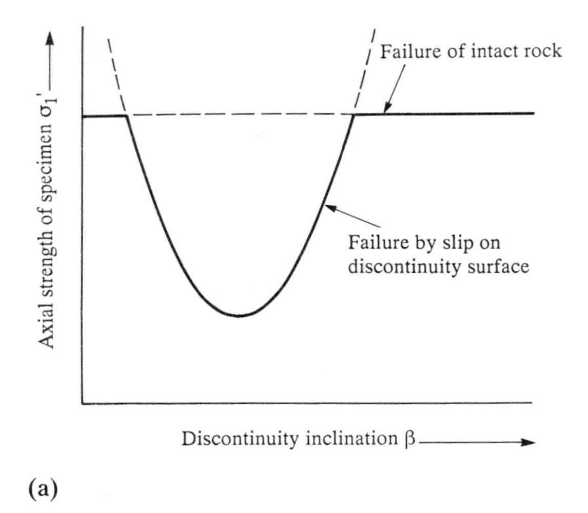

(a)

Confining pressure
σ_3': MN/m^2
● 34.5
■ 69
▲ 138
○ 207

Major principal stress at failure σ_1' (MN/m^2)
0
200
400
600
800
20
40
60
80
Experimental values
Predicted strength values
Angle β between failure plane and major principal stress direction

(b)

Figure 7.9 (a) *Strength of specimen predicted by means of Equations (7.4) and (7.7)* (b) *triaxial test results for slate with different failure plane indications, obtained by McLamore and Gray (1967), compared with strength predictions from Equations (7.4) and (7.7) (after Hoek, 1983)*

values of β whilst values of β which are near 90° yield negative values of σ_1' which are meaningless. Such very high or negative values mean that slip cannot occur along a plane of weakness and that failure will take place through the intact rock. None the less the two-strength model represented in Figure 7.8 gives an oversimplified view of the variation in strength which occurs in anisotropic rocks. After carrying out a series of triaxial tests on slate at a range of confining pressures and cleavage orientations, McLamore and Gray (1967) attempted a fuller explanation by proposing that both cohesion, c_i', and tan ϕ_i' vary according to orientation in relation to the following expressions:

$$c_i' = A - B[\cos 2(\alpha - \alpha_c)]^n \tag{7.8}$$

and

$$\tan \phi_i' = C - D[\cos 2(\alpha - \alpha_\phi)]^m \tag{7.9}$$

where *A, B, C, D, m* and *n* are constants, and α_c and α_ϕ are values of α ($\alpha = \pi/2 - \beta$) at which c_i' and ϕ_i' have minimum values respectively.

Barton (1976) proposed the following empirical expression for deriving the shear strength (τ) along joint surfaces:

$$\tau = \sigma_n \tan (\text{JRC} \log_{10}(\text{JCS}/\sigma_n) + \phi_b) \tag{7.10}$$

where σ_n is the effective normal stress, JRC is the joint roughness coefficient, JCS is the joint wall compressive strength and ϕ_b is the basic friction angle. According to Barton, the values of the joint roughness coefficient range from 0 to 20, from the smoothest to the roughest surface (Figure 7.10). The joint wall compressive strength is equal to the unconfined compressive strength of the rock if the joint is unweathered. This may be reduced by up to 75% when the walls of the joints are weathered. Both these factors are related as smooth-walled joints are less affected by the value of JCS, since failure of asperities plays a less important role. The smoother the walls of the joints, the more significant is the part played by their mineralogy (ϕ_b). The experienced gained from rock mechanics indicates that under low effective normal stress levels, such as occur in engineering, the shear strength of joints can vary within relatively wide limits. The maximum effective normal stress acting across joints considered critical for stability lies, according to Barton, in the range 0.1–2.0 MN/m^2.

In practice it is found that JRC is only a constant for a fixed joint length. Generally longer profiles (of the same joint) have lower JRC values. Indeed Barton and Bandis (1980) suggested that mobilization of peak strength along a joint surface seems to be a measure of the distance the joint has to be displaced in order that asperities are brought into contact. This distance increases with increasing joint length. Consequently when testing, longer samples tend to give lower values of peak shear strength.

Hoek (1983) recommended the use of Equation (7.10) for estimation of shear strength in the field. He went on, however, to point out that this equation was not

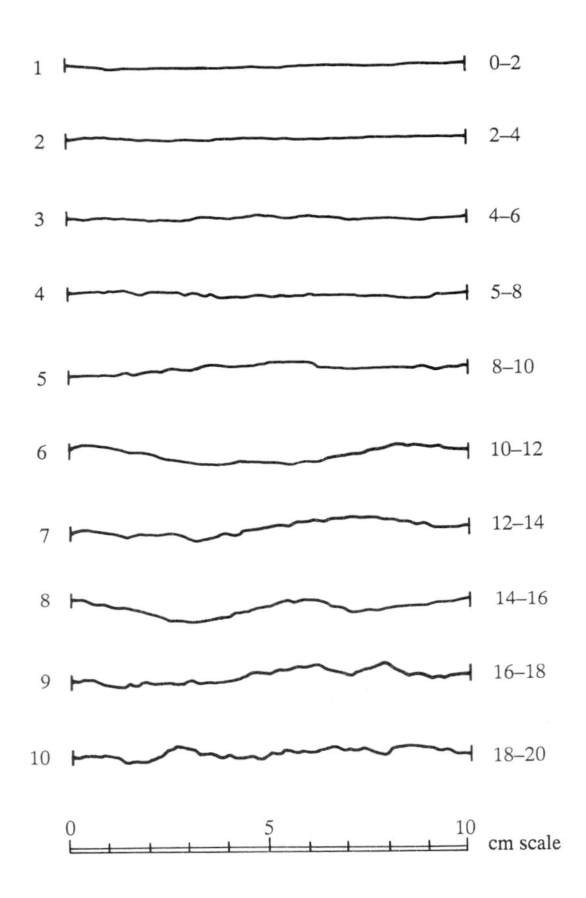

Figure 7.10 *Roughness profiles and corresponding range of JRC values associated with each one (after Barton, 1976)*

the only one which could be used for fitting to shear test data obtained in the laboratory. For example, he maintained that the equations for τ and ϕ_i', given in Figure 7.8 provide a reasonably accurate estimation of the shear strength along rough discontinuities in rock masses under a wide range of effective normal stress conditions. Nevertheless Equation (7.10) suggests that there are three components of shear strength, namely, a basic frictional component (ϕ_b), a geometrical component which is governed by surface roughness (JRC) and an asperity failure component which depends upon the ratio JCS/σ_n. From Figure 7.11 it can be seen that the geometrical and asperity failure components together give the net roughness component, i°. Accordingly the total frictional resistance can be derived from $(\phi_b + i)^\circ$. The shear strength developed along a rough discontinuity depends

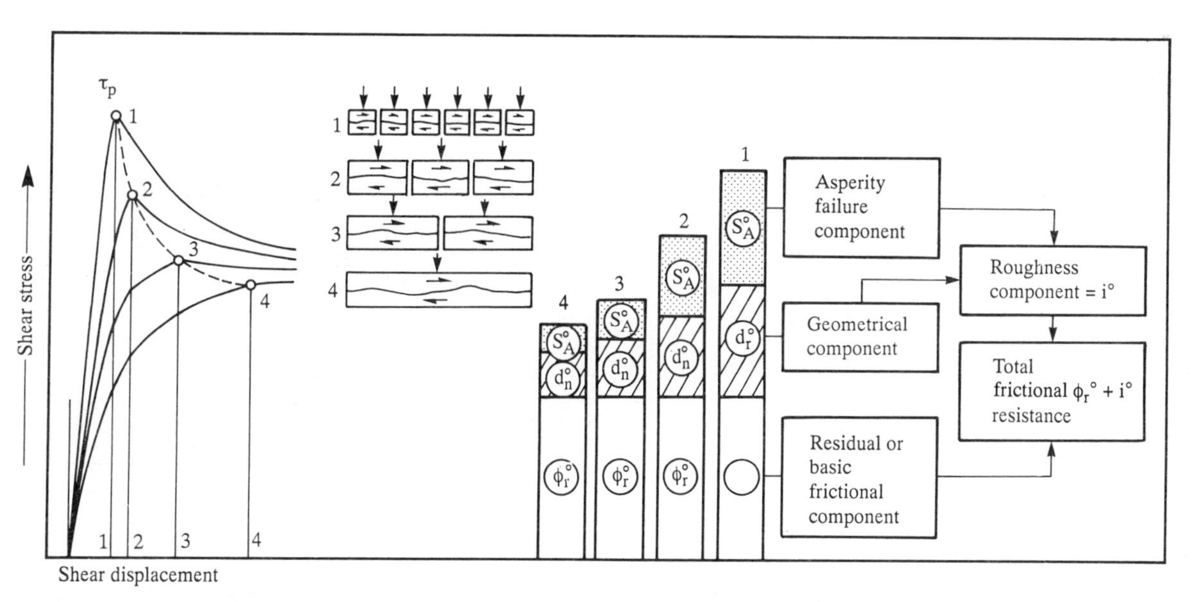

Figure 7.11 *An illustration of the size dependence of shear stress–deformation behaviour for non-planar joints (after Bandis et al.,* 1983)

upon the scale and amount of stress involved. As the effective normal stress, σ_n, increases, so the term $\log_{10}$ (JCS/σ_n') decreases, as does the net apparent friction angle. The steeper asperities are sheared off and the inclination of the controlling roughness decreases with increasing scale. Increasing scale also means that the asperity failure component decreases since compressive strength of the rock, JCS, declines in significance with increasing size. Hence the shear force-displacement curves change with increasing scale in that the behaviour along a discontinuity on shearing changes from brittle to plastic as the shear stiffness is reduced (Barton and Bandis, 1980).

Barton and Choubey (1977) had suggested that tilt and push tests provided a more reliable means of estimating the joint roughness coefficient than comparison with typical profiles. Barton and Bandis (1980) also supported the use of such tests, particularly in heavily jointed rock masses, when three joint sets are present. In a tilt test, two immediately adjacent blocks are extracted from an exposure and the upper is laid upon the lower in the exact same position as it was in the rock mass. Both are then tilted and the angle (α) at which sliding occurs is recorded (Figure 7.12(a)). The JRC is estimated from

$$\mathrm{JRC} = \frac{\alpha - \phi_r}{\log_{10}(\mathrm{JCS}/\sigma_{no})} \tag{7.11}$$

where $\sigma_{no} = \gamma H \cos^2 \alpha$ (i.e. normal stress induced by self-weight of block), γ is the unit weight, H is the thickness of upper block, and ϕ_r is the residual friction angle.

In a pull test an external shearing force (T_2) is applied via a bolt grouted into the block in question (Figure 7.12b). The value of JRC is given by

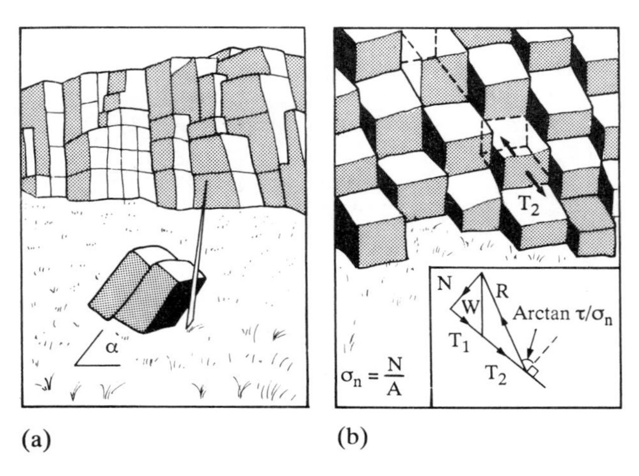

Figure 7.12 *Two extremely simple and inexpensive ways of determining an accurate scale-free value of JRC:* (a) *tilt test;* (b) *pull test (after Barton and Bandis, 1980)*

$$\text{JRC} = \frac{\arctan\left[\dfrac{T_1 + T_2}{N}\right] - \phi_r}{\log_{10}(\text{JCS.A/N})} \qquad (7.12)$$

where A is the joint area and N is the normal and tangential components of the self-weight of the upper block. In both cases the joint wall compression strength (JCS) and the residual friction angle (ϕ_r) can be estimated by using a Schmidt hammer (Barton and Choubey, 1977) where:

$$\log_{10} \text{JCS} = 0.00088\, \gamma_d \text{R} + 1.01 \qquad (7.13)$$

where γ_d is the dry unit weight, R is the Schmidt hammer rebound number and

$$\phi_r = (\phi_b - 20°) + 20\,(r/R) \qquad (7.14)$$

where ϕ_b is the basic friction angle, r is the Schmidt hammer rebound number on wet joint surface, and R is the Schmidt hammer rebound number on a dry unweathered sawn surface.

Probably the most thorough approach to the assessment of the strength of discontinuous rock masses has been made by Hoek and Brown (1980) and again by Hoek (1983). Their method is summarized in Table 7.1 and involves estimating the values of the empirical constants m and s from a description of the rock mass. In other words, the appropriate box in Table 7.1 is determined from a description of the rock mass or preferably from the Q (NGI) or Geomechanics (CSIR) systems of rock mass classification (see Chapter 10). However, the size of the structure which is to be constructed influences the spacing of the discontinuities in that the larger the structure the greater the number of discontinuities that are going to have an effect on its design. Hence the type of model chosen to represent rock mass behaviour also depends on scale. These estimates, together with an estimate of the unconfined compressive strength, can then be used to construct an approximate Mohr failure envelope for the discontinuous rock mass. Hoek (1983) suggested that the values listed in Table 7.1, as far as practical engineering design is concerned, are somewhat conservative and strength estimates derived therefrom can be regarded as lower bound values for design purposes.

References

Bandis, S., Lumsden, A. C. and Barton, N. (1983). 'Fundamentals of rock joint deformation', *Int. J. Rock Mech. Min. Sci. and Geomech. Abstr.*, **20**, 249–68.

Barton, N. (1976). 'The shear strength of rock and rock joints', *Int. J. Rock Mech. Min. Sci and Geomech. Abstr.*, **13**, 255–79.

Barton, N. and Bandis, S. (1980). 'Some effects of scale on the shear strength of joints', *Int. J. Rock Mech. Min. Sci and Geomech. Abstr.*, **17**, 69–76.

Barton, N. and Choubey, V. (1977). 'The shear strength of rock joints in theory and practice', *Rock Mechanics,* **10**, 1–54.

Brady, B. H. G. and Brown, E. T. (1985). *Rock Mechanics for Underground Mining*, George Allen and Unwin, London.

Brown, E. T., Richards, L. R. and Barr, M. V. (1977) 'Shear strength characteristics of the Delabole Slates', *Proc. Conf. Rock Engg*, Newcastle University, **1**, 33–51.

Colback, P. S. B. and Wiid, B. L. (1965). 'Influence of moisture content on the compressive strength of rock', *Symp. Can. Dept. Min. Tech. Survey*, Ottawa, 65–83.

Coulomb, C. A. (1773). 'Sur une application des regles de maximus et minimus a quelques problemes de statique relatifs a l'architecture', *Acad. Roy. des Sci., Mem de Math, et de Phys. par divers Sovans,* **7**, 343–82.

Deere, D. U. and Miller, R. P. (1966). *Engineering Classification and Index Properties for Intact Rock*, Technical Report No. AFWL-TR-65-115, Air Force Weapons Laboratory, Kirtland Air Base, New Mexico.

Dobereiner, L. and De Freitas, M. H. (1986). 'Geotechnical properties of weak sandstones', *Geotechnique,* **36**, 79–94.

Griffith, A. A. (1920). 'The phenomenon of rupture and flows in solids', *Phil. Trans. Roy. Soc. London,* **A221**, 163–98.

Griggs, D. T. (1936). 'Deformation of rocks under high confining pressures', *J. Geol.,* **44**, 541–77.

Griggs, D. T. (1940). 'Experimental flow of rocks under conditions favoring recrystallization', *Bull. Geol. Soc. Am.,* **51**, 1001–22.

Hawkes, I., Mellor, M. and Gariepy, S. (1973). 'Deformation of rocks under uniaxial tension', *Int. J. Rock Mech. Min. Sci. and Geomech. Abstr.,* **10**, 493–507.

Hoek, E. (1968). 'Brittle fracture of rocks', in *Rock Mechanics in Engineering Practice*, Stagg, K. G. and Zienkiewicz, O. C. (eds.), Wiley, London, 99–124.

Hoek, E. (1983). 'Strength of jointed rock masses', Rankine Lecture, *Geotechnique,* **33**, 187–223.

Hoek, E. and Brown, E. T. (1980). 'Empirical strength criterion for rock mass', *Proc. ASCE, J. Geot. Engg Div.,* **106**, GT9, 1013–35.

John, K. W. (1965). 'Civil engineering approach to evaluate strength and deformability of regularly jointed rock', *Rock Mech.,* **1**, 69–80.

McLamore, R. and Gray, K. E. (1967). 'The mechanical behavior of anisotropic sedimentary rocks', *J. Engg for Industry, Trans. Am. Soc. Mech. Eng.*, Ser B, **89**, 62–73.

Mohr, O. (1882). *Abhandlungen aus dem Gebiete der Technische Meckanik*, Ernst und Sohn, Berlin.

Priest, S. D. and Selvakumar, S. (1986). *The Failure Characteristics for Selected British Rocks*, Report for the Transport and Road Research Laboratory, Imperial College, University of London.

Schwartz, A. E. (1964). 'Failure of rock in triaxial shear test', *Proc. 6th Rock Mechanics Symp.*, Rolla, Missouri, 109–35.

Vutukuri, V. S. (1974). 'The effects of liquid on the tensile strength of limestone', *Int. J. Rock Mech. Min. Sci. and Geomech. Abstr.,* **11**, 27–9.

Walsh, J. B. (1965). 'The effects of cracks on the uniaxial elastic compression of rocks', *J. Geophys. Res.,* **70**, 399–411.

Wawersik W. R. and Fairhurst, C. (1970). 'A study of brittle rock fracture in laboratory compression experiments', *Int. J. Rock Mech. Min. Sci.,* **7**, 561–75.

Chapter 8

Discontinuities in rock masses

A discontinuity represents a plane of weakness within a rock mass across which the rock material is structurally discontinuous. Although discontinuities are not necessarily planes of separation, most in fact are and they possess little or no tensile strength. Discontinuities vary in size from small fissures on the one hand to huge faults on the other. The most common discontinuities are joints and bedding planes (Figure 8.1). Other important discontinuities are planes of cleavage and schistosity.

Figure 8.1

8.1 Nomenclature of joints

Joints are fractures along which little or no displacement has occurred and are present within all types of rock. At the surface, joints may open as a consequence of stress release and weathering.

A group of joints which run parallel to each other is termed a *joint set* whilst two or more joint sets which intersect at a more or less constant angle are referred to as a *joint system*. If joints are planar and parallel or sub-parallel they are described as systematic; conversely, when they are irregular they are termed non-systematic. If one set of joints is dominant, then these joints are known as primary joints, the other set of joints being termed secondary.

On a basis of size, joints can be divided into master joints which penetrate several rock horizons and persist for hundreds of metres; major joints which are smaller joints but which are still well-defined structures and minor joints which do not transcend bedding planes. Lastly, minute fractures occasionally occur in finely bedded sediments and such micro-joints may only be a few millimetres in size.

Joints may be associated with folds and faults, having developed towards the end of an active tectonic phase or when such a phase has subsided. However, joints do not appear to form parallel to other planes of shear failure such as normal and thrust faults. The orientation of joint sets in relation to folds depends upon their size, the type and size of the fold, and the thickness and competence of the rocks involved. At times the orientation of the joint sets can be directly related to the folding and may be defined in terms of the *a*, *b* and *c* axes of the 'tectonic cross' (Figure 8.2). Those joints which cut the fold at right angles to the axis are called *ac*

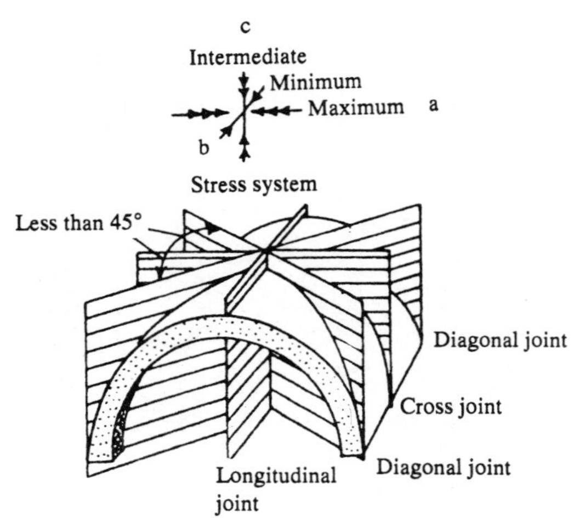

Figure 8.2 *Geometric orientation of longitudinal, cross and diagonal joints relative to fold axis and to principal stress axes (from Willis and Willis, 1934)*

or cross joints. The *bc* or longitudinal joints are perpendicular to the latter joints and diagonal or oblique joints make an angle with both the *ac* and the *bc* joints. Diagonal joints are classified as shear joints whereas *ac* and *bc* joints are regarded as tension joints.

8.2 Origins of joints

Joints are formed through failure in tension, in shear, or through some combination of both. Rupture surfaces formed by extension tend to be clean and rough with little detritus. They tend to follow minor lithological variations. If such surfaces are sheared, the resulting load deformation curve displays a peak, rising well above the residual strength which is only reached at large values of displacement. Simple surfaces of shearing are generally smooth and contain considerable detritus. They are unaffected by local lithological changes. Shearing along this type of discontinuity does not yield as great a contrast between peak and residual strength as does a tension joint. The material along shear joints is commonly much more susceptible to alteration than that along tension joints.

Price (1966) contended that the majority of joints are post-compressional structures, formed as a result of the dissipation of residual stress after folding has occurred. Some spatially restricted small joints associated with folds, such as radial tension joints, are probably initiated during folding. The shearing stresses at the time of joint formation probably measure a few hundred megapascals. Such stresses usually can be dissipated by a movement amounting to a millimetre or so along a shear plane. But such movement can only dissipate the residual stresses in the immediate neighbourhood of a joint plane so that a very large number of joints need to form in order to dissipate the stresses throughout a large area. Price (1966) suggested that joint frequency was related to the lithology of the rock type, the dimensions of the rock unit and the degree of tectonic deformation. On the last point he quoted the work of Harris *et al.* (1960) who found that the highest joint frequencies were associated with areas where the structures exhibited maximum curvature.

Joints are also formed in other ways. For example, joints develop within igneous rocks when they initially cool down, and in wet sediments when they dry out. The most familiar of these are the columnar joints in lava flows, sills and some dykes. Cross joints, longitudinal joints, diagonal joints and flat-lying joints are associated with large granitic intrusions (Figure 8.3).

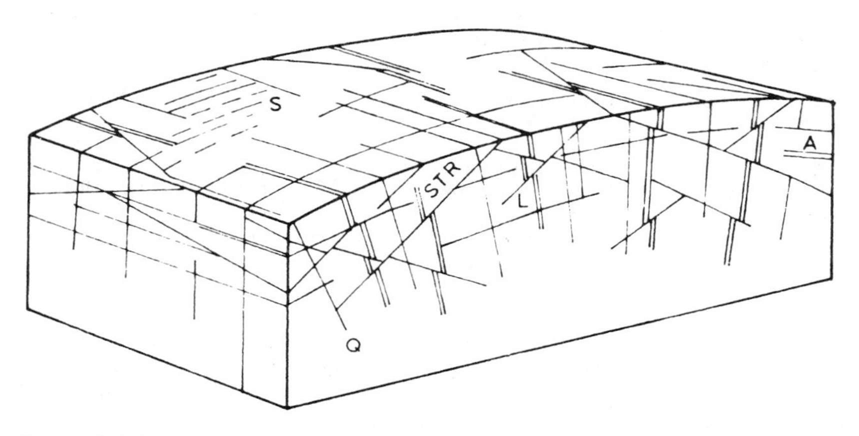

Figure 8.3 *Block diagram showing the types of structures in a batholith. Q = cross joints; S = longitudinal joints; L = flat-lying joints; STR = planes of stretching; F = linear flow structures; A = aplite dykes (after Balk, 1938)*

Sheet or mural joints have a similar orientation to flay-lying joints. When they are closely spaced and well developed they impart a pseudo-stratification to the host rock. It has been noted that the frequency of sheet jointing is related to the depth of overburden, in other words, the thinner the rock cover the more pronounced the sheeting. This suggests a connection between removal of overburden by denudation and the development of sheeting. It may well be that some granitic intrusions contain considerable residual strain energy and that with the gradual removal of load the associated residual stresses are dissipated by the formation of sheet joints.

8.3 Description of jointed rock masses

8.3.1 Incidence of discontinuities

The shear strength of a rock mass and its deformability are very much influenced by the discontinuity pattern, its geometry and how well it is developed. Observation of discontinuity spacing, whether in a field exposure or in a core stick, aids appraisal of rock mass structure. In sedimentary rocks bedding planes are usually the dominant discontinuity and the rock mass can be described as shown in Table 8.1. The same boundaries can be used to describe the spacing of joints (Anon, 1977).

Table 8.1 *Description of bedding plane and joint spacing*

Description of bedding plane spacing	*Description of joint spacing (cf. Barton 1978)*	*Limits of spacing*	*Mass factor (j)*
Very thickly bedded	Extremely wide	Over 2 m	0.8–1.0
Thickly bedded	Very wide	0.6–2 m	0.5–0.8
Medium bedded	Wide	0.2–0.6 m	0.2–0.5
Thinly bedded	Moderately wide	60 mm–0.2 m	0.1–0.2
Very thinly bedded	Moderately narrow	20–60 mm	Less than 0.1
Laminated	Narrow	6–20 mm	
Thinly laminated	Very narrow	Under 6 mm	

The mechanical behaviour of a rock mass is strongly influenced by the number of sets of discontinuities which intersect it, since this influences the amount of deformation that the rock mass will undergo. The number of sets also affects the degree of overbreak which occurs on excavation, they therefore may be an important factor in rock slope stability.

Systematic sets should be distinguished from non-systematic sets when recording the discontinuities in the field. Barton (1978) suggested that the number of sets of discontinuities at any particular location could be described in the following manner:

(1) massive, occasional random joints,
(2) one discontinuity set,
(3) one discontinuity set plus random,
(4) two discontinuity sets,
(5) two discontinuity sets plus random,
(6) three discontinuity sets,
(7) three discontinuity sets plus random,
(8) four or more discontinuity sets,
(9) crushed rock, earth-like.

8.3.2 Geometry of discontinuities

As joints represent surfaces of weakness, the larger and more closely spaced they are, the more influential they become in reducing the effective strength of the rock mass. The persistence of a joint plane refers to its continuity. This is one of the most difficult properties to quantify since joints frequently continue beyond the rock exposure and consequently in such instances it is impossible to estimate their continuity. Nevertheless Barton (1978) suggested that the modal trace lengths measured for each discontinuity set can be described as set out in Table 8.2. Simple sketches and block diagrams help to indicate the relative persistence of the various sets of discontinuities.

Table 8.2 *Persistence of joints*

Very low persistence	Less than 1 m
Low persistence	1–3 m
Medium persistence	3–10 m
High persistence	10–20 m
Very high persistence	Greater than 20 m

Block size provides an indication of how a rock mass is likely to behave, since block size and interblock shear strength determine the mechanical performance of a rock mass under given conditions of stress. The following descriptive terms have been recommended for the description of rock masses in order to convey an impression of the shape and size of blocks (Barton, 1978):

(1) massive – few joints or very wide spacing,
(2) blocky – approximately equidimensional,
(3) tabular – one dimension considerably shorter than the other two,
(4) columnar – one dimension considerably larger than the other two,
(5) irregular – wide variations of block size and shape,
(6) crushed – heavily jointed to ‘sugar cube’.

The orientation of the short or long dimensions should be specified in the columnar and tabular blocks respectively. In addition it may be useful to note the ratio of the orthogonal dimensions, for example, 1 vertical: 2 north: 6 east. The block size may be described by using the terms given in Table 8.3 (Anon, 1977).

Table 8.3 *Block size*

Term	*Block size*	*Equivalent discontinuity spacings in blocky rock*	*Volumetric joint count (J_v)* (joints/m^3)*
Very large	Over 8 m^3	Extremely wide	Less than 1
Large	0.2–8 m^3	Very wide	1–3
Medium	0.008–0.2 m^3	Wide	3–10
Small	0.0002–0.008 m^3	Moderately wide	10–30
Very small	Less than 0.0002 m^3	Less than moderately wide	Over 30

* After Barton (1978)

Discontinuities, especially joints, may be open or closed. How open they are (Table 8.4) is of importance in relation to the overall strength and permeability of a rock mass and this often depends largely on the amount of stress release and/or weathering which the rocks have undergone. Furthermore, where the effects of weathering have penetrated deeply into a joint, a wide weak zone may be present. Some joints may be partially or completely filled. The type and amount of filling not only influence the effectiveness with which the opposing joint surfaces are bound together, thereby affecting the strength of the rock mass, but also influence

Table 8.4 *Description of the aperture of discontinuity surfaces*

Anon (1977)			*Barton (1978)*	
Description	*Width of aperture*		*Description*	*Width of aperture*
Tight	Zero		Very tight	Less than 0.1 mm
Extremely narrow	Less than 2 mm	Closed	Tight	0.1–0.25 mm
Very narrow	2–6 mm		Partly open	0.25–0.5 mm
Narrow	6–20 mm		Open	0.5–2.5 mm
Moderately narrow	20–60 mm	Gapped	Moderately wide	2.5–10 mm
Moderately wide	60–200 mm		Wide	Over 10 mm
Wide	Over 200 mm		Very wide	10–100 mm
		Open	Extremely wide	100–1000 mm
			Cavernous	Over 1 m

permeability. If the infilling is sufficiently thick, for example, over 100 mm, the walls of the joint will not be in contact and hence the strength of the joint plane will be that of the infill material. Materials such as clay or sand may have been introduced into a joint opening. Mineralization is frequently associated with joints. This may effectively cement a joint; however, in other cases the mineralizing agent may have altered and weakened the rocks along the joint conduit.

Infill occupying discontinuities may possess a wide range of physical properties, especially with regard to its shear strength, deformability and permeability. Its short-term and long-term behaviour may differ appreciably. The range of behaviour is influenced by the mineralogy of the infill, its particle size distribution, its water content and permeability, its over-consolidation ratio, any previous shear displacement, width of aperture, and roughness and state of wall rock. The infill may be assessed by using the same method(s) as used to assess the wall rock (see below).

8.3.3 Surfaces of discontinuities

The nature of the opposing joint surfaces also influences rock mass behaviour as the smoother they are, the more easily can movement take place along them. However, joint surfaces are usually rough and may be slickensided. Hence the nature of a joint surface may be considered in relation to its waviness, roughness and the condition of the walls. Waviness and roughness differ in terms of scale and their effect on the shear strength of the joint. Waviness refers to first-order asperities which appear as undulations of the joint surface and are not likely to shear off during movement. Therefore the effects of waviness do not change with displacements along the joint surface. Waviness modifies the apparent angle of dip but not the frictional properties of the discontinuity. On the other hand, roughness refers to second-order asperities which are sufficiently small to be sheared off during movement. Increased roughness of the discontinuity walls results in an increased effective friction angle along the joint surface. These effects diminish or disappear when infill matter is present. The procedure for measuring joint roughness in the field has been given by Barton (1978).

The visual classification of roughness shown in Table 8.5 can be used when

Table 8.5 *Joint roughness*

Category	*Degree of roughness*
1	Polished
2	Slickensided
3	Smooth
4	Rough
5	Defined ridges
6	Small steps
7	Very rough

quantitative measurements are not made (Anon, 1977). This classification has meaning only when the direction of the irregularities on the surface is in the least favourable direction to resist sliding. As a consequence it is necessary to record the trend of the lineation on the joint surface in relation to the direction of shearing. Uniformity of assessment may be obtained by identifying and photographing each category at the site in question.

An alternative set of descriptive terms has been suggested by Barton (1978) and these should be based upon two scales of observation, namely, small scale (several centimetres) and intermediate scale (several metres). The intermediate scale of roughness is divided into stepped, undulating and planar, and the small scale of roughness, superimposed upon the former, includes rough (or irregular), smooth and slickensided categories. The direction of the slickensides should be noted as shear strength may vary with direction. Barton recognized the following classes (Figure 8.4):

Class	↑
(1) rough (or irregular), stepped	
(2) smooth, stepped	
(3) slickensided, stepped	
(4) rough (irregular), undulating	Increasing
(5) smooth, undulating	shear
(6) slickensided, undulating	strength
(7) rough (irregular) planar	
(8) smooth, planar	
(9) slickensided, planar	

The compressive strength of the rock comprising the walls of a discontinuity is a very important component of shear strength and deformability, especially if the walls are in direct rock to rock contact, as in the case of unfilled joints. Weathering and alteration frequently are concentrated along the walls of discontinuities, thereby reducing their strength. The weathered material can be assessed in terms of its grade (Table 9.3) and manual index tests (Table 1.2). Alternatively, a Schmidt hammer can be used to obtain an idea of the compressive strength of the material concerned (Barton, 1978). Samples of wall rock can be tested in the laboratory, not just for strength, but if they are highly weathered, also for swelling and durability.

8.3.4 Flow of water and discontinuities

Seepage of water through rock masses usually takes place via the discontinuities, although in some sedimentary rocks seepage through the pores may also play an important role. The prediction of groundwater levels, probable seepage paths and approximate water pressures frequently provides an indication of stability or construction problems. Barton (1978) suggested that seepage from open or filled discontinuities could be assessed according to the descriptive scheme given in Table 8.6.

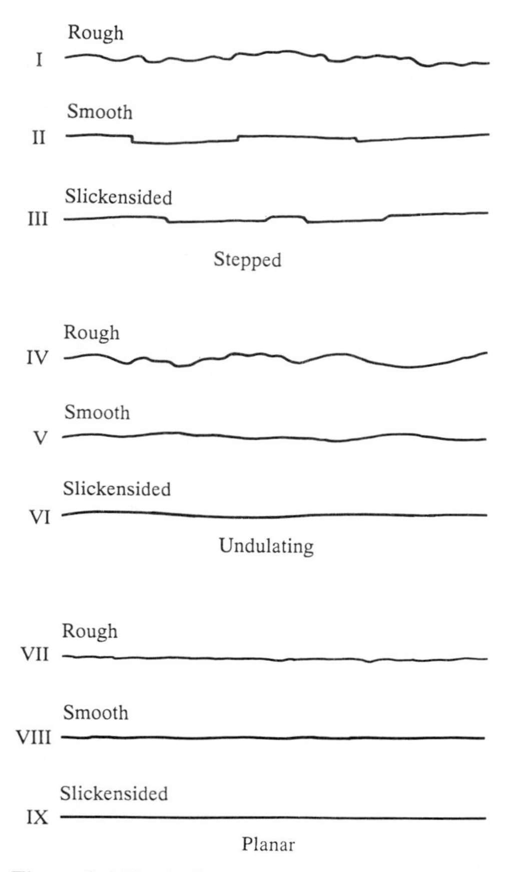

Figure 8.4 *Typical roughness profiles and suggested nomenclature. The length of each profile is in the range 1–10 m. The vertical and horizontal scales are equal (after Barton, 1978)*

8.4 Discontinuities and rock quality indices

Several attempts have been made to relate the numerical intensity of fractures to the quality of unweathered rock masses and to quantify their effect on deformability. For example, the concept of rock quality designation (RQD) was introduced by Deere (1964). It is based on the percentage core recovery when drilling rock with NX (57.2 mm) or larger diameter diamond core drills. Assuming that a consistent standard of drilling can be maintained, the percentage of solid core obtained depends on the strength and degree of discontinuities in the rock mass concerned. The RQD is the sum of the core sticks in excess of 100 mm expressed as

Table 8.6 *Seepage from discontinuities*

	A. Open discontinuities	*B. Filled discontinuities*
Seepage rating	*Description*	*Description*
(1)	The discontinuity is very tight and dry, water flow along it does not appear possible.	The filling material is heavily consolidated and dry, significant flow appears unlikely due to very low permeability.
(2)	The discontinuity is dry with no evidence of water flow.	The filling materials are damp but no free water is present.
(3)	The discontinuity is dry but shows evidence of water flow, i.e. rust staining, etc.	The filling materials are wet, occasional drops of water.
(4)	The discontinuity is damp but no free water is present.	The filling materials show signs of outwash, continuous flow of water (estimate l/min).
(5)	The discontinuity shows seepage, occasional drops of water but no continuous flow.	The filling materials are washed out locally, considerable water flow along outwash channels (estimate l/min and describe pressure, i.e. low, medium, high).
(6)	The discontinuity shows a continuous flow of water (estimate l/min and describe pressure, i.e. low, medium, high).	The filling materials are washed out completely, very high water pressures are experienced, especially on first exposure (estimate l/min and describe pressure).

a percentage of the total length of core drilled. However, the RQD does not take account of the joint opening and condition, a further disadvantage being that with fracture spacings greater than 100 mm the quality is excellent irrespective of the actual spacing (Table 8.7). This particular difficulty can be overcome by using the fracture spacing index as suggested by Franklin *et al.* (1971). This simply refers to the frequency with which fractures occur within a rock mass (Table 8.7).

The concept of the rock mass factor (j) was introduced by Hobbs (1975). He defined the rock mass factor as the ratio of the deformability of a rock mass within any readily identifiable lithological and structural component to that of the deformability of the intact rock comprising the component. Consequently it reflects the effect of discontinuities on the expected performance of the intact rock (Table 8.7). The value of j depends upon the method of assessing the deformability of the rock mass, and the value beneath an actual foundation will not necessarily be the same as that determined even from a large-scale field test. According to Hobbs, the greatest difficulties which occur in a jointed rock mass in relation to foundation design are experienced when the fracture spacing falls within a range of about 100–500 mm, in as much as small variations in fracture spacing and condition result in exceptionally large changes in *j*-value.

The effect of discontinuities in a rock mass can be estimated by comparing the *in situ* compressional wave velocity with the laboratory sonic velocity of an intact core sample obtained from the rock mass. The difference in these two velocities is caused by the structural discontinuities which exist in the field. The velocity ratio, V_{cf}/V_{cl}, where V_{cf} and V_{cl} are the compressional wave velocities of the rock mass *in situ* and of the intact specimen respectively, was first proposed by Onodera (1963). For a high-quality massive rock with only a few tight joints, the velocity ratio approaches unity. As the degree of jointing and fracturing becomes more severe, the velocity ratio is reduced (Table 8.7). The sonic velocity is determined for the core sample in the laboratory under an axial stress equal to the computed overburden stress at the depth from which the rock material was taken, and at a moisture content equivalent to that assumed for the *in situ* rock. The field seismic velocity preferably is determined by uphole or crosshole seismic measurements in drillholes or test adits, since by using these measurements it is possible to explore individual homogeneous zones more precisely than by surface refraction surveys.

Table 8.7 *Classification of rock quality in relation to the incidence of discontinuities*

Quality classification	*RQD (%)*	*Fracture frequency per metre*	*Mass factor (j)*	*Velocity ratio* (V_{ef}/V_{cl})
Very poor	0–25	Over 15		0.0–0.2
Poor	25–50	15–8	Less than 0.2	0.2–0.4
Fair	50–75	8–5	0.2–0.5	0.4–0.6
Good	75–90	5–1	0.5–0.8	0.6–0.8
Excellent	90–100	Less than 1	0.8–1.0	0.8–1.0

8.5 Recording discontinuity data

8.5.1 Direct discontinuity surveys

Before a discontinuity survey commences the area in question must be mapped geologically to determine rock types and delineate major structures. It is only after becoming familiar with the geology that the most efficient and accurate way of conducting a discontinuity survey can be devised. A comprehensive review of the procedure to be followed in a discontinuity survey has been provided by Barton (1978).

One of the most widely used methods of collecting discontinuity data is simply by direct measurement on the ground. A direct survey can be carried out subjectively in that only those structures which appear to be important are measured and recorded. In a subjective survey the effort can be concentrated on the apparently significant joint sets. Nevertheless, there is a risk of overlooking sets which might be important. Conversely, in an objective survey all structures intersecting a fixed line or area of the rock face are measured and recorded.

Several methods have been used for carrying out direct discontinuity surveys. Halstead *et al.* (1968) used the fracture set mapping technique by which all discontinuities occurring in 6 by 2 m zones, spaced at 30 m intervals along the face, were recorded. On the other hand, Piteau (1971) maintained that using a series of line scans provides a satisfactory method of joint surveying. The technique involves extending a metric tape across an exposure, levelling the tape and then securing it to the face. Two other scanlines are set out as near as possible at right angles to the first, one more or less vertical, the other horizontal. The distance along a tape at which each discontinuity intersects is noted, as is the direction of the pole to each discontinuity (this provides an indication of the dip direction). The dip of the pole from the vertical is recorded as this is equivalent to the dip of the plane from the horizontal. The strike and dip directions of discontinuities in the field can be measured with a compass and the amount of dip with a clinometer. Measurement of the length of a discontinuity provides information on its continuity. It has been suggested that measurements should be taken over distances of about 30 m, and to ensure that the survey is representative the measurements should be continuous over that distance. The line scanning technique yields more detail on the incidence of discontinuities and their attitude than other methods (Priest and Hudson, 1981). A minimum of at least 200 readings per locality is recommended to ensure statistical reliability.

Hudson and Priest (1979) pointed out that where discontinuities occur in sets, the discontinuity frequency along a scanline is a function of scanline orientation. They showed that the spacing distributions of discontinuities is a negative exponential distribution with the mean spacing of discontinuities being the reciprocal of the average number of discontinuities per metre, This value can simply be calculated by dividing the number of scanline intersections by the total scanline length. In addition, Hudson and Priest showed how the distributions of block areas, for most locations, can be predicted adequately from discontinuity frequency measurements made along scanlines, and how to derive cumulative frequency curves for block volumes from scanline data.

In their joint survey of sediments of Cretaceous age in south-east England, Fookes and Denness (1969) used the cavity technique, that is, they excavated blocks of material from the faces of exposures, and measurements were taken in the area vacated by the blocks. This meant that only fresh discontinuities were recorded. In addition to recording the dip and strike data for the joints and fissures they also measured the height, and, where possible, the length of each joint. A large orientated block was excavated for examination in the laboratory.

8.5.2 Drill holes and discontinuity surveys

The information gathered by any of the above methods can be supplemented with data from orientated cores from drill holes. The value of the data depends in part on the quality of the rock concerned, in that poor quality rock is likely to be lost during drilling. However, it is impossible to assess the persistence, degree of separation or the nature of the joint surfaces. What is more, infill material, especially if it is soft, is not recovered by the drilling operations.

Core orientation can be achieved by a core orientator or by integral sampling (Rocha, 1971 and Figure 8.5).

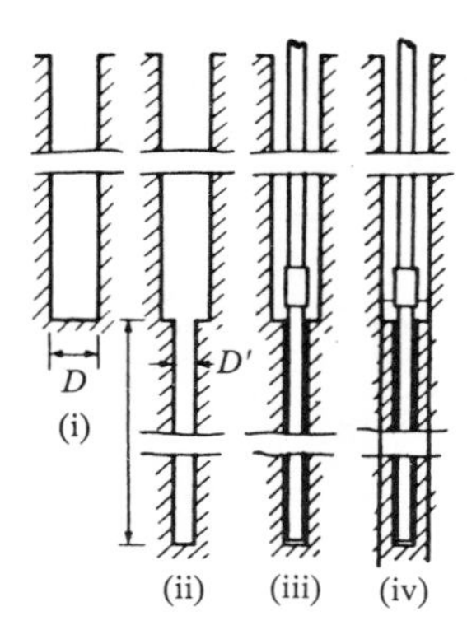

Figure 8.5 *Stages of integral sampling. A drill hole, diameter* D, *is drilled to a depth where the integral sample is to be obtained. Then another hole (diameter* D'*) coaxial with the former is drilled from the bottom of the hole, into which a reinforcing bar is placed and bonded to the rock mass. Overdrilling then occurs to obtain the integral sample*

Drill hole inspection techniques include the use of drill hole periscopes, drill hole cameras or closed-circuit television. The drill hole periscope affords direct inspection and can be orientated from outside the hole. However, its effective use is limited to about 30 m. The drill hole camera can also be orientated prior to photographing a section of the wall of a drill hole. The television camera provides a direct view of the drill hole and a recording can be made on videotape. These three systems are limited in that they require relatively clear conditions and so may be of little use below the water table, particularly if the water in the drill hole is murky.

The televiewer produces an acoustic picture of the drill hole wall. One of its advantages is that drill holes need not be flushed prior to its use.

8.5.3 Photographs and discontinuity surveys

Many data relating to discontinuities can be obtained from photographs of exposures. Photographs may be taken looking horizontally at the rock mass from the ground or they may be taken from the air looking vertically, or occasionally

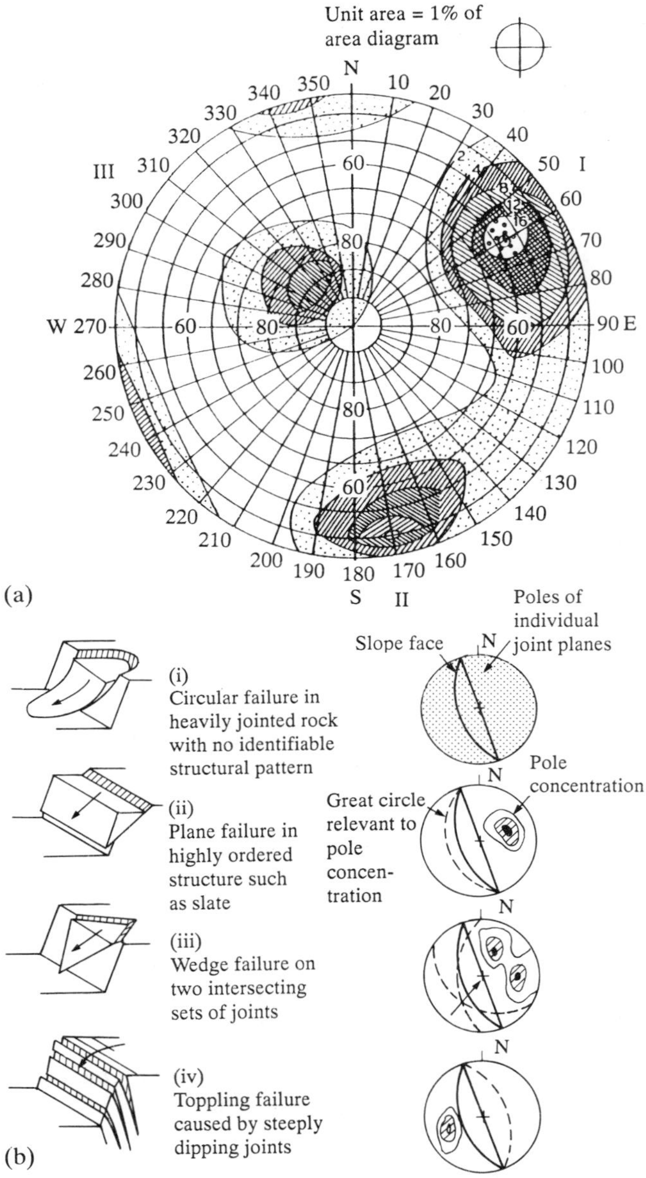

Figure 8.6 (a) *Schmidt contour diagram representing the orientation of three sets of joints plotted on a polar equal-area net. The main sets I and II are approximately normal to each other and the minor set III is nearly horizontal (after Barton, 1978).* (b) *Representation of structural data concerning four possible slope failure modes plotted on equal-area nets as poles and great circles (after Hoek and Bray, 1977)*

obliquely, down at the outcrop. These photographs may or may not have survey control. Uncontrolled photographs are taken using hand-held cameras, stereo-pairs being obtained by taking two photographs of the same face from positions about 5% of the distance of the face apart, along a line parallel to the face. Delineation of major discontinuity patterns and preliminary subdivision of the face into structural zones can be made from these photographs. Unfortunately, data cannot be transferred with accuracy from them onto maps and plans. Conversely discontinuity data can be accurately located on maps and plans by using controlled photographs. Controlled photographs are obtained by aerial photography with complementary ground control or by ground-based phototheodolite surveys. Aerial and ground-based photography are usually done with panchromatic film but the use of colour and infrared techniques is becoming more popular. Aerial photographs, with a suitable scale, have proved useful in the investigation of discontinuities. Photographs taken with a phototheodolite can also be used with a stereo-comparator which produces a stereoscopic model. Measurements of the locations or points in the model can be made with an accuracy of approximately 1 in 5000 of the mean object distance. As a consequence, a point on a face photographed from 50 m can be located to an accuracy of 10 mm. In this way the frequency, orientation and continuity of discontinuities can be assessed. Such techniques prove particularly useful when faces which are inaccessible or unsafe have to be investigated.

8.5.4 Recording discontinuity data

The simplest method of recording discontinuity data is by using a histogram on which the frequency is plotted along one axis and the strike direction along the other. Directional information, however, is more effectively represented on a rose diagram. This provides a graphical illustration of the angular relationships between joint sets. The strikes of the joints and their frequencies are represented by the directions on each rose diagram, the lengths of the vectors being plotted either on a half or full-circle. Directions are usually plotted for data contained in 5° arcs, while magnitudes are plotted to scale.

Data from a discontinuity survey are now, however, usually plotted on a stereographic projection. The use of spherical projections, commonly the Schmidt or Wulf net, means that traces of the planes on the surface of the 'reference sphere' can be used to define the dips and dip directions of discontinuity planes. In other words the inclination and orientation of a particular plane is represented by a great circle or a pole, normal to the plane, which are traced on an overlay placed over the stereonet. The method whereby great circles or poles are plotted on a stereogram has been explained by Hoek and Bray (1981). When recording field observations of the amount and direction of dip of discontinuities it is convenient to plot the poles rather than the great circles. The poles can then be contoured in order to provide an expression of orientation concentration. This affords a qualitative appraisal of the influence of the discontinuities on the engineering behaviour of the rock mass concerned (Figure 8.6).

References

Anon (1977). 'The description of rock masses for engineering purposes. Working Party Report', *Q. J. Engg Geol.*, **10**, 355–88.

Balk, R. (1938). *Structural Behaviour of Igneous Rocks*, Memoir 6, Geol. Soc. Am., New York.

Barton, N. (1978). 'Suggested methods for the quantitative description of discontinuities in rock masses', ISRM Commission on Standardization of Laboratory and Field Tests, *Int. J. Rock Mech. Min. Sci and Geomech. Abstr.*, **15**, 319–68.

Deere, D. U. (1964). 'Technical description of cores for engineering purposes', *Rock Mech. Engg Geol.*, **1**, 18–22.

Fookes, P. G. and Denness, B. (1969). 'Observational studies on fissure patterns in Cretaceous sediments of south-east England', *Geotechnique,* **19**, 453–77.

Franklin, J. L., Broch, E. and Walton, G. (1971). 'Logging the mechanical character of rock', *Trans. Inst. Min. Metall.*, **81**, Mining Section, A1–9.

Halstead, P. N., Call, P. D. and Rippere, K. H. (1968). 'Geological structural analysis for open pit slope design, Kimberley pit, Ely, Nevada', Reprint: *Annual AIME Meeting*, New York.

Harris, J. F., Taylor, G. L. and Walper, J. L. (1960). 'Relation of deformational features in sedimentary rocks and regional and local structure', *Bull. Am. Assoc. Petrol. Geol.*, **44**, 1853–73.

Hobbs, D. W. (1968). 'The formation of tension joints in sedimentary rocks', *Geol. Mag.*, **104**, 550–6.

Hobbs, N. B. (1975). 'Factors affecting the prediction of settlement of structures on rocks with particular reference to the Chalk and Trias', in *Settlement of Structures*, Brit. Geot. Soc., Pentech Press, London, 579–610.

Hoek, E. and Bray, J. W. (1981). *Rock Slope Engineering*, Inst. Min. Metall., London.

Hudson, J. A. and Priest, S. D. (1979). 'Discontinuities and rock mass geometry', *Int. J. Rock Mech. Min. Sci and Geomech. Abstr.*, **16**, 339–62.

Onodera, T. F. (1963). 'Dynamic investigation of foundation rocks', *Proc. 5th Symp. Rock Mech., Minnesota*, Pergamon Press, New York, 517–33.

Piteau, D. R. (1971). 'Geological factors significant to the stability of slopes cut in rock', *Symp. Planning Open Pit Mines*, Johannesburg, A. A. Balkema, Rotterdam, 43–53.

Price, N. L. (1966). *Fault and Joint Development in Brittle and Semi-Brittle Rock*, Pergamon Press, London.

Priest, S. D. and Hudson, J. A. (1981). 'Estimation of discontinuity spacing and trace length using scanline surveys', *Int. J. Rock Mech. Min. Sci. and Geomech. Abstr.*, **18**, 183–97.

Rocha, M. (1971). 'Method of integral sampling', *Rock Mech.*, **3**, 1–12.

Willis, B. and Willis, R. (1934). *Geologic Structures*, McGraw-Hill, New York.

Chapter 9

Weathering of rocks and rock masses

Weathering of rocks is brought about by physical disintegration, chemical decomposition and biological activity. The weathering process is primarily controlled by the presence of discontinuities in that they provide access for the agents of weathering. Some of the earliest effects of weathering are seen along discontinuity surfaces. Weathering then proceeds inwards until the whole of a discontinuity bounded block is affected. The agents of weathering, unlike those of erosion, do not themselves provide for the transportation of debris from a rock surface. Therefore unless this rock waste is otherwise removed it eventually acts as a protective blanket, preventing further weathering taking place. If weathering is to be continuous, fresh rock exposures must be constantly revealed, which means that the weathered debris must be removed by the action of gravity, running water, wind or moving ice.

9.1 Rate of weathering

The rate at which weathering proceeds depends not only upon the vigour of the weathering agent but also on the durability of the rock mass concerned. This, in turn, is governed by the mineralogical composition, texture and porosity of the rock on the one hand, and the incidence of discontinuities within the rock mass on the other.

Most studies regarding the rate at which weathering occurs have been made upon stone used for construction purposes and have involved measuring the rate at which the surfaces of stones have been removed (Table 9.1). In addition, a number of tests have been used to simulate and accelerate the rate of weathering of construction stone. The problem with such tests is the difficulty in relating the results obtained to the natural performance of the rocks concerned. Nevertheless according to Fookes *et al.* (1988) the rate of weathering declines with time where a residual layer is developed at the surface of the rock and, if surface reaction is involved, then the rate of weathering is linear.

Table 9.1 *Records of rates of surface lowering reported for various rock types (From Fookes et al., 1988)*

Rock type	*Surface studied*	*Period* (years)	*Remarks*	*Average* (mm^{a-1})	*Location*
Granite	Ancient structures	4 000	Fresh	0.0009	Aswan
		5 400	Flaking	0.0015	Giza
	Glacial surface	10 000	–	0.00105	Narvik
Slate	Tombstones	90	Engraving clear	–	Edinburgh
Mica schist	Glacial surface	8	–	0.04–0.15	Karkevagge
Marble	Tombstones	90	Crumbling	–	Edinburgh
	Tombstones	500	–	0.051	Yorkshire
	Tombstones	300	–	0.085	Yorkshire
	Tombstones	250	–	0.102	Yorkshire
	Tombstones	240	–	0.106	Yorkshire
	Great Pyramid	1 000	Hard grey	–	Giza
	Great Pyramid	1 000	Soft grey	0.01–0.02	Giza
Limestone	Kammetz fortress	230	–	1.32	Ukraine
	Bare surface	1 000	–	0.009–0.0125	Austrian Alps
	Covered surface	1 000	Under acid soil	0.028	Austrian Alps
	Jetty	16	Intertidal	0.5–1.0	Norfolk Island
	Coastal notch		–	1.0	Point-Perm, Australia
	Inscriptions		–	0.5	La Jolla, California
	Inter-tidal notch	155	–	1.0	Puerto Rico
Carb. limestone	Erratic block	12 000	–	0.025–0.042	N. England
Carb. limestone	Glacial striae	13	–	2.2–3.8	N. England
Carb. limestone	Runnels in glacial surface	13	–	11.5	N. England
Limestone	Great Pyramid	1 000	Hard grey	Little	Giza
	Great Pyramid	1 000	Soft grey	0.01–0.02	Giza
Kirkby-Stephen limestone	Tombstones	500	–	0.051	Yorkshire
Tailbrig limestone	Tombstones	300	–	0.085	Yorkshire
Penrith limestone	Tombstones	250	–	0.102	Yorkshire
Askrigg limestone	Tombstones	240	–	0.106	Yorkshire
Algal limestone	Tombstones	<1	Microerosion meter	0.11	Aldabra
Limestone (poorly cemented)	Limestone pavement		–	0.003–6.3	Co. Clare
Limestone chert	Glacial surface	70	–	0.02–0.2	Spitzbergen
Limey shale	Great Pyramid	1 000	Rubble	0.2	Giza
Grey shale	Great Pyramid	1 000	–	0.2	Giza
Sandstone	Tombstones	200	Little	–	Edinburgh

9.2 Mechanical weathering

Mechanical or physical weathering is particularly effective in climatic regions which experience significant diurnal changes of temperature. This does not necessarily imply a large range of temperature, as frost and thaw action can proceed where the range is limited.

As far as frost susceptibility is concerned the porosity, pore size and degree of saturation all play an important role. When water turns to ice it increases in volume by up to 9%, thus giving rise to an increase in pressure within the pores. This action is further enhanced by the displacement of pore water away from the developing ice front. Once ice has formed the ice pressures rapidly increase with decreasing temperature, so that at approximately −22°C ice can exert a pressure of 200 MN/m^2 (Winkler, 1973). Usually coarse-grained rocks withstand freezing better than fine-grained types. Indeed the critial pore size for freeze–thaw durability appears to be about 0.005 mm. In other words rocks with larger mean pore diameters allow outward drainage and escape of fluid from the frontal advance of the ice line and are therefore less frost susceptible. Fine-grained rocks which have 5% sorbed water are often very susceptible to frost damage whilst those containing less than 1% are very durable. Alternate freeze–thaw action causes cracks, fissures, joints and some pore spaces to be widened. As the process advances, angular rock debris is gradually broken from the parent body.

The mechanical effects of weathering are well displayed in hot deserts, where wide diurnal ranges of temperature cause rocks to expand and contract. Because rocks are poor conductors of heat these effects are mainly localized in their outer layers where alternate expansion and contraction create stresses which eventually rupture the rock. In this way flakes of rock break away from the parent material, the process being termed exfoliation. The effects of exfoliation are concentrated at the corners and edges of rocks so that their outcrops gradually become rounded. Furthermore minerals possess different coefficients of expansion and differential expansions within a polymineralic rock fabric generate stresses at grain contacts and can lead to granular disintegration.

The repeated action of wetting and drying together with the expansive force of water can represent a disruptive influence, especially in pores of a rock with a low tensile strength. For instance, when the temperature of water is raised from 0°C to 60°C, then it expands by about 1.5% exerting a pressure of up to 52 MN/m^2 within the pores and a diurnal temperature difference of 40°C can develop pressures of around 26 MN/m^2. Hence the expansion and contraction of water in narrow capillaries can produce sufficient pressures to disrupt many of the weaker rock types within a year.

Soluble salts which occur in the pores of rock tend to move towards the surface. The more soluble salts, however, such as some chlorides and sulphates, move back and forth in the rock with changes in weather and moisture gradient. The less soluble salts crystallize at or near the surface. The development of subfluorescences just below the surface can lead to the outer skin of a rock losing its support and exfoliating. Calcium and magnesium sulphates are among the most common salts involved. Resistance to crystallization damage as with frost damage, is strongly dependent on the internal structure of the rock and decreases as the proportion of fine pores increases. The pressures produced on crystallization in small pores are appreciable, for example, gypsum ($CaSO_4.nH_2O$) exerts a pressure of up to 100 MN/m^2; anhydrite ($CaSO_4$), 120 MN/m^2; kieserite ($MgSO_4.nH_2O$), 100 MN/m^2; and halite (NaCl), 200 MN/m^2; and these are sufficient to cause disruption.

Disruption in rock also may take place due to the considerable contrasts in thermal expansion of salts in the pores. For instance, halite expands by some 0.5% from 0°C to 60°C, which may aid rock decay.

9.3 Chemical and biological weathering

Chemical weathering leads to mineral alteration and the solution of rocks. Alteration is principally effected by oxidation, hydration, hydrolysis and carbonation whilst solution is brought about by acidified or alkalized waters. Chemical weathering also aids rock disintegration by weakening the rock fabric and by emphasizing any structural weaknesses, however slight, that it possesses. When decomposition occurs within a rock the altered material frequently occupies a greater volume than that from which it was derived and in the process internal stresses are generated. If this swelling occurs in the outer layers of a rock, then it causes them to peel off from the parent body.

In dry air rocks decay very slowly. The presence of moisture hastens the rate tremendously, first, because water is itself an effective agent of weathering and, second, because it holds in solution substances which react with the component minerals of the rock. The most important of these substances are free oxygen, carbon dioxide, organic acids and nitrogen acids.

Free oxygen is an important agent in the decay of all rocks which contain oxidizable substances, iron and sulphur being especially suspect. The rate of oxidation is quickened by the presence of water; indeed it may enter into the reaction itself, as for example, in the formation of hydrates. However, its role is chiefly that of a catalyst.

Carbonic acid is produced when carbon dioxide is dissolved in water and it may possess a pH value of about 5.7. The principal source of carbon dioxide is not the atmosphere but the air contained in the pore spaces in the soil where its proportion may be a hundred or so times greater than it is in the atmosphere. An abnormal concentration of carbon dioxide is released when organic material decays. Furthermore humic acids are formed by the decay of humus in soil waters; they ordinarily have pH values between 4.5 and 5.0 but occasionally they may be under 4.0. The nitrogen acids, HNO_3 and HNO_2, are formed by organic decay or bacterial action in soils. They play only a minor part in weathering.

The simplest reactions which take place on weathering are the solution of soluble minerals and the addition of water to substances to form hydrates. Solution commonly involves ionization; for example, this takes place when salt and gypsum deposits and carbonate rocks are weathered. Hydration and dehydration take place amongst some substances, a common example being gypsum and anhydrite:

$$\underset{\text{(anhydrite)}}{CaSO_4} + 2H_2O = \underset{\text{(gypsum)}}{CaSO_4.2H_2O}$$

The above reaction produces an increase in volume (Table 9.2) and accordingly causes the enclosing rocks to be wedged further apart. These reactions are slow but those involving ferric oxides and hydrates are even slower. Iron oxides and hydrates are conspicuous products of weathering, usually the oxides are a shade of red and the hydrates yellow to dark brown.

Table 9.2 *Crystalline solid expansion due to mineral alteration (From Taylor, 1988)*

Original mineral	*Mineral formed*	*Volume increase (%)*
Pyrite, FeS_2	Jarosite	115
	Melanterite	536
	Anhydrous ferrous sulphate	350
Calcite, $CaCO_3$	Gypsum	103
	Bassanite	189
Illite, $KAl_2Si_3O_8(OH)_2$	Jarosite	10
	Alunite	8

Suplhur compounds are readily oxidized by weathering. Because of the hydrolysis of the dissolved metal ion, solutions forming from the oxidation of sulphides are acidic. For instance, when pyrite is initially oxidized, ferrous sulphate and sulphuric acid are formed. Further oxidation leads to the formation of ferric sulphate. Very insoluble ferric oxide or hydrated oxide is formed if highly acidic conditions are produced:

$$\underset{\text{(pyrite)}}{FeS_2} + H_2O + 7(O) = FeSO_4 + H_2SO_4$$
$$2FeSO_4 + (O) + H_2SO_4 = Fe_2(SO)_4 + H_2O$$
$$2FeS_2 + 15(O) + 4H_2O = F_2O_3 + 4H_2SO_4$$

Ferrous sulphate may react with illite to form jarosite, which also involves an expansion in volume (Table 9.2).

Perhaps the most familiar examples of rock prone to chemical attack are limestones and dolostones. Limestones are chiefly composed of calcium carbonate and they are suspect to acid attack because CO_3 readily combines with H to form the stable bicarbonate HCO_3:

$$CaCO_3 + H_2CO_3 = Ca(HCO_3)_2$$

In water with a temperature of 25°C the solubility of calcium carbonate ranges from 0.01–0.05 g/l, depending upon the degree of saturation with carbon dioxide.

Dolostone is somewhat less soluble than limestone. When a carbonate rock is subject to dissolution any insoluble material present in it remains behind.

Weathering of the silicate minerals is primarily a process of hydrolysis. Much of the silica which is released by weathering forms silicic acid but where it is liberated in large quantities some of it may form colloidal or amorphous silica. As noted above, mafic silicates usually decay more rapidly than felsic silicates and in the process they release magnesium, iron and lesser amounts of calcium and alkalis. Olivine is particularly unstable, decomposing to form serpentine, which on further weathering forms talc and carbonates. Chlorite is the commonest alteration product of augite (the principal pyroxene) and of hornblende (the principal amphibole).

When subjected to chemical weathering feldspars decompose to form clay minerals, the latter are consequently the most abundant residual products. The process is effected by the hydrolysing action of weakly carbonated waters which leach the bases out of the feldspars and produce clays in colloidal form. The alkalis are removed in solution as carbonates from orthoclase (K_2CO_3) and albite (Na_2CO_3), and as bicarbonate from anorthite ($Ca(HCO_3)_2$). Some silica is hydrolysed to form silicic acid. Although the exact mechanism of the process is not fully understood the equation given below is an approximation towards the truth:

$$\underset{\text{(orthoclase)}}{2KAlSi_3O_6} + 6H_2O + CO_2 = \underset{\text{(kaolinite)}}{Al_2Si_2O_5(OH)_4} + 4H_2SiO_4 + K_2CO_3$$

The colloidal clay eventually crystallizes as an aggregate of minute clay minerals. Deposits of kaolin are formed when percolating acidified waters decompose the feldspars contained in granitic rocks.

Clays are hydrated aluminium silicates and when they are subjected to severe chemical weathering in humid tropical regimes they break down to form laterite or bauxite. The process involves the removal of siliceous material and this is again brought about by the action of carbonated waters. Intensive leaching of soluble mineral matter from surface rocks takes place during the wet season. During the subsequent dry season groundwater is drawn to the surface by capillary action and minerals are precipitated there as the water evaporates. The minerals generally consist of hydrated peroxides of iron, and sometimes of aluminium, and very occasionally of manganese. The precipitation of these insoluble hydroxides gives rise to an impermeable lateritic soil. When this point is reached the formation of laterite ceases as no further leaching can occur. As a consequence lateritic deposits are usually less than 7 m thick.

Plants and animals play an important role in the breakdown and decay of rocks, indeed their part in soil formation is of major significance. Tree roots penetrate cracks in rocks and gradually wedge the sides apart whilst the adventitious root system of grasses breaks down small rock fragments to particles of soil size. Burrowing rodents also bring about mechanical disintegration of rocks. The action of bacteria and fungi is largely responsible for the decay of dead organic matter. Other bacteria are responsible, for example, for the reduction of iron or sulphur compounds. It has also been suggested that bacterial action plays an important part in the formation of residual deposits such as laterites and bauxites.

9.4 Slaking and swelling of mudrocks

Physical disintegration of mudrocks is a much more important breakdown process than chemical weathering. The two principal controls on breakdown are slaking and the expansion of mixed-layer clay minerals. Indeed some freshly exposed mudrocks belonging to the Coal Measures in Britain can commence breakdown within a few days, weeks or months, depending on the character of the parent rock.

Slaking refers to the breakdown of rocks, especially mudrocks, by alternate wetting and drying. If a fragment of mudrock is allowed to dry out, air is drawn into the outer pores and high suction pressures develop. When the mudrock is next saturated the entrapped air is pressurized as water is drawn into the rock by capillary action (indurated mudrocks with small expandable clay mineral contents require suction pressures in excess of 10 kN/m^2 before air entry commences, whereas air entry occurs over the whole suction range in mudrocks with high expandable clay mineral contents). This slaking process causes the internal arrangement of grains to be stressed. Given enough cycles of drying and wetting, breakdown can occur as a result of air breakage, the process ultimately reducing the mudrock involved to tabular-shaped gravel-size particles.

In order to assess the durability of the mudrocks, samples are subjected to slake-durability testing. The various categories of the slake-durability index are shown in Table 9.3. However, Taylor (1988) suggested that durable mudrocks can be better distinguished from non-durable types on a basis of compressive strength and three-cycle slake-durability index (i.e. those mudrocks with a compressive strength of over 3.6 MN/m^2 and a three-cycle slake-durability index in excess of 60% are regarded as durable).

On the other hand Morgenstern and Eigenbrod (1974) used a water absorption test to assess the amount of slaking undergone by argillaceous material. This test measures the increase in water content in relation to the number of wetting and drying cycles undergone. They found that the maximum slaking water content

Table 9.3 *Description of the degree of slaking*

Amount of slaking	*Slake-durability index (%)**	*Liquid limit (%)†*
Very low	0–25	0–20
Low	25–50	20–50
Medium	50–75	50–90
High	75–95	90–140
Very high	over 95	over 140

* After Franklin and Chandra (1972)
† After Mergenstern and Eigenbrod (1974)

increased linearly with increasing liquid limit and that during slaking all materials eventually reached a final water content equal to their liquid limit. Materials with medium to high liquid limits exhibited very substantial volume changes during each wetting stage, which caused large differential strains, resulting in complete destruction of the original structure. Thus materials characterized by high liquid limits are more severely weakened during slaking than materials with low liquid limits. Classes of slaking were therefore defined in terms of the value of liquid limit (Table 9.3). The rate of slaking was grouped into three classes according to the change in liquidity index after immersion in water for two hours, that is, slow – less than 0.75; fast – 0.75 to 1.25; and very fast – over 1.25.

Intraparticle swelling (i.e. swelling due to the take up of water not only between particles of clay minerals but also within them – into the weakly bonded layers between molecular units) of clay minerals on saturation can cause mudrocks to break down where the proportion of such minerals constitutes more than 50% of the rock. The expansive clay minerals such as montmorillonite can expand many times their original volume.

It was concluded by Taylor and Spears (1970) that intraparticle swelling of mixed-layer clay during periods of saturation followed by desiccation in the near surface zone was a major control on the breakdown of those British Coal Measures mudrocks which contained significant amounts of expandable mixed-layer clay. Furthermore the exchangeable Na ion was most abundant in those mudrocks which underwent the greatest breakdown. In fact the exchangeable sodium percentage (ESP = exchangeable Na/cation exchange capacity × 100%) has been used as a guide to mudrock behaviour in that it is related to the amount of dispersion in shales.

Failure of consolidated and poorly cemented rocks occurs during saturation when the swelling pressure (or internal saturation swelling stress, σ_2), developed by capillary suction pressures, exceeds their tensile strength. An estimate of σ_s can be obtained from the modulus of deformation (E):

$$E = \sigma_s/\epsilon_D \tag{9.1}$$

where ϵ_D is the free swelling coefficient. The latter is determined by a sensitive dial gauge recording the amount of swelling of an oven-dried core specimen per unit height, along the vertical axis during saturation in water for 12 hours, ϵ_D being obtained as follows:

$$\epsilon_D = \frac{\text{Change in length after swelling}}{\text{Initial length}} \tag{9.2}$$

Sulphur compounds are frequently present in shales and mudstones. An expansion in volume large enough to cause structural damage can occur when sulphide minerals such as pyrite and marcasite suffer oxidation to give anhydrous and hydrous sulphates. According to Fasiska *et al.* (1974) the pyrite structure may be regarded as a stacking of almost close-packed hexagonal sheets of sulphide ions with iron ions in the interstices between the sulphide layers. The packing density is related to the radius of the sulphide ion which is 1.85 Å (volume = 26.1 $Å^3$). In the sulphate structure each atom of sulphur is surrounded by four atoms of oxygen in tetrahedral coordination. The packing density is related to the radius of the sulphate ion which is 2.8 Å, giving a volume of 92.4 Å. This represents an increase in volume per packing unit of approximately 350%. Hydration involves a further increase in volume. In fact such a reaction is electrolytic, that is, water is required and the sulphate ion exists in solution. Any cation in the system may cause the precipitation of sulphate crystals. If calcium carbonate is present, gypsum may be formed, which may give rise to an eight-fold increase in volume over the original sulphide, exerting pressures of up to about 0.5 MN/m^2. This leads to further disruption and weakening of the rocks involved.

Swelling can also occur as a result of hydration. For example, when anhydrite is hydrated to form gypsum there is a volume increase of between 30 and 58% which exerts pressures that have been variously estimated between 2 and 69 MN/m^2. It is thought that no great length of time is required to bring about such hydration. When it occurs at shallow depths it causes expansion but the process is gradual and is usually accompanied by the removal of gypsum in solution. At greater depths anhydrite is effectively confined during the process. This results in a gradual build-up of pressure and the stress is finally liberated in an explosive manner. According to Brune (1965) such uplifts in the United States have taken place beneath reservoirs, these bodies of water providing a constant supply for the hydration process, percolation taking place via cracks and fissures. Examples are known where the ground surface has been elevated by about 6 m. The rapid explosive movement causes strata to fold, buckle and shear.

9.5 Engineering classification of weathering

The early stages of weathering are usually represented by discoloration of the rock material which increases from slightly to highly discoloured as the degree of weathering increases. Because weathering brings about changes in engineering properties, in particular it commonly leads to an increase in bulk (and so in porosity) with a corresponding reduction in density and strength, these changes are reflected in the amount of discoloration. In other words the engineering properties of a slightly discoloured rock may differ notably from those of the same rock which is highly discoloured (Dearman, 1986). As weathering proceeds the rock material becomes increasingly decomposed and/or disintegrated until ultimately a soil is formed. Hence various stages in the reduction process of a rock to a soil can be recognized and can be used to form the basis of an engineering classification of weathering. Between five and seven grades of weathering have been recognized in the various schemes which have been proposed, the identification of grades being based upon the presence or absence of discoloration in rock material; the rock to soil ratio; and the presence or absence of relict rock fabric in the groups which are predominantly soil. It is almost inevitable that the boundaries between grades are gradational.

Table 9.4 *Engineering grade classifications of weathered rock and their relation to engineering behaviour*

Grade	*Degree of decomposition*	*Field recognition (after Little (1969); Fookes et al (1972); Dearman (1974))*			*Engineering behaviour*		
		Rocks (mainly chemical decomposition)	*Rocks (physical disintegration)**	*Carbonate rocks (solution)*	*After Little (1969)**	*After Hobbs (1975)†*	*After Hencher & Martin (1982)**
VI	Residual soil	The rock is discoloured and is completely changed to a soil in which the original fabric of the rock is completely destroyed. There is a large volume change	The rock is changed to a soil by granular disintegration and/or grain fracture. The structure of the rock is destroyed and the soil is a residuum of minerals unaltered from the original rock		Unsuitable for important foundations. Unstable on slopes when vegetation cover is destroyed and may erode easily unless hard cap is present. Requires selection before use as fill	In completely weathered rock and residual soil it may be possible to obtain fair quality samples depending upon the parent rock type and the consistency of the product. Generally the samples will tend to be less disturbed than when taken in the same rock in the highly weathered state. The bearing capacity and settlement characteristics of rock in these extreme states can be assessed using the usual methods for testing soils	A soil mixture with the original texture of the rock completely destroyed
V	Extremely weathered	The rock is discoloured and is wholly decomposed and friable, but the original fabric is mainly preserved. The properties of the rock mass depend in part on the nature of the parent rock. In granitic rocks feldspars are completely kaolinized	The rock is changed to a soil by granular disintegration and/or grain fracture. The structure of the rock is preserved	Grades V and VI cannot occur. These grades can be applied to interbedded soluble and insoluble rocks. Void size should be recorded	Cannot be recovered as cores by ordinary rotary drilling methods. Can be excavated by hand or ripping without the use of explosives. Unsuitable for foundations of concrete dams or large structures. May be suitable for foundation of earth dams and for fill. Unstable in high cuttings at steep angles. New joint patterns may have formed. Requires erosion protection		No rebound from N. Schmidt hammer; slakes readily in water; geological pick easily indents when pushed into surface; rock is wholly decomposed but rock texture preserved

IV	Highly weathered†	The rock is discoloured: discontinuities may be open and have discoloured surfaces (e.g. stained by limonite) and the original fabric of the rock near the discontinuities is altered: alteration penetrates deeply inwards, but corestones are still present. The rock mass is partially friable. Less than 50% rock	More than 50% and less than 100% of the rock is disintegrated by open discontinuities or spheroidal scaling spaced at 60 mm or less and/or by granular disintegration. The structure of the rock is preserved	More than 50% of the rock has been removed by solution. A small residuum may be present in the voids	Similar to Grade V. Sometimes recovered as core by careful rotary drilling. Unlikely to be suitable for foundations of concrete dams. Erratic presence of boulders makes it an unreliable foundation for large structures	In highly weathered rock difficulties will generally be encountered in obtaining undisturbed samples for testing. If samples are obtained the strength and modulus will generally be underestimated, frequently by large margins, even with apparently undisturbed samples. In such rocks *in situ* tests with either the Menard pressuremeter or the plate should be carried out to determine the bearing capacity and settlement characteristics. The greatest difficulties in assessing bearing capacity and settlement are likely to be encountered in highly weathered rocks, in which the rock fabric becomes increasingly disintegrated or increasingly more plastic	N. Schmidt rebound value 0 to 25; does not slake readily in water, geological pick cannot be pushed into surface; hand penetrometer strength index >250 kN/m^2; rock weakened so that large pieces broken by hand; individual grains plucked from surface

Table 9.4 *continued*

Grade	*Degree of decomposition*	*Field recognition (after Little (1969); Fookes et al (1972); Dearman (1974))*			*Engineering behaviour*		
		Rocks (mainly chemical decomposition)	*Rocks (physical disintegration)**	*Carbonate rocks (solution)*	*After Little (1969)**	*After Hobbs (1975)†*	*After Hencher & Martin (1982)**
III	Moderately weathered	The rock is discoloured, discontinuities may be open and have greater discoloration with the alteration penetrating inwards, the intact rock is noticeably weaker, as determined in the field than the fresh rock. The rock mass is not friable. 50–90% rock	Up to 50% of the rock is disintegrated by open discontinuities or by spheroidal scaling spaced at 60 mm or less and/or by granular disintegration. The structure of the rock is preserved	Up to 50% of the rock has been removed by solution. A small residuum may be present in the voids. The structure of the rock is preserved	Possessing some strength–large pieces (e.g. NX drill core) cannot be broken by hand. Excavated with difficulty without the use of explosives. Mostly crushes under bulldozer tracks. Suitable for foundations of small concrete structures and rock fill dams. May be suitable for semipervious fill. Stability in cuttings depends on structural features especially joint attitudes	In moderately weathered rock the intact modulus and strength can be very much lower than in the fresh rock and thus the j-value will be higher than in the fresh state, unless the joints and fractures have been opened by erosion or softened by the accumulation of weathered products. The intact modulus and strength can be measured in the laboratory and the bearing capacity assessed, in the same way as for fresh rock. Triaxial tests may be more appropriate than uniaxial tests, and it would be advisable to adopt conservative values for the factor of safety	N. Schmidt rebound value 25 to 45; considerably weathered but possessing strength such that pieces 55 mm diameter cannot be broken by hand; rock material not friable

II	Slightly weathered	The rock may be slightly discoloured, particularly adjacent to discontinuities which may be open and have slightly discoloured surfaces: the intact rock is not noticeably weaker than the fresh rock. Some decomposed feldspar in granites. Over 90% rock	100% rock: discontinuities open and spaced at more than 60 mm	100% rock: discontinuity surfaces open. Very slight solution etching of discontinuity surfaces may be present	Requires explosives for excavation. Suitable for concrete dam foundations. Highly permeable through open joints. Often more permeable than the zones above or below. Questionable as concrete aggregate	In faintly and slightly weathered rock it is possible that the *j*-value, owing to the reduction in stiffness of the joints as a result of penetrative weathering alone, will show a fairly sharp decrease compared with that of the same rock in the fresh state. The intact modulus, by definition, is unaffected by penetrative weathering. The safe bearing capacity is not therefore affected by faint weathering, and may be only slightly affected by slight weathering	N. Schmidt rebound value >45; more than one blow of geological hammer to break specimen; strength approaches that of fresh rock
I	Fresh rock	The parent rock shows no discoloration, loss of strength or other effects due to weathering	100% rock; discontinuities closed	100% rock; discontinuities closed	Staining indicates water percolation along joints; individual pieces may be loosened by blasting or stress relief and support may be required in tunnels and shafts		No visible signs of weathering

* Discontinuity spacing should be recorded.
† The ratio of the original rock to altered material should be estimated where possible.

However, as was pointed out by Martin and Hencher (1986) weathering processes are rarely sufficiently uniform to give gradual and predictable changes in the engineering properties of a weathered profile. In fact such profiles usually consist of heterogenous materials at various stages of decomposition and/or disintegration. Nonetheless an indication of the degree of weathering required to describe rock specimens or drill core is provided by the alteration of individual minerals, the loss of bonding between grains and the development and growth of microfractures. Martin and Hencher suggested that such descriptions should be made in terms of material grades which are uniform and so can be defined within quite precise limits. At a larger scale, for example, when describing rock masses for mapping purposes, then it frequently is necessary to group grades of rock weathering into mass zones which, for engineering purposes, may be regarded as having distinctive characteristics. In addition zonal classifications of weathered rock masses may be of value in terms of general design. Even so, where highly complex ground conditions exist it may be impossible to apply a rigid zonal classification. In such conditions Martin and Hencher recommended that the weathered material should be carefully described and combined with detailed lithological and structural mapping.

Several attempts have been made to devise engineering classifications of weathered rock and rock masses. Some schemes have involved quantification of the amount of mineralogical alteration and structural defects with the aid of the petrological microscope. Others have resorted to some combination of simple index tests to provide a quantifiable grade of weathering. Some of the earliest methods of assessing the degree of weathering were based on a description of the character of the rock mass concerned as seen in the field. Such descriptions recognized different grades of weathering and attempted to relate them to engineering performance. However, grading based on description of the degree of weathering is subjective and accordingly such grading systems now are being coupled with assessments made by index tests to provide better precision and quantification.

9.5.1 Descriptive classifications

Moye (1955) was one of the first to propose a grading system for the degree of weathering as found in granite at the Snowy Mountains scheme in Australia. Similar classifications were advanced by Kiersch and Treacher (1955), Ruxton and Berry (1957), Little (1969), Knill and Jones (1965), Fookes and Horswill (1970) and Fookes *et al.* (1972). These classifications were mainly based on the degree of decomposition exhibited by a rock mass and were primarily directed towards weathering in granitic rocks. Subsequently Dearman (1974) suggested descriptions which could be used to establish the grade of mechanical weathering, and that of solution weathering of relatively pure carbonate rock (Table 9.4). Others, working on different rock types, have proposed modified classifications of weathering grade. For example, Lovegrove and Fookes (1972) made slight variations from the above in their identification of grades of weathering of volcanic tuffs and associated sediments in Fiji. Classifications of weathered chalk and weathered marl have been developed by Ward *et al.* (1968) and Chandler (1969) respectively (see Chapter 11).

Nevertheless Dearman (1974) maintained that an ideal profile of weathering, which was irrespective of rock type, would be worthwhile in that each grade would provide an indication of the general engineering properties of the material concerned. Usually the grades will lie one above the other in a weathered profile developed from a single rock type, the highest grade being at the surface, but this is not necessarily the case in complex geological conditions (Knill and Jones, 1965).

In a review of recent classifications of weathering advanced by the Geological Society (Anon, 1977), the International Association of Engineering Geology (Anon, 1981a), the International Society for Rock Mechanics (Anon, 1981b) and the British Standard (Anon, 1981c), Martin and Hencher (1986) were critical of the way in which the term 'grade' was used both as a type rather than a scale of weathering of rock and to classify a zone of heterogeneous rock mass, which they argued led to confusion. Furthermore Martin and Hencher noticed a lack of definition or guidance for the description of rock material grades, in particular they referred to the inadequate definition of the terms 'rock' and 'soil' and pointed out that not all rocks are characterized by the presence of corestones when they weather. They also maintained that boundaries of mass weathering zones were on occasions arbitrarily drawn. Accordingly they proposed the following guidelines to provide a means of describing uniform grades of weathered rock material:

(1) Grade descriptions should apply to uniform materials.
(2) Index tests should be used whenever practical to define boundaries between grades more precisely.
(3) Wherever possible boundaries between grades should be established according to engineering relevance.
(4) Constant grade numbering and nomenclature should be used.
(5) A six-fold division of grades should be used in accordance with common practice.
(6) A single classification should be used whenever possible to cover all types (decomposition, disintegration) and degrees of material weathering.

They further commented that the end terms 'fresh rock' and 'residual soil' should be adopted as standard and that intermediate grades should be described as slightly, moderately, highly and extremely decomposed or disintegrated according to the dominant type of weathering. Martin and Hencher went on to propose that zonal schemes should be developed and used for the description of weathered rock masses which could be used for mapping purposes and in general engineering design. Any such scheme should take account of the following:

(1) Zones must be recognizable in naturally occurring profiles.
(2) The complete range of expected materials must be accounted for.
(3) Boundaries must be defined so that they separate zones with significantly different engineering properties.

They then presented a general zonal scheme for the description of rock masses whose weathering profiles exhibit a heterogeneous mixture of rock materials (Figure 9.1).

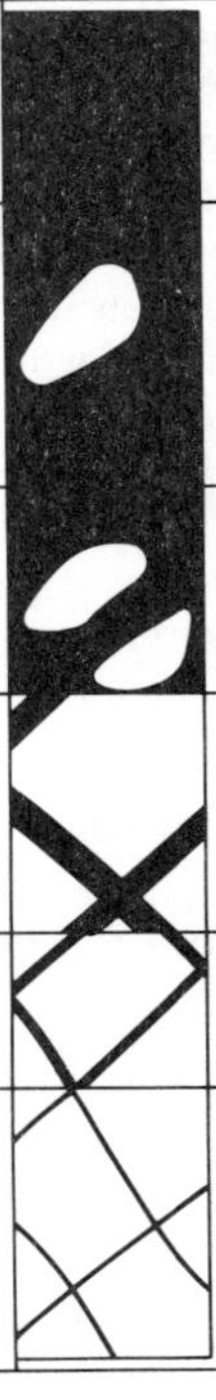

Zone symbol	Zonal characteristics
6	• Soil derived from *in situ* weathering: 100% soil (grades IV, V or VI) • May or may not have lost rock mass features completely
5	• Soil with corestones: less than 30% rock (grades I, II or III) • Shearing can be effected through matrix • Rock content significant for investigation and construction
4	• Poor quality rock mass: 30% to 50% rock (grades I, II or III) • Corestones affect shear behaviour of mass
3	• Moderate quality rock mass: 50% to 90% rock (grades I, II or III) • Severe weathering along discontinuities • Locked structure
2	• Good quality rock mass: greater than 90% rock (grades I, II or III) • Weathering along discontinuities
1	• Excellent quality rock mass: 100% rock (grades I, II) • No visible signs of rock weathering apart from slight discoloration along joints • Joint surfaces strongly interlocking

Figure 9.1 *Proposed zonal scheme for the classification of heterogeneous weathered rock masses (after Martin and Hencher, 1986). The zonal sequence is given in Arabic numerals to avoid confusion with grading systems. The scheme is based on varying proportions of rock and soil. Rock is generally defined as material grades I, II or III. Soil comprises grades IV, V and VI*

9.5.2 Classifications based on index tests and petrographic techniques

As pointed out above, the mineralogical composition and texture, as well as the engineering properties of rock, alter as it undergoes progressive weathering. Hence petrographic techniques and simple index tests can be used, either separately or in combination, to assess the degree of weathering.

Hamrol (1961) devised a quantitative classification of weatherability in which he first distinguished two weathering types. Type I weathering excluded cracking of any kind, whilst Type II weathering consisted entirely of cracking. This represents a division between chemical and physical weathering respectively, but such distinction can be extremely difficult to make. In Type I the void ratio increases as weathering progresses which means that the saturation moisture content increases and the dry density decreases. These two parameters therefore were used as the basis of an index test, the numerical value (i_I) of which was expressed as the weight of water absorbed by an oven-dried rock when it is saturated for a limited period, divided by its dry weight and expressed as a percentage. This is referred to as a quick absorption test. When considering Type II weathering, Hamrol distinguished between unfilled and filled cracks, the index being

$$i_{II} = (x+y+z)\times 100 \qquad (9.3)$$

where x, y and z were the dimensions of the crack along three orthogonal axes. Further indices could be obtained by relating the change in the degree of weathering (j) to a given time (Δt), hence

$$j_I = \Delta i_I/\Delta t \quad \text{and} \quad j_{II} = \Delta i_{II}/\Delta t \qquad (9.4a,b)$$

Unfortunately Hamrol gave no scale to his indices so that their meaning in terms of engineering performance is lacking (he did mention that with an $i_I = 10$, a weathered granite would crumble in the fingers, but little else).

Onodera *et al.* (1974) also used the number and width of microcracks (cracks less than 1 mm in width which occur in the rock fabric) as an index of the physical weathering of granite. They found a linear relationship between effective porosity (n_c) and density of microcracks (determined with the aid of a petrological microscope), defined as

$$n_c = 100 \times (\text{total width of cracks/length of measured line}) \qquad (9.5)$$

They also found that the mechanical strength of granite decreased rapidly as the density of microcracks increased from about 1.5 to 4%.

Lumb (1962) defined a quantitative index, Xd, related to the weight ratio of quartz and feldspar in decomposed granite from Hong Kong, as follows:

$$Xd = (N_q - N_{qo})/(1 - N_{qo}) \qquad (9.6)$$

where N_q is the weight ratio of quartz and feldspar in the soil sample, and N_{qo} is the weight ratio of quartz and feldspar in the original rock. For fresh rock $Xd = 0$, whilst for completely decomposed rock $Xd = 1$.

Yet another index or coefficient of weathering (K) for granitic rock was developed by Iliev (1967). This coefficient was based upon the ultrasonic velocities of the rock material according to the expression

$$K = (V_u - V_w)/V_u \qquad (9.7)$$

where V_u and V_w are the ultrasonic velocities of the fresh and weathered rock respectively. A quantitative index indicating the grade of weathering as determined from the ultrasonic velocity and the corresponding coefficient of weathering is as follows.

Grade of weathering	*Ultrasonic velocity* (m/s)	*Coefficient of weathering*
Fresh	Over 5000	0
Slightly weathered	4000–5000	0–0.2
Moderately weathered	3000–4000	0.2–0.4
Strongly weathered	2000–3000	0.4–0.6
Very strongly weathered	Under 2000	0.6–1.0

After an extensive testing programme Irfan and Dearman (1978a) concluded that the quick absorption, Schmidt hammer, and point load strength tests prove reliable field tests for the determination of a quantitative weathering index for granite (Table 9.5). This index can be related to the various grades of weathering recognized by visual determination and given in Table 9.4.

Table 9.5 *Weathering indices for granite (after Irfan and Dearman, 1978a)*

Type of weathering	*Quick absorption* (%)	*Bulk density* (Mg/m^3)	*Point load strength* (MN/m^2)	*Unconfined compressive strength* (MN/m^2)
Fresh	Less than 0.2	2.61	Over 10	Over 250
*Partially stained	0.2–1.0	2.56–2.61	6–10	150–250
*Completely stained	1.0–2.0	2.51–2.56	4–6	100–150
Moderately weathered	2.0–10.0	2.05–2.51	0.1–4	2.5–100
Highly/completely weathered	Over 10	Less than 2.05	Less than 0.1	Less than 2.5

* Slightly weathered

Irfan and Dearman (1978b) also developed a quantitative method of assessing the grade of weathering of granite in terms of its megascopic and microscopic petrography. The megascopic factors included an evaluation of the amount of discoloration, decomposition and disintegration shown by the rock. The microscopic analysis involved assessment of mineral composition and degree of alteration by modal analysis and a microfracture analysis. The latter involved

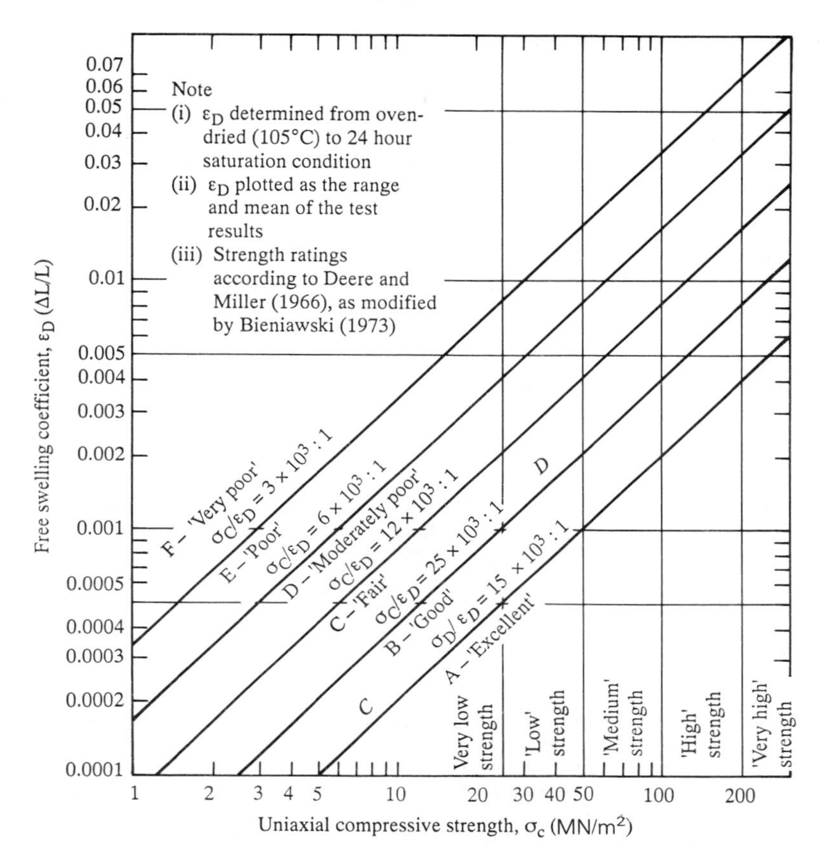

Figure 9.2 *Geodurability classification of the intact rock material (after Olivier, 1979)*

counting the number of clean and strained microcracks and voids under the microscope in a 10 mm traverse across a thin section. The types of microfracture recognized included stained grain boundaries, open grain boundaries, stained microcracks in quartz and feldspar, infilled microcracks in quartz and feldspar, clean transgranular microcracks crossing the grains, filled or partially infilled microcracks, and pores in plagioclase. The data are used to derive the micropetrographic index, I_p, as follows:

$$I_p = \frac{\%\ \text{sound or primary minerals}}{\%\ \text{unsound constituents}} \tag{9.8}$$

Table 9.6 *Stages of weathering of rock material in terms of microscopical properties (after Irfan and Dearman 1978b)*

Stage 1:	No penetration of brown iron-staining. Microcracks are very short, fine intragranular and structural. The centres of plagioclases are clouded and slightly sericitized. Altered minerals <6%; microcrack intensity <0.5%; micropetrographic index >12.
Stage 2:	Three substages are recognized depending on the amount of discoloration and type and amount of microfracturing. (a) The rock is iron-stained only along the joint faces. No penetration of iron-staining. (b) Penetration of iron-staining (brown) inwards from the joint faces along the microcracks. Formation of simple, branched microcracks; tight and partially stained. Slight alteration of the centres of plagioclases. Occasional staining along quartz–quartz and quartz–feldspar grain boundaries. Grain boundaries are sharp. (c) More inward penetration of brown iron-staining along microcracks and partial staining of plagioclases. Microfracturing of feldspars and quartz by mainly intragranular, but some transgranular, microcracks. Unstained core: Altered minerals 6–9%; microcrack intensity 0.5–1.0%; micropetrographic index 9–12. Stained rims: Altered minerals 9–12%; microcrack intensity 1.0–2.0%; micropetrographic index 6–9.
Stage 3:	Complete discoloration of rock by deep brown iron-staining. Partial alteration of plagioclases to sericite and gibbsite(?). Formation of single pores in plagioclases due to leaching. Potash feldspars are unaltered. Slight loss of pleochroism and bleaching of biotite. Grain boundaries are tight but stained brown by iron-oxide. The rock fabric is highly microfractured by complex branched, transgranular microcracks. Altered minerals 12–15%; microcrack intensity 2.0–5.0%; micropetrographic index 4–6.
Stage 4:	Nearly complete alteration of plagioclase to sericite and gibbsite and formation of nearly opaque areas in plagioclases. Very slight alteration of potash feldspar. Interconnected pores are formed in plagioclase feldspars due to removal and leaching of alteration products. Some solution of silica forming diffused quartz grain boundaries. Intense microfracturing of the rock fabric by a complex branched and dendritic pattern of microcracks. The whole of the rock is iron stained. Altered minerals 15–20%; microcrack intensity 5.0–10.0%; micropetrographic index 2–4.
Stage 5:	Complete alteration of plagioclases. Potash feldspar is partially altered, but highly microfractured. Biotite is partially and muscovite slightly altered; expansion of biotite. Quartz is reduced in grain size and amount by microfracturing and solution. Almost all the grain boundaries are open. The fabric is intensely microfractured by a dendritic pattern of micro- and macrocracks. Parallel sided, partially filled or clean macrocracks are formed. Highly bleached, highly porous. Rock texture is intact. Altered minerals 20%; microcrack intensity 10%; micropetrographic index 2.

The unsound constituents are the secondary minerals together with microcracks and voids. Irfan and Dearman were thus able to identify five stages and three substages of weathering in granite (Table 9.6).

Olivier (1979) proposed geodurability classification which was based on the free-swelling coefficient and uniaxial compressive strength (Figure 9.2). This classification was developed primarily to assess the durability of mudrocks and poorly cemented sandstones during tunnelling operations, since the tendency of such rocks to deteriorate after exposure governs the stand-up time of tunnels. Olivier suggested that the classification could be used in decisions relating to primary tunnel support, particularly where and when shotcrete should be applied.

References

Anon (1977). 'The description of rock masses for engineering purposes', Geological Society Engineering Group Working Party Report. *Q. J. Engg Geol.,* **10**, 355–88.

Anon (1981a). 'Rock and soil description and classification for engineering geological mapping', Int. Assoc. Engg Geol. Commission on Engineering Geological Mapping, *Bull. Int. Assoc. Engg Geol.,* **24**, 235–74.

Anon (1981b). 'Basic geotechnical description of rock masses', International Society for Rock Mechanics Commission on the Classification of Rocks and Rock Masses', *Int. J. Rock. Mech. Min. Sci. Geomech. Abstr.,* **18**, 85–110.

Anon (1981c). *Code of Practice for Site Investigation*, BS 5930. British Standards Institution, London.

Bieniawski, Z. T. (1973). 'Engineering classification of jointed rock masses', *Trans. S. Afr. Inst. Civ. Engrs,* **15**, 335–43.

Brune, G. (1965). 'Anhydrite and gypsum problems in engineering geology', *Bull. Assoc. Engg Geol.,* **3**, 26–38.

Chandler, R. J. (1969). 'The effect of weathering on the shear strength properties of the Keuper Marl', *Geotechnique,* **19**, 321–34.

Dearman, W. R. (1974). 'Weathering classification in the characterisation of rock for engineering purposes in British practice', *Bull. Int. Assoc. Engg Geol.,* **9**, 33–42.

Dearman, W. R. (1986). 'State of weathering: the search for a natural approach', in *Site Investigation Practice: Assessing BS 5930*. Engineering Geology Special Publication No 2, Hawkins, A. B. (ed.) Geological Society, London, 193–8.

Deere, D. U. and Miller, R. P. (1966). *Engineering Classification and Index Properties for Intact Rock*. Technical Report AFWL-TR-65-116, Air Force Weapons Laboratory, Kirtland Air Base, New Mexico.

Fasiska, E., Wagenblast, N. and Dougherty, M. T. (1974). 'The oxidation mechanism of sulphide minerals', *Bull. Assoc. Engg Geol.,* **11**, 75–82.

Fookes, P. G. and Horswill, P. (1970). 'Discussion on the load deformation behaviour of the Middle Chalk at Mundford, Norfolk, in *In Situ Investigations in Soils and Rocks*, British Geotechnical Society, London, 53–7.

Fookes, P. G., Dearman, W. R. and Franklin, J. A. (1972). 'Some engineering aspects of weathering with field examples from Dartmoor and elsewhere', *Q. J. Engg Geol.,* **3**, 1–24.

Fookes, P. G., Gourley, C. S. and Ohikere, E. (1988). 'Rock weathering in engineering time', *Q. J. Engg Geol.,* **21**, 33–57.

Franklin, J. A. and Chandra, A. (1972). 'The slake durability test', *Int. J. Rock Mech. Min. Sci.,* **9**, 325–41.

Hamrol, A. (1961). 'A quantitative classification of weathering and weatherability of rocks', *Proc. 5th Int. Conf. Soil Mech. Found. Engg, Paris,* **2**, 771–3.

Hencher, S. R. and Martin, R. P. (1982). 'The description and classification of weathered rocks in Hong Kong for engineering purposes', *Proc. 7th South East Asian Geotechnical Conference*, Hong Kong, **1**, 143–9.

Hobbs, N. B. (1975). *Foundations on Rock*, Soil Mechanics Ltd, Bracknell.

Iliev, I. G. (1966). 'An attempt to estimate the degree of weathering of intrusive rocks from their physico-mechanical properties', *Proc. 1st Int. Cong. Rock Mech.*, Lisbon, 109–14.

Irfan, T. Y. and Dearman, W. R. (1978). 'Engineering classification and index properties of weathered granite', *Bull. Int. Assoc. Engg Geol.,* **17**, 79–90.

Irfan, T. Y. and Dearman, W. R. (1978). 'The engineering petrography of a weathered granite in Cornwall, England', *Q. J. Engg Geol.,* **11**, 233–44.

Kiersch, G. A. and Treacher, R. C. (1955). 'Investigations, areal and engineering geology – Folsam dam project, central California', *Econ. Geol.,* **50**, 271–310.

Knill, J. L. and Jones, K. S. (1965). 'The recording and interpretation of geological conditions in the foundations of the Rosieres, Kariba and Latiyan dams', *Geotechnique,* **15**, 94–124.

Little, A. L. (1969). 'The engineering classification of residual tropical soils', *Proc. 7th Int. Conf. Soil Mech. Found. Engg*, Mexico, **1**, 1–10.

Lovegrove, C. W. and Fookes, P. G. (1972). 'The planning and implementation of a site investigation for a highway in tropical conditions in Fiji', *Q. J. Engg Geol.,* **5**, 43–68.

Lumb, P. 'The properties of decomposed granite', *Geotechnique,* **12**, 226–43.

Martin, C. P. and Hencher, S. R. (1986). 'Principles for description and classification of weathered rock for engineering purposes', in *Site Investigation Practice: Assessing BS 5930*, Engineering Geology Special Publication No 2, Hawkins, A. B. (ed.), Geological Society, London, 299–308.

Morgenstern, N. R. and Eigenbron, K. D. (1974). 'Classification of argillaceous shales and rocks', *Proc. ASCE, J. Geot. Engg Div.,* **100**, GT10, 1137–56.

Moye, D. G. (1955). 'Engineering geology for the Snowy Mountains scheme', *J. Inst. Eng. Aust.,* **27**, 287–98.

Olivier, H. J. (1979). 'A new engineering-geological rock durability classification', *Engg Geol.,* **14**, 255–79.

Onodera, T. F., Yoshinaka, R. and Oda, M. (1974). 'Weathering and its relation to mechanical properties of granite', *Proc. 3rd Int. Cong. Soc. Rock Mech.*, Denver, **2A**, 71–98.

Ruxton, B. P. and Berry, L. (1957). 'Weathering of granite and associated erosional features in Hong Kong', *Bull. Geol. Soc. Am.,* **68**, 1263–92.

Taylor, R. K. (1988). 'Coal Measures mudrocks: composition classification and weathering processes', *Q. J. Engg Geol.,* **21**, 85–100.

Taylor, R. K. and Spears, D. A. (1970). 'The breakdown of British Coal Measures rocks', *Int. J. Rock Mech. Min. Sci.,* **7**, 481–501.

Ward, W. H., Burland, J. B. and Gallois, R. W. (1968). 'Geotechnical assessment of a site at Mundford, Norfolk, for a large proton accelerator', *Geotechnique,* **18**, 399–431.

Winkler, E. M. (1973). *Stone Properties, Durability in Man's Environment*, Springer-Verlag, New York.

Chapter 10

Description and classification of rocks and rock masses

10.1 Description of rocks and rock masses

Description is the initial step in an engineering assessment of rocks and rock masses. It should therefore be both uniform and consistent in order to gain acceptance.

The complete specification of a rock mass requires descriptive information on the nature and distribution in space of both the materials that constitute the mass (rock, soil, water and air-filled voids) and the discontinuities which divide it (Anon, 1977). The intact rock may be considered as a continuum or polycrystalline solid consisting of an aggregate of minerals or grains whereas a rock mass may be looked upon as a discontinuum of rock material transected by discontinuities. The properties of the intact rock are governed by the physical properties of the materials of which it is composed and the manner in which they are bonded to each other. The parameters which may be used in a description of intact rock therefore include petrological name, mineral composition, colour, texture, minor lithological characteristics, degree of weathering or alteration (see Chapter 9), density, porosity, strength, hardness, intrinsic or primary permeability, seismic velocity and modulus of elasticity. Swelling and slake durability (see Chapter 9) can be taken into account where appropriate, such as in the case of argillaceous rocks.

The behaviour of a rock mass is, to a large extent, determined by the type, spacing, orientation and characteristics of the discontinuities present (see Chapter 8). As a consequence, the parameters which ought to be used in a description of a rock mass include the nature and geometry of the discontinuities as well as its overall strength, deformation modulus, secondary permeability and seismic velocity. It is not necessary, however, to describe all the parameters for either a rock mass or intact rock.

The data collected should be recorded on data sheets for subsequent automatic processing. A data sheet for the description of rock masses and another for discontinuity surveys have been recommended by the Geological Society (Anon, 1977; Figures 10.1, 10.2). Because geological data often have strong spatial interrelationships they are usually presented in cartographic, graphical or tabulated form which facilitates assessment.

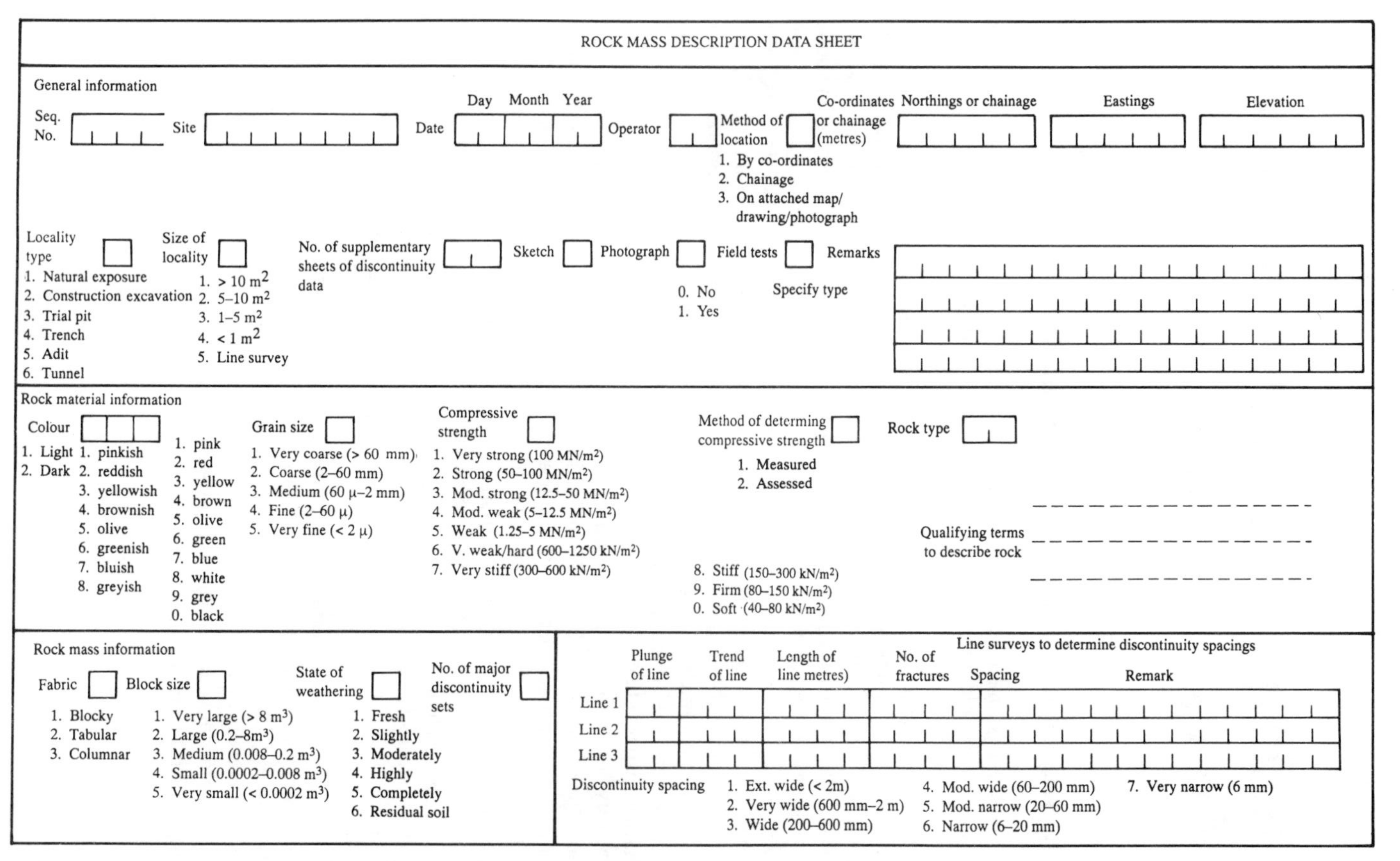

ROCK MASS DESCRIPTION DATA SHEET

General information

Seq. No. | Site | Date (Day, Month, Year) | Operator | Method of location | Co-ordinates or chainage (metres): Northings or chainage, Eastings, Elevation

Method of location:
1. By co-ordinates
2. Chainage
3. On attached map/drawing/photograph

Locality type
1. Natural exposure
2. Construction excavation
3. Trial pit
4. Trench
5. Adit
6. Tunnel

Size of locality
1. > 10 m²
2. 5–10 m²
3. 1–5 m²
4. < 1 m²
5. Line survey

No. of supplementary sheets of discontinuity data

Sketch | Photograph | Field tests | Remarks

0. No
1. Yes

Specify type

Rock material information

Colour
1. Light
2. Dark

1. pinkish
2. reddish
3. yellowish
4. brownish
5. olive
6. greenish
7. bluish
8. greyish

1. pink
2. red
3. yellow
4. brown
5. olive
6. green
7. blue
8. white
9. grey
0. black

Grain size
1. Very coarse (> 60 mm)
2. Coarse (2–60 mm)
3. Medium (60 μ–2 mm)
4. Fine (2–60 μ)
5. Very fine (< 2 μ)

Compressive strength
1. Very strong (100 MN/m²)
2. Strong (50–100 MN/m²)
3. Mod. strong (12.5–50 MN/m²)
4. Mod. weak (5–12.5 MN/m²)
5. Weak (1.25–5 MN/m²)
6. V. weak/hard (600–1250 kN/m²)
7. Very stiff (300–600 kN/m²)
8. Stiff (150–300 kN/m²)
9. Firm (80–150 kN/m²)
0. Soft (40–80 kN/m²)

Method of determing compressive strength
1. Measured
2. Assessed

Rock type

Qualifying terms to describe rock

Rock mass information

Fabric
1. Blocky
2. Tabular
3. Columnar

Block size
1. Very large (> 8 m³)
2. Large (0.2–8m³)
3. Medium (0.008–0.2 m³)
4. Small (0.0002–0.008 m³)
5. Very small (< 0.0002 m³)

State of weathering
1. Fresh
2. Slightly
3. Moderately
4. Highly
5. Completely
6. Residual soil

No. of major discontinuity sets

Line surveys to determine discontinuity spacings

	Plunge of line	Trend of line	Length of line metres)	No. of fractures	Spacing	Remark
Line 1						
Line 2						
Line 3						

Discontinuity spacing
1. Ext. wide (< 2m)
2. Very wide (600 mm–2 m)
3. Wide (200–600 mm)
4. Mod. wide (60–200 mm)
5. Mod. narrow (20–60 mm)
6. Narrow (6–20 mm)
7. Very narrow (6 mm)

Figure 10.1 *Rock mass data description sheet (after Anon, 1977)*

DISCONTINUITY SURVEY DATA SHEET

GENERAL INFORMATION

Seq. no. | Site | Date (Day Month Year) | Operator | Discontinuity data sheet No. of

NATURE AND ORIENTATION OF DISCONTINUITY

Chainage or No.	Type	Dip	Dip direction	Persistence	Aperture	Nature of infilling	Consistency of infilling	Surface roughness	Trend of lineation	Waviness wavelength	Waviness amplitude	Water/flow	Remarks

Type
0. Fault zone
1. Fault
2. Joint
3. Cleavage
4. Schistosity
5. Shear
6. Fissure
7. Tension crack
8. Foliation
9. Bedding

Dip, dip direction and trend of lineation

(Expressed in degrees)

Persistence (Expressed in metres)

Aperture
1. Wide (> 200 mm)
2. Mod. wide (60–200 mm)
3. Mod. narrow (20–60 mm)
4. Narrow (6–20 mm)
5. Very narrow (2–6 mm)
6. Ext. narrow (< 2mm)
7. Tight

Nature of infilling
1. **Clean**
2. **Surface staining**
3. **Non-cohesive**
4. **Inactive clay or clay matrix**
5. **Swelling clay or clay matrix**
6. **Cemented**
7. **Chlorite, talc or gypsum**
8. Others-specify

Compressive strength on infilling
1. Very soft (< 40 kN/m^2)
2. Soft (40–80 kN/m^2)
3. Firm (80–150 kN/m^2)
4. Stiff (150–300 kN/m^2)
5. Very stiff (300–600 kN/m^2)
6. Hard/v. weak (600–1250 kN/m^2)
7. Weak (1.25 MN/m^2)
8. Mod. weak (5.12.5 MN/m^2)
9. Mod. strong (12.5.50 MN/m^2)
10. Strong (50–100 MN/m^2)
11. Very strong (100–200 MN/m^2)
12. Ext. strong (> 200 MN/m^2)

Roughness
1. Polished
2. Slickensided
3. Smooth
4. Rough
5. Defined ridges
6. Small steps
7. Very rough

Waviness

Express wavelength and amplitude in metres

Water
1. Dry
2. Seepage

Flow

3. < 10 ml/s
4. 10–100 ml/s
5. 0.1–1 l/s
6. 1–10 l/s
7. 10–100 l/s
8. >100 l/s

Figure 10.2 *Discontinuity survey data sheet (after Anon, 1977)*

10.2 Properties of rocks and rock masses

10.2.1 Geological properties

Intact rock may be described from a geological or engineering point of view. In the first case the origin and mineral content of a rock are of prime importance, as is its texture and any change which has occurred since its formation. In this respect the name of a rock provides an indication of its origin, mineralogical composition and texture (Figure 10.3). Only a basic petrographical description of the rock is required when describing a rock mass. A useful system of petrographical description has been provided by Dearman (1974) and was further developed by the International Asociation of Engineering Geology (Anon, 1979).

The colour of a rock has a composite character attributable to the different minerals of which it is formed, to the size of these minerals, and, in the case of sedimentary rocks, to the type and amount of cement present. Hence the overall colour should be assessed by reference to a colour system or chart since it is difficult to make a quantitative assessment by the eye alone. For example, the Munsell colour system evaluates colour in terms of hue, value and chroma. Hue refers to the basic colour or a mixture of basic colours. The chroma indicates the intensity, strength or degree of departure of a particular hue from a neutral grey of the same value. Value indicates the degree of lightness or darkness of a colour in relation to a neutral grey.

A simple subjective scheme has been suggested by the Geological Society (Anon, 1977) which involves the choice of colour from column 3 below, supplemented if necessary by a term from column 2 and/or column 1:

1	2	3
light	pinkish	pink
dark	reddish	red
	yellowish	yellow
	brownish	brown
	olive	olive
	greenish	green
	bluish	blue
		white
	greyish	grey
		black

The texture of a rock refers to its component grains and their mutual arrangement or fabric. It is dependent upon the relative sizes and shapes of the grains and their positions with respect to one another and the groundmass or matrix, when present. Grain size, in particular, is one of the most important aspects of texture, in that it exerts an influence on the physical properties of a rock. It is now generally accepted that the same descriptive terms for grain size ranges should be applicable to all rock types and should be the same as those used to describe soils (Table 10.1).

Table 10.1 *Description of grain size*

Term	*Particle size*	*Equivalent soil grade*
Very coarse grained	Over 60 mm	Boulders and cobbles
Coarse grained	2–60 mm	Gravel
Medium grained	0.06–2 mm	Sand
Fine grained	0.002–0.06 mm	Silt
Very fine grained	Less than 0.002	Clay

Other aspects of texture include the relative grain size and the grain shape. The IAEG (Anon, 1979) suggested three types of relative grain size, namely, uniform, non-uniform and porphyritic. Grain shape was described in terms of angularity (angular, subangular, subrounded and rounded); form (equidimensional flat, elongated, flat and elongated, and irregular) and surface texture (rough and smooth).

10.2.2 Rock composition and texture in relation to physical properties

The micropetrographic description of rocks for engineering purposes includes the determination of all parameters which cannot be obtained from a macroscopic examination of a rock sample, such as mineral content, grain size and texture, and which have a bearing on the mechanical behaviour of the rock or rock mass (Hallbauer *et al.*, 1978). In particular a microscopic examination should include a modal analysis, determination of microfractures and secondary alteration, determination of grain size and, where necessary, fabric analysis. The ISRM recommends that the report of a petrographic examination should be confined to a short statement on the origin, classification and details relevant to the mechanical properties of the rock concerned. Wherever possible this should be combined with a report on the mechanical parameters (Figure 10.4).

10.2.3 Physical properties of rock

The IAEG (Anon, 1979) grouped the dry density and porosity of rocks into five classes as shown in Table 10.2.

Table 10.2 *Classification of dry density and porosity (after Anon, 1979)*

Class	*Dry density* (Mg/m^3)	*Description*	*Porosity* (%)	*Description*
1	Less than 1.8	Very low	Over 30	Very high
2	1.8–2.2	Low	30–15	High
3	2.2–2.55	Moderate	15–5	Medium
4	2.55–2.75	High	5–1	Low
5	Over 2.75	Very high	Less than 1	Very low

GRAIN SIZE (mm)	GENETIC/GROUP			DETRITIAL SEDIMENTARY						PYROCLASTIC		CHEMICAL/ ORGANIC
	Usual structure			BEDDED								
	Composition		Grains of rock, quartz, feldspar and clay minerals				At least 50% of grains are of carbonate			At least 50% of grains are of fine-grained igneous rock		
			Grains are of rock fragments						LIMESTONE and DOLOMITE (undifferentiated)			
60	Very coarse-grained	RUDACEOUS	BOULDERS COBBLES —	Rounded grains: CONGLOMERATE			CARBONATE GRAVEL	CALCI-RUDITE		Rounded grains AGGLOMERATE		SALINE ROCKS Halite Anhydrite
2	Coarse-grained	RUDACEOUS	GRAVEL	Angular grains: BRECCIA						Angular grains VOLCANIC BRECCIA LAPILLI TUFF		Gypsum
			Grains are mainly mineral fragments									
	Medium-grained	ARENACEOUS	SAND	SANDSTONE: Grains are mainly mineral fragments QUARTZ SANDSTONE: 95% quartz, voids empty or cemented ARKOSE: 75% quartz, up to 25% feldspar: voids empty or cemented GREYWACKE: 75% quartz, 15% fine detrital material: rock and feldspar fragments			CARBONATE SAND	CALC-ARENITE		TUFF	VOLCANIC ASH	LIMESTONE DOLOMITE CHERT FLINT
0.06	Fine-grained	ARGILLACEOUS or LUTACEOUS	SILT	SILTSTONE: 50% fine-grained particles	MUDSTONE SHALE: fissile mudstone	MARLSTONE	CARBONATE SILT	CALCI-SILTITE CHALK		Fine-grained TUFF		
0.002	Very fine-grained	ARGILLACEOUS or LUTACEOUS	CLAY	CLAYSTONE: 50% very fine grained particles			CARBONATE MUD	CALCI-LUTITE		Very fine-grained TUFF		PEAT LIGNITE COAL

METAMORPHIC		IGNEOUS				GENETIC GROUP	GRAIN SIZE (mm)
FOLIATED		MASSIVE				Usual structure	
Quartz. feldspars, micas, acicular dark minerals		Light coloured minerals are quartz, feldspar, mica		Dark and light minerals	Dark minerals	Composition	
		Acid rocks	Intermediate	Basic rocks	Ultrabasic		
GNEISS (ortho-, para-, alternate layers of granular and flaky minerals)	MARBLE	PEGMATITE			PYROXENITE and PERIDOTITE	Very coarse-grained	60
		GRANITE	DIORITE	GABBRO		Coarse-grained	2
	GRANULITE				SERPENTINITE		
MIGMATITE		MICROGRANITE	MICRODIORITE	DOLERITE		Medium-grained	
SCHIST	QUARTZITE HORNFELS AMPHIBOLITE						
PHYLLITE							0.06
SLATE MYLONITE		RHYOLITE	ANDESITE	BASALT		Fine-grained	0.002
						Very fine-grained	
		OBSIDIAN and PITCHSTONE		TACHYLYTE		GLASSY AMORPHOUS	
		VOLCANIC GLASSES					

Figure 10.3 *Rock type classification (after Anon, 1979)*

Project:
Location:
Co-ordinates: Collected by:
Specimen No:
Description of sampling point:
Thin section No: Date:

Geological description

Rock name:

Petrographic classification:

Geological formation:

Photo-micrograph of typical features of thin section

Macroscopic description of sample

Degree of weathering:

Structure (incl. bedding):

Discontinuities:

Qualitative description

Texture:

Fracturing:

Alteration:

Matrix:

Mineral composition (modal analysis)

Major components	Vol. %	Minor components	Vol. %	Accessories	Vol. %

Results of rock property tests

Point load index: Porosity ....

......MN/m^2, wet/dry normal/parallel to foliation

Density......Mg/m^3

Water absorption:

Any other results:

General remarks

Significance of results for rock engineering

Grain size and distribution

Microns	%

Figure 10.4 *ISRM suggested form of petrogrpahic report (after Hallbauer* et al., *1978)*

Determination of the strength and deformability of intact rock is achieved with the aid of some type of laboratory test. If the strength of rock is not measured then it can be estimated as shown in Table 10.3. Obviously such estimates can only be very approximate.

Table 10.3 *Estimation of the strength of intact rock (after Anon, 1977)*

Description	*Approximate unconfined compressive stength* (MN/m^2)	*Field estimation*
Very strong	Over 100	Very hard rock – more than one blow of geological hammer required to break specimen
Strong	50–100	Hard rock – hand-held specimen can be broken with a single blow of hammer
Moderately strong	12.5–50	Soft rock – 5 mm indentations with sharp end of pick
Moderately weak	5.0–12.5	Too hard to cut by hand
Weak	1.25–5.0	Very soft rock – material crumbles under firm blows with the sharp end of a geological hammer

The uniaxial compressive strength of a rock may be regarded as the highest stress that a specimen can carry when a unidirectional stress is applied, normally in an axial direction, to its ends. Although its application is limited, the unconfined compressive strength does allow comparisons to be made between rocks and affords some indication of rock behaviour under more complex stress systems. The grades of unconfined compressive strength shown in Table 10.3 have been suggested by the Geological Society (Anon, 1977).

Rocks have a much lower tensile than compressive strength. Unfortunately, however, the determination of the direct tensile strength has frequently proved

Table 10.4 *Classification of uniaxial compressive strength (after Anon, 1977)*

Term	*Strength* (MN/m^2)
Very weak	Less than 1.25
Weak	1.25–5.00
Moderately weak	5.00–12.50
Moderately strong	12.50–50
Strong	50–100
Very strong	100–200
Extremely strong	Over 200

difficult since it is not easy to grip the specimen without introducing bending stresses. Hence most values of tensile strength quoted have been obtained by indirect methods of testing. One of the most popular of these methods is the point load test. It has been recommended that this test should be limited to rocks with unconfined compressive strengths above 25 MN/m^2 (i.e. point load index above 1 MN/m^2). Even so the classification of point load strength index shown in Table 10.5, devised by Franklin and Broch (1972), is commonly used.

Table 10.5 *Classification of point load strength (after Franklin and Broch, 1972)*

	Point load strength index (MN/m^2)
Extrmely high strength	Over 10
Very high strength	3–10
High strength	1–3
Medium strength	0.3–1
Low strength	0.1–0.3
Very low strength	0.03–0.1
Extremely low strength	Less than 0.03

As far as deformability is concerned the five classes shown in Table 10.6 have been proposed by the IAEG (Anon, 1979).

Hardness can be defined as the mechanical competence of the intact rock. In the strict sense it is a surface property which is measured by using abrasion, indentation or rebound tests. These tests tend to reflect the hardnesses of individual grains rather than the intergranular bond or coherence of the rock.

Table 10.6 *Classification of deformability (after Anon, 1979)*

Class	*Deformability* (GN/m^2)	*Description*
1	Less than 5	Very high
2	5–15	High
3	15–30	Moderate
4	30–60	Low
5	Over 60	Very low

The permeability of intact rock (primary permeability) is usually several orders less than the *in situ* permeability (secondary permeability), as most water normally flows via discontinuities in rock masses. Although the secondary permeability is affected by the openness of discontinuities on the one hand and the amount of infilling on the other, a rough estimate of the permeability can be obtained from the frequency of discontinuities (Table 10.7; see also Figure 12.3). Admittedly such estimates must be treated with caution and cannot be applied to rocks which are susceptible to solution.

Table 10.7 *Estimation of secondary permeability from discontinuity frequency (Anon, 1977)*

Rock mass description	*Term*	*Permeability* k (m/s)
Very closely to extremely closely spaced discontinuities	Highly permeable	10^{-2}–1
Closely to moderately widely spaced discontinuities	Moderately permeable	10^{-5}–10^{-2}
Widely to very widely spaced discontinuities	Slightly permeable	10^{-9}–10^{-5}
No discontinuities	Effectively impermeable	Less than 10^{-9}

The seismic velocity refers to the velocity of propagation of shock waves through a rock mass. Its value is governed by the mineral composition, density, porosity, elasticity and degree of fracturing within a rock mass. The IAEG (Anon, 1979) recognized five classes of sonic velocity for rocks (Table 10.8).

Table 10.8 *Classification of sonic velocity (after Anon, 1979)*

Class	*Sonic velocity* (m/s)	Description
1	Less than 2500	Very low
2	2500–3500	Low
3	3500–4000	Moderate
4	4000–5000	High
5	Over 5000	Very high

Igneous rocks generally possess values of sonic velocity above 5000 m/s, those of metamorphic rocks range upwards from 3500 m/s and those of sedimentary rocks tend to vary between 1500 m/s and 4500 m/s. The latter range does not include unconsolidated deposits. The dynamic value of Young's modulus and Poisson's ratio can be derived from the seismic velocity (Onodera, 1963) and both can be correlated with the degree of fracturing (Grainger *et al.*, 1973).

10.3 Basic geotechnical description of ISRM

The basic geotechnical description (BGD) of rock masses proposed by the ISRM (Anon, 1981) considered the following characteristics:

(1) rock name with a simplified geological description,
(2) the layer thickness and fracture (discontinuity) intercept of the rock mass,
(3) the unconfined compressive strength of the rock material and the angle of friction of the fractures.

It was suggested that, where necessary, the rock mass should be divided into geotechnical units or zones. The division of the rock mass should be made in relation to the project concerned and the BGD should then be applied to each unit. The rock name is given in accordance with Figure 10.2. Although the simplified geological description depends upon the character of the rock masses involved, together with the requirements of the proposed scheme, it usually takes account of the mineralogical composition, texture and colour of the rock on the one hand and the degree of weathering, the nature of the discontinuities and the geological structure of the rock mass on the other. In addition the ISRM recommended that it would be advisable to provide a general geological description for the rock mass as well as one for each geotechnical unit.

The same class limits are used to describe the layer thickness of a geotechnical unit as are used for fracture intercept (Table 10.9; *cf.* Table 8.1). The ISRM defined the fracture intercept as the mean distance between successive fractures as measured along a straight line. When the fracture spacing changes with direction, the value adopted in the BGD is that corresponding to the direction along the smallest mean intercept. Fractures or discontinuities can be grouped into sets. The average fracture intercept, measured perpendicular to the fractures, is recorded for each set and given as supplementary information.

The unconfined compressive strength of the intact rock within a geotechnical unit of a rock mass represents the mean strength of rock samples taken from the zone.

Table 10.9 *Classification of layer thickness and fracture intercept (after Anon, 1981)**

Interval (m)	*Layer thickness*		*Fracture intercept*	
	Symbol†	*Description*	*Symbol*	*Description*
Over 2.0	L_1	Very large	F_1	Very large
0.6–2.0	L_2	Large	F_2	Wide
0.2–0.6	L_3	Moderate	F_3	Moderate
0.06–0.2	L_4	Small	F_4	Close
Less than 0.06	L_5	Very small	F_5	Very close

* The IAEG has proposed the same class limits.
† If a unit is not layered then it is given the symbol L_0.

The BGD includes the groupings shown in Table 10.10; those suggested by the IAEG (Anon, 1979) are given for comparison (See also Table 10.4).

If the rock material is notably anisotropic, the mean strengths obtained in different directions should be recorded, and special note should be made of that direction along which the lowest mean strength occurs.

Table 10.10 *ISRM and IAEG classifications of unconfined compressive strength*

BGD strength (MN/m^2)	*Symbol*	*Description*	*IAEG strength* (MN/m^2)	*Description*
Over 200	S_1	Very high	Over 230	Extremely strong
60–200	S_2	High	120–230	Very strong
20–60	S_3	Moderate	50–120	Strong
6–20	S_4	Low	15–150	Moderately strong
Under 6	S_5	Very low	1.5–15	Weak

The angle of friction of fractures as defined by the ISRM refers to the slope of the tangent to the peak strength envelope at a normal stress of 1 MN/m^2. The smallest mean value of the angle of friction is recorded when fracture sets differ in their shear strength. Table 10.11 shows the class limits for the angle of friction of fractures adopted by the BGD.

Table 10.11 *Angle of friction of fractures*

Interval	*Symbol*	*Description*
Over 45°	A_1	Very high
35–45°	A_2	High
25–35°	A_3	Moderate
15–25°	A_4	Low
Less than 15°	A_5	Very low

The data sheet given in Figure 10.5 is used for the BGD. Each zone is characterized by its rock name, followed by the class symbols corresponding to the parameter values, e.g. Granite L_0, F_3, S_2, A_3. Supplementary information is incorporated in the BGD when the rock mass concerned exhibits special features or if the requirements of the project so demand.

10.4 Principles of classification

Classifications of rocks devised by geologists usually have a genetic basis. Unfortunately, however, such classifications may provide little information relating to the engineering behaviour of the rocks concerned.

Example of application of BGD

Type of work:[1] Concrete Dam

Investigation stage:[2] Preliminary	Exposure:[3] Outcrop	
Location: Rocha da Galé Portugal	Observer:[4] Gomes Coelho	Date: June 77

5

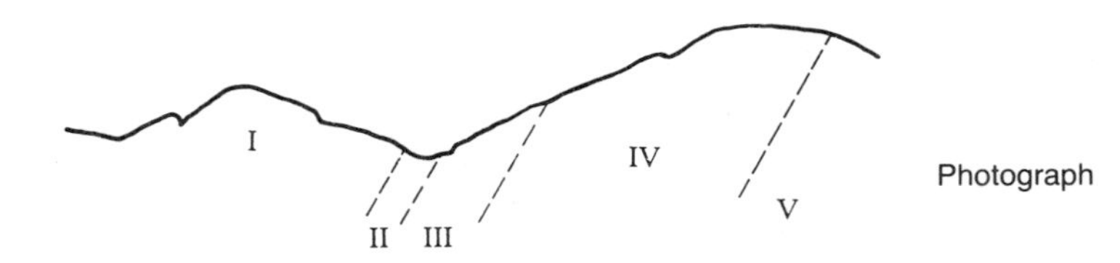

Photograph

Rock name and general geological description:[6]

Isoclinical sequence of metasedimentary and meta-volcanic rocks composed by interbedded siliceous schists and greywackes (I), pyroclastic rocks like tuff and breccia (II), agglomerate with rhyolitic matrix (III), rhyolite (IV), and porphyritic quartz-diorite (V).

Supplementary geological description:[7]

Zone 1: Rock mass formed of grey to red siliceous schist and interbedded greywacke thinly bedded (2–20 cm), and very often thinly laminated (0.6–2 cm). Rock mass is crossed by widely to very widely spaced joints, open, without filling material. Rock is fresh (W_1) and strong.

Zone II: Interbedded zone 30 m thick, composed of pyroclastic tuff and breccia, moderately to highly weathered (W_3–W_4) and moderately weak.

Zone III: Rock mass formed by agglomerate with a matrix of rhyolitic composition; rock is fresh to slightly weathered (W_1–W_2) and strong to very strong.

Zone IV: Rock mass formed by rhyolite massive, fresh to slightly weathered (W_1–W_2) and strong to very strong.

Zone V: Rock mass formed by porphyritic quartz-diorite, with slight foliation and widely spaced bedding (0.6–2 m); rock is fresh to moderately weathered (W_1-W_2) and moderately strong.

Zones	*Occurrence* (%)[8]	*Characterization*[9]	*Zones*	*Occurrence* (%)[8]	*Characterization*[9]
I	20	Siliceous schist L_4; $F_{4.5}$; S_2; A_2	V	45	Quartz-diorite L_2; F_4; S_3; A_2
II	2	Breccia and Tuff L_4; F_4; S_4; A_3	VI		
III	8	Agglomerate L_0; F_4; S_2; A_2	VII		
IV	25	Rhyolite L_0; F_3; S_2; A_2	VIII		

Computation of parameters

Zone	Parameters	Samples 1	2	3	4	Average	Std dev	BGD symbols
I	Layer thickness (cm)	4	10	8	6	7		L_4
	Fracture interc. (cm)							$F_{4.5}$
	U. comp. strength (MN/m^2)	66	65	150*	80	70		S_2
	Angle of friction (°)					35		A_2
(II)	Layer thickness (cm)	10	12	16	15	13		L_4
	Fracture interc. (cm)	15				15		F_4
	U. comp. strength (MN/m^2)	15	20	12	15	15		S_4
	Angle of friction (°)					30		A_3
(III)	Layer thickness (cm)	–	–	–	–	–		L_o
	Fracture interc. (cm)	4	12	6	8	7		F_4
	U. comp. strength (MN/m^2)	236	250	150	170	200		S_2
	Angle of friction (°)					40		A_2
(IV)	Layer thickness (cm)	–	–	–	–	–		L_o
	Fracture interc. (cm)	22	25	50	45	36		F_3
	U. comp. strength (MN/m^2)	210	140	180	220	185		S_2
	Angle of friction (°)					40		A_2
(V)	Layer thickness (cm)	80	210	120	160	140		L_2
	Fracture interc. (cm)	4	7	12	18	10		F_4
	U. comp. strength (MN/m^2)	92*	55	60	50	55		S_4
	Angle of friction (°)					40		A_2

Remarks(10)

Layer thickness: measured on outcrops
Fracture interc.: measured and estimated
U comp. strength: lab. test and estimated
Angle of friction: estimated

Supplementary information

* Normal to layering

(1) Main characteristics of the structure. (2) Preliminary, final, . . . (3) Outcrop, trench, cores . . . (4) Name and qualification. (5) Stereo pair of photographs, with the zones outlined. Other stereo pairs may be added. Ordinary photographs and/or sketches can be resorted to. (6) Rock name, structure (folds, faults). Fracturing (fracture sets, fracture characteristics); weathering. (7) Specific aspects should be considered for each zone. (8) Estimated proportion, by volume, of the occurrence of each zone relative to the observed rock mass. (9) Rock name followed by the interval symbols of the parameter values. (10) Methods followed in the determination of the parameters and difficulties encountered

Figure 10.5 *Basic geotechnical description (after Anon, 1981)*

According to Coates (1964) classification is needed in geotechnical engineering in order to assist in making an initial assessment of a problem and to point to areas where additional information must be sought in order to obtain the required answer. Any classification to some extent should be tailored to suit the application but if classification criteria are carefully selected initially, then a change of emphasis from one application to the next, rather than a complete reorganization, should suffice. Broad terms of reference therefore are necessary in designing a classification system.

Tests which are used for the engineering classification of rocks are termed index tests. If the right index tests are chosen then rocks having similar index properties, irrespective of their origin, will probably exhibit similar engineering performance. In order for an index test to be useful it must satisfy certain criteria. It should be simple to carry out, inexpensive and rapidly performed. The test results must be reproducible and the index property must be relevant to the engineering requirement.

Classification systems of intact rock may be developed either by selecting individual index properties to represent others that are closely related, or by summing the values of closely related properties to derive a single score (see Cottiss *et al.*, 1971). Obviously if two types of test are closely related, then the values of one may be used to predict the values of the other. Where economy of effort is important there is little to be gained by performing both tests.

Deere and Miller (1966) maintained that information concerning the physical properties of rocks and the nature of the discontinuities within rock masses is required in order to make rational predictions about their engineering behaviour under superimposed stresses. They further stated that the properties of intact rock should be investigated initially in an attempt to develop a meaningful system of evaluation of the *in situ* behaviour of rock. Appropriate reduction factors, attributable to the discontinuities, should then be determined for application to the intact rock data.

In fact two types of classification have been developed. First, there are those based upon some selected properties of the intact rock, as mentioned above. Secondly, and more importantly, there are those which take account of the properties of the rock mass, especially the nature of the discontinuities. The specific purpose for which a classification is developed obviously plays an important role in determining whether the emphasis is placed on the physical properties of the intact rock or on the continuity of the rock mass. The object of both types of classification is to provide a reliable basis for assessing rock quality.

10.5 Review of classifications

Any classification of intact rock for engineering purposes should be relatively simple, being based on significant physical properties so that it has a wide application. For example, Deere and Miller (1966) based their engineering classification of intact rock on the unconfined compressive strength (Table 10.12) and the modulus ratio (Table 10.13).

Strength

Class	*Description*	*Unconfined compressive strength* (MN/m^2)
A	Very high strength	Over 224
B	High strength	112–224
C	Medium strength	56–112
D	Low strength	28–56
E	Very low strength	Less than 28

Modulus ratio

Class	*Description*	*Modulus ratio*
H	High modulus ratio	Over 500
M	Medium modulus ratio	200–500
L	Low modulus ratio	Less than 200

The strength categories follow a geometric progression and the dividing line between categories A and B was chosen at 224 MN/m^2 since it is about the upper limit of the strength of most rocks. A rock may be classified as CH, BH, DL, etc.

Deere and Miller (1966) found that different rock types, when plotted on Figure 10.6, occupied different positions. For instance, the envelope enclosing sandstones and siltstones indicates that they have a unique position with respect to other rocks. It shows that they are more compressible in relation to their strength than most rock types. Granites also occupy a rather special position in the centre of the zone of average modulus ratio. Deere and Miller suggested that specific rock types fall within certain areas on the classification chart because it is sensitive to mineralogy, fabric and direction of anisotropy.

10.6 The rating concept

As far as the classification of rock masses for engineering purposes is concerned most work has been done in relation to tunnelling and the construction of underground chambers. Engineers have been especially concerned with the determination of rock mass quality in relation to the time the rock mass can remain unsupported, and the type and amount of support necessary.

Wickham *et al.* (1972) introduced the concept of rock structure rating (RSR) which refers to the quality of rock structure in relation to ground support in tunnelling. Although their classification is specifically related to tunnels it did introduce the rating principle which has been adopted subsequently in other

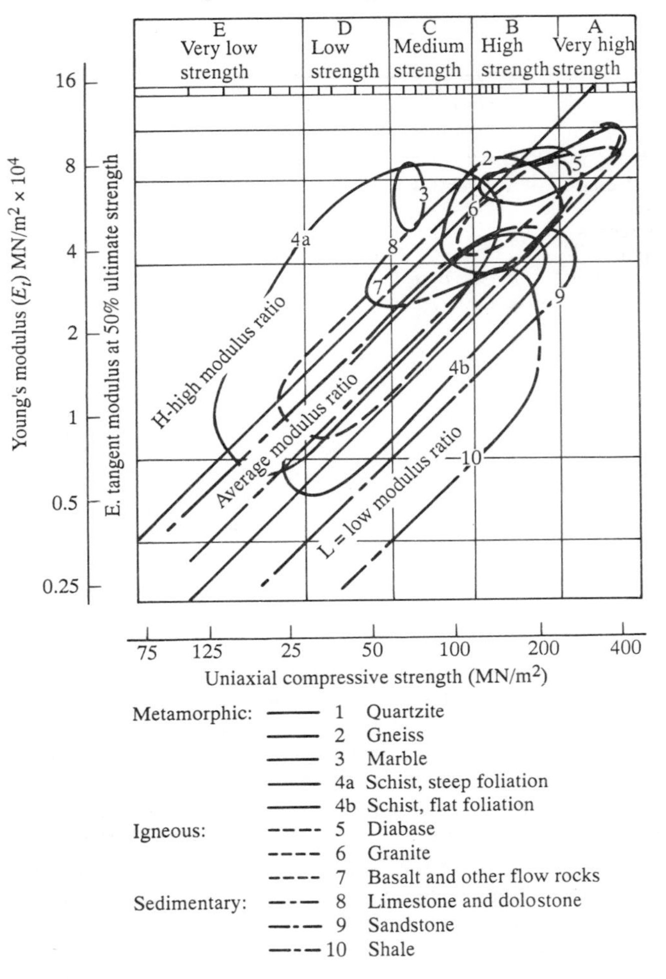

Figure 10.6 *Engineering classification of intact rock based on uniaxial compressive strength and modulus ratio. Fields are shown for igneous, sedimentary and metamorphic rocks (after Deere and Miller, 1966)*

classifications. The rating principle allows several parameters to be taken into account and their influence on the rock mass is collectively assessed.

One of the more sophisticated systems of classification of rock masses was advanced by Bieniawski (1973). This system initially incorporated the following properties: the RQD; the unconfined compressive strength; the degree of weathering; the spacing, orientation, separation and continuity of the discontinuities; and the groundwater flow. Unfortunately, no account was taken of the roughness of joint surfaces or the character of the infill material. The unconfined compressive strength of the intact rock has an important bearing on the engineering performance of the rock mass when the discontinuities are widely spaced or the rock mass is weak. It is also important if the joints are not continuous. Because the unconfined compressive strength and the degree of weathering are two interdependent factors, Bieniawski (1974) subsequently revised his views and suggested that both factors should be regarded as one parameter, namely, the strength of the rock material. He chose a somewhat modified version of Deere's (1964) classification of intact strength for his classification. The point load test was also used to determine the intact strength on site.

Special caution should be exercised in using this classification in the case of shales and other swelling materials. For instance, Bieniawski (1973) suggested that their behaviour under conditions of alternate wetting and drying should be assessed by a slake-durability test.

The presence of discontinuities reduces the overall strength of a rock mass and their spacing and orientation govern the degree of such reduction. Hence the spacing and orientation of the discontinuities are of paramount importance as far as the stability of structures in jointed rock masses is concerned. Bieniawski (1973) accepted the classification of discontinuity spacing which was proposed by Deere (1968).

Another revision of Bieniawski's ideas included the continuity and separation of discontinuities which he later (1974) grouped together under the heading, condition of discontinuities. This parameter also took account of the surface roughness of discontinuities and the quality of the wall rock.

Groundwater has an important effect on the behaviour of a jointed rock mass. However, as the pore water pressures are of greater significance in foundations than groundwater inflow, Pells (1974) suggested that the pore water pressure ratio (r_u), where r_u is defined as the ratio of the pore pressure to the major principal stress, should be included within the classification. This view was accepted by Bieniawski (1974) and the pore water pressure ratio was incorporated into his classification.

Bieniawski (1976, 1983) grouped each of the chosen rock mass parameters into five classes (Table 10.12). He considered five classes to be sufficient to provide a meaningful discrimination in all the chosen properties. Because these parameters vary in their relative importance from rock mass to rock mass and can contribute individually or collectively to its engineering performance, Bieniawski used a rating system. In other words a weighted numerical value was given to each class in each parameter. The total rock mass rating is the sum of the weighted values of the individual parameters, the higher the total rating, the better the rock mass condition (Table 10.12c). In describing a rock mass the class rating should be quoted with the class number, for example, class 3, rating 68.

Laubscher (1977) developed a geomechanics classification for jointed rock masses which was based on Bieniawski's classification. This classification also was designed principally for use in relation to subsurface excavation. Laubscher's classification similarly involves the rating concept, the rating extending from 0 to 100, and is supposed to cover all variations of jointed rock masses from very poor to very good.

Table 10.12 *The rock mass rating system (geomechanics classification of rock masses) (After Bieniawski, 1989)*

(a) Classification parameters and their ratings

	Parameter		Range of values						
1	Strength of intact rock material	Pointed load strength index (MN/m^2)	>10	4–10	2–4	1–2	For this low range, uniaxial compressive test is preferred		
		Uniaxial compressive strength (MN/m^2)	>250	100–250	50–100	25–50	5–25	1–5	<1
		Rating	15	12	7	4	2	1	0
2	Drill core quality RQD (%)		90–100	75–90	50–75	25–50	<25		
		Rating	20	17	13	8	3		
3	Spacing of discontinuities		>2 m	0.6–2 m	200–600 mm	200–600 mm	<60 mm		
		Rating	20	15	10	8	5		
4	Condition of discontinuities		Very rough surfaces Not continuous No separation Unweathered wall rock	Slightly rough surfaces Separation <1 mm Slightly weathered walls	Slightly rough surface Separation <1 mm Highly weathered wall	Slickensided surfaces or Gouge <5 mm thick or Separation 1–5 mm Continuous	Soft gouge >5 mm thick or Separation >5 mm Continuous		
		Rating	30	25	20	10	0		

5	Groundwater	Inflow per 10 m annual length (L/min)	None	<10	10–25	25–125	>125
			or	or	or	or	or
		Ratio $\frac{\text{Joint water pressure}}{\text{Major principal stress}}$	0	<0.1	0.1–0.2	0.2–0.5	>0.5
			or	or	or	or	or
		General conditions	Completely dry	Damp	Wet	Dripping	Flowing
	Rating		15	10	7	4	0

(b) Rating adjustment for discontinuity orientations

Strike and dip orientations of discontinuities		Very favourable	Favourable	Fair	Unfavourable	Very unfavourable
Ratings	Tunnels and mines	0	−2	−5	−10	−12
	Foundations	0	−2	−7	−15	−25
	Slopes	0	−5	−25	−50	−60

(c) Rock mass classes determined from total ratings

Rating	100 ← 81	80 ← 61	60 ← 41	40 ← 21	<20
Class no.	I	II	III	IV	V
Description	Very good rock	Good rock	Fair rock	Poor rock	Very poor rock

(c) Meaning of rock mass classes

Class no.	I	II	III	IV	V
Average stand-up time	20 yr for 15 m span	1 yr for 10 m span	1 wk for 5 m span	10 h for 2.5 m span	30 min for 1 m span
Cohesion of the rock mass (kN/m^2)	>400	300–400	200–300	100–200	<100
Friction angle of the rock mass (deg)	>45	35–45	25–35	15–25	<15

Table 10.13 *Descriptions and ratings for the six parameters used to describe rock mass quality in Q system (from Barton et al., 1975)*

	Description	*Value*	*Notes*
(1)	ROCK QUALITY DESIGNATION	RQD	
(A)	Very poor	0–25	(1) Where RQD is reported or measured as ≤10 (including 0), a nominal value of 10 is used to evaluate Q.
(B)	Poor	25–50	
(C)	Fair	50–75	
(D)	Good	75–90	(2) RQD intervals of 5, i.e. 100, 95, 90 etc. are sufficiently accurate.
(E)	Excellent	90–100	
(2)	JOINT SET NUMBER	J_n	
(A)	Massive, no or few joints	0.5–1.0	
(B)	One joint set	2	
(C)	One joint set plus random	3	
(D)	Two joint sets	4	
(E)	Two joint sets plus random	6	
(M)	Three joint sets	9	(1) For intersections use ($3.0 \times J_n$)
(G)	Three joint sets plus random	12	
(H)	Four or more joint sets, random, heavily jointed 'sugar cube', etc	15	(2) For portals use ($2.0 \times J_n$)
(J)	Crushed rock, earthlike	20	
(3)	JOINT ROUGHNESS NUMBER	J_r	
	(a) Rock wall contact and *(b) Rock wall contact before 10 cm shear*		
(A)	Discontinuous joints	4	
(B)	Rough or irregular, undulating	3	
(C)	Smooth, undulating	2	
(D)	Slickensided, undulating	1.5	(1) Add 1.0 if the mean spacing of the relevant joint set is greater than 3 m.
(E)	Rough or irregular, planar	1.5	
(F)	Smooth, planar	1.0	(2) $J_r = 0.5$ can be used for planar, slickensided joints having lineations, provided the lineations are orientated for minimum strength.
(G)	Slickensided, planar	0.5	
	(c) No rock wall contact when sheared		

(H)	Zone containing clay minerals thick enough to prevent rock wall contact.	1.0		
(J)	Sandy, gravelly or crushed zone thick enough to prevent rock wall contact.	1.0		
(4)	JOINT ALTERATION NUMBER	J_a	ϕ_r (approx.)	
	(a) Rock wall contact			
(A)	Tightly healed, hard, non-softening, impermeable filling, e.g. quartz	0.75	–	
(B)	Unaltered joint walls, surface staining only	1.0	(25–35°)	
(C)	Slightly altered joint walls, non-softening mineral coatings, sandy particles, clay-free disintegrated rock, etc.	2.0	(25–30°)	(1) Values of ϕ_r, the residual friction angle, are intended as an approximate guide to the mineralogical properties of the alteration products, if present.
(D)	Silty-, or sandy-clay coatings, small clay-fraction (non-softening)	3.0	(20–25°)	
(E)	Softening or low friction clay mineral coatings, i.e. kaolinite, mica. Also chlorite, talc, gypsum and graphite etc., and small quantities of swelling clays. (Discontinuous coatings, 1–2 mm or less in thickness.).	4.0	(8–16°)	
	(b) Rock wall contact before 10 cm shear			
(F)	Sandy particles, clay-free disintegrated rock etc.	4.0	(25–30°)	
(G)	Strongly over-consolidated, non-softening clay mineral fillings (continuous, <5 mm thick).	6.0	(16–24°)	
(H)	Medium or low over-consolidation, softening, clay mineral fillings, (continuous, <5 mm thick)	8.0	(12–16°)	
(J)	Swelling clay fillings, i.e. montmorillonite (continuous, <5 mm thick). Values of J_a depend on percentage of swelling clay-size particles, and access to water	8.0–12.0	(6–12°)	
	(c) No rock wall contact when sheared			
(K) (L) (M)	Zones or bands of disintegrated or crushed rock and clay (see G, H and J for clay conditions)	6.0 8.0 8.0–12.0	(6–24°)	
(N)	Zones or bands of silty- or sandy clay, small clay fraction, (non-softening)	5.0		
(Q) (P) (R)	Thick, continuous zones or bands of clay (see G, H and J for clay conditions)	10.0–13.0 13.0–20.0	(6–24°)	

Table 10.13 *continued*

	Description	*Value*			*Notes*
(5)	JOINT WATER REDUCTION FACTOR		J_w	approx. water pressure (kgf/cm^2)	
(A)	Dry excavations or minor inflow, i.e. <5 l/min locally		1.0	<1.0	
(B)	Medium inflow or pressure, occasional outwash of joint fillings		0.66	1.0–2.5	
(C)	Large inflow or high pressure in competent rock with unfilled joints		0.5	2.5–10.0	(1) Factors (C)–(F) are crude estimates. Increase J_w if drainage measures are installed.
(D)	Large inflow or high pressure, considerable outwash of fillings		0.33	2.5–10.0	
(E)	Exceptionally high inflow or pressure at blasting, decaying with time		0.2–0.1	>10	(2) Special problems caused by ice formation are not considered.
(F)	Exceptionally high inflow or pressure continuing without decay		0.1–0.05	>10	
(6)	STRESS REDUCTION FACTOR				
	(*a*) *Weakness zones intersecting excavation, which may cause loosening of rock mass when tunnel is excavated.*			SRF	
(A)	Multiple occurrences of weakness zones containing clay or chemically disintegrated rock, very loose surrounding rock (any depth)			10.0	(1) Reduce these values of SRF by 25–50% if the relevant shear zones only influence but do not intersect the excavation
(B)	Single weakness zones containing clay, or chemically disintegrated rock (excavation depth < 50 m)			5.0	
(C)	Single weakness zones containing clay, or chemically disintegrated rock (excavation depth >50 m)			2.5	
(D)	Multiple shear zones in competent rock (clay free), loose surrounding rock (any depth)			7.5	
(E)	Single shear zones in competent rock (clay free), (depth of excavation < 50 m)			5.0	
(F)	Single shear zones in competent rock (clay free), (depth of excavation >50 m)			2.5	(2) For strongly anisotropic virgin stress field (if measured): when $5 \leq \sigma_1/\sigma_3 \leq 10$, reduce σ_c to $0.8\sigma_c$ and σ_t to $0.8\sigma_t$. When $\sigma_1/\sigma_3 > 10$, reduce σ_c and σ_t to $0.6\sigma_c$ and $0.6\sigma_t$, where σ_c = unconfined compressive strength, and σ_t = tensile strength (point load) and σ_1 and σ_3 are the major and minor principal stresses.
(G)	Loose open joints, heavily jointed or 'sugar cube' (any depth)			5.0	
	(*b*) *Competent rock, rock stress problems*	σ_c/σ_1	σ_t/σ_1	SRF	
(H)	Low stress, near surface	>200	>13	2.5	
(J)	Medium stress	200–10	13–0.66	1.0	
(K)	High stress, very tight structure (usually favourable to stability, may be unfavourable for wall stability)	10–5	0.66–0.33	0.5–2	

(L)	Mild rock burst (massive rock)	5–2.5	0.33–0.16	5–10	(3) Few case records available where depth of crown below surface is less than span width. Suggest SRF increase from 2.5 to 5 for such cases (see H).
(M)	Heavy rock burst (massive rock)	<2.5	<0.16	10–20	
	(c) Squeezing rock, plastic flow of incompetent rock under the influence of high rock pressure				
(N)	Mild squeezing rock pressure			5–10	
(O)	Heavy squeezing rock pressure			10–20	
	(d) Swelling rock, chemical swelling activity depending upon the presence of water				
(P)	Mild swelling rock pressure			5–10	
(R)	Heavy swelling rock pressure			10–20	

ADDITIONAL NOTES ON THE USE OF THIS TABLE

When making estimates of the rock mass quality (Q) the following guidelines should be followed, in addition to the notes listed in the tables:

(1) When drill hole core is unavailable, RQD can be estimated from the number of joints per unit volume, in which the number of joints per metre for each joint set are added. A simple relation can be used to convert this number to RQD for the case of clay free rock masses:
RQD = 115 − 3.3 J_v (approx.) where J_v = total number of joints per m^3
(RQD = 100 for $J_v < 4.5$)

(2) The parameter J_n representing the number of joint sets will often be affected by foliation, schistosity, slatey cleavage or bedding etc. If strongly developed these parallel 'joints' should obviously be counted as a complete joint set. However, if there are few 'joints' visible, or only occasional breaks in the core due to these features, then it will be more appropriate to count them as 'random joints' when evaluating J_n.

(3) The parameters J_r and J_a (representing shear strength) should be relevant to the *weakest significant joint set or clay-filled discontinuity* in the given zone. However, if the joint set or discontinuity with the minimum value of (J_r/J_a) is favourably oriented for stability, then a second, less favourably oriented joint set or discontinuity may sometimes be more significant, and its higher value of J_r/J_a should be used when evaluating Q. *The value of J_r/J_a should in fact relate to the surface most likely to allow failure to initiate.*

(4) When a rock mass contains clay, the factor SRF appropriate to *loosening loads* should be evaluated. In such cases the strength of the intact rock is of little interest. However, when jointing is minimal and clay is completely absent the strength of the intact rock may become the weakest link, and the stability will then depend on the ratio rock-stress/rock-strength. A strongly anisotropic stress field is unfavourable for stability and is roughly accounted for as in note (2) in the table for stress reduction factor evaluation.

(5) The compressive and tensile strengths (σ_c and σ_t) of the intact rock should be evaluated in the saturated condition if this is appropriate to present or future *in situ* conditions. A very conservative estimate of strength should be made for those rocks that deteriorate when exposed to moist or saturated conditions.

Barton *et al.* (1975) proposed the concept of rock mass quality (Q) which they defined in terms of six parameters:

(1) The RQD or an equivalent system of joint density.
(2) The number of joint sets (J_n), which is an important indication of the degree of freedom of a rock mass. The RQD and the number of joint sets provide a crude measure of relative block size (RQD/J_n) with two extreme values (100/0.6 and 10/20) which differ by a factor of 400.
(3) The roughness of the most unfavourable joint set (J_r). The joint roughness and the number of joint sets determine the dilatancy of the rock mass.
(4) The degree of alteration or filling of the most unfavourable joint set (J_a). The roughness and degree of alteration of the joint walls or the filling materials provide an approximation of the shear strength of the rock mass (J_r/J_a). The quotient is weighted in favour of rough, unaltered joints in direct contact; such surfaces are close to peak strength and dilate strongly when sheared.
(5) The degree of water seepage or the joint water reduction factor (J_w).
(6) The stress reduction factor (SRF). Reduction of load due to excavation, stress in competent rock, and squeezing and swelling are taken account of in the stress reduction factor. The active stress is defined as J_w/SRF.

Descriptions and ratings of the six parameters are given in Table 10.15. The rock-mass quality (Q) is then derived from

$$Q = (\mathrm{RQD}/J_n) \times (J_r/J_n) \times (J_w/\mathrm{SRF}) \qquad (10.1)$$

This is the most sophisticated method of classifying rocks so far devised. The numerical value of Q ranges from 0.001 for exceptionally poor quality squeezing ground, up to 1000 for exceptionally good quality rock which is practically unjointed.

References

Anon (1977). 'The description of rock masses for engineering purposes'. Working Party Report, *Q. J. Engg Geol.*, **10**, 355–88.
Anon (1979). 'Classification of rocks and soils for engineering geological mapping. Part 1 – Rock and soil materials', *Bull. Int. Ass. Engg Geol.*, No. 19, 364–71.
Anon (1981). 'Basic geotechnical description of rock masses'. ISRM Commission on Classification of Rocks and Rock Masses. *Int. J. Rock Mech. Min. Sci. and Geomech. Abstr.*, **18**, 85–110.
Barton, N., Lien, R. and Lunde, J. (1975). 'Engineering classification of rock masses for design of tunnel support' *Norwegian Geot. Inst., Publ.*, **106** and *Rock Mech.*, **6**, 189–236, 1974.
Bieniawski, Z. T. (1973). 'Engineering classification of jointed rock masses', *Trans. S. Afr. Inst. Civil Engrs*, **15**, 335–43.
Bieniawski, Z. T. (1974). 'Geomechanics classification of rock masses and its application in tunnelling', *Proc. 3rd Int. Cong. Rock Mech., Denver,* **2**, 27–32.

Bieniawski, Z. T. (ed.) (1976). 'Rock mass classification in rock engineering', *Proc. Symp. on Exploration for Rock Engineering*, A. A. Balkema, Cape Town, **1**, 97–106.

Bieniawski, Z. T. (1983). 'The Geomechanics classification (RMR system) in design applications to underground excavations', *Int. Symp. on Engineering Geology and Underground Construction*, Lisbon (IAEG/SPG/LNEC), **2**, 11.33–47.

Bieniawski, Z. T. (1989). *Engineering Rock Mass Classifications*, Wiley-Interscience, New York.

Coates, D. F. (1964). 'Classification of rocks for rock mechanics', *Int. J. Rock Mech. Min. Sci.*, **1**, 421–9.

Cottiss, G. I., Dowell, R. W. and Franklin, J. A. (1971). 'A rock classification system applied to civil engineering', *Civ. Engg Pub. Works Rev.*, **66**, Part 1 – No. 777, 611–714; Part 2 – No. 780, 736–43.

Dearman, W. R. (1974). 'The characterisation of rock for civil engineering practice in Britain'. Centenaire de la Société Geologique de Bélgique, Colloque', *Géologie de L'Ingénieur*, Liège, 1–75.

Deere, D. U. (1964). 'Technical description of cores for engineering purposes', *Rock Mech. Engg Geol.*, **1**, 17–22.

Deere, D. U. (1968). 'Geologic considerations', In *Rock Mechanics in Engineering Practice*, Stagg, K. G. and Zienkiewicz, O. C. (eds.), Wiley, London, 1–19.

Deere, D. U. and Miller, R. P. (1966). *Engineering Classification and Index Properties for Intact Rock*, Tech. Rep. No. AFWL-TR-65-116, Air Force Weapons Laboratory, Kirtland Air Base, New Mexico.

Franklin, J. A. and Broch, E. (1972). 'The point load strength test', *Int. J. Rock Mech. Min. Sci.*, **9**, 669–97.

Grainger, P., McCann, D. M. and Gallois, R. W. (1973). 'Application of seismic refraction techniques for the study of fracturing in the Middle Chalk at Mundford, Norfolk', *Geotechnique*, **23**, 219–32.

Hallbauer, D. K., Nieble, C., Berard, J., Rummel, F., Houghton, A., Broch, E. and Szlavin, J. (1978). 'Suggested methods for petrographic description', ISRM Commission on Standardization of Laboratory and Field Tests, *Int. J. Rock Mech. Min. Sci. and Geomech. Abstr.*, **15**, 41–5.

Laubscher, D. H. (1977). 'Geomechanics classification of jointed rock masses – mining applications', *Trans. Inst. Min. Metall.*, Section A – Mining Industry, **86**, A1–A8.

Onodera, T. F. (1963). 'Dynamic investigation of rocks *in situ*', *Proc. 5th Symp. Rock Mech.*, Minnesota University, Rolla, Pergamon Press, New York, 517–33.

Pells, P. J. H. (1974). 'Discussion: engineering classification of jointed rock masses', *Trans. S. Afr. Inst. Civil Engrs*, **16**, 242.

Wickham, G. E., Tiedemann, H. R. and Skinner, E. H. (1972). 'Support determination based on geologic predictions', *Proc. 1st N. Am. Tunneling Conf.*, AIME, New York, 43–64.

Chapter 11

Engineering properties of rock

11.1 Engineering character of igneous and metamorphic rocks

The plutonic igneous rocks are characterized by granular texture, massive structure and relatively homogeneous composition. In their unaltered state they are essentially sound and durable with adequate strength for any engineering requirement (Table 11.1). In some instances, however, intrusives may be highly altered, by weathering or hydrothermal attack. Furthermore fissure zones are by no means uncommon in granites. The rock mass may be very much fragmented along such zones, indeed it may be reduced to sand-size material (Terzaghi, 1946), and it may have undergone varying degrees of kaolinization.

In humid regions valleys carved in granite may be covered with residual soils which extend to depths often in excess of 30 m. Fresh rock may only be exposed in valley bottoms which have actively degrading streams. At such sites it is necessary to determine the extent of weathering and the engineering properties of the weathered products. Generally the weathered product of plutonic rocks has a large clay content although that of granitic rocks is sometimes porous with a permeability comparable to that of medium-grained sand.

Joints in plutonic rocks are often quite regular, steeply dipping structures in two or more intersecting sets. Sheet joints tend to be approximately parallel to the topographic surface. Consequently they may introduce a dangerous element of weakness into valley slopes. For example, in a consideration of Mammoth Pool Dam foundations on sheeted granite, Terzaghi (1962) observed that the most objectionable feature was the sheet joints orientated parallel to the rock surface. In the case of dam foundations such joints, if they remain untreated, may allow the escape of large quantities of water from the reservoir. This, in turn, may lead to the development of hydrostatic pressures in the rock downstream which are high enough to dislodge sheets of granite.

Generally speaking the older volcanic deposits do not prove a problem in foundation engineering, ancient lavas having strengths frequently in excess of 200 MN/m^2. But volcanic deposits of geologically recent age at times can prove treacherous, particularly if they have to carry heavy loads such as concrete dams. This is because they often represent markedly anisotropic sequences in which lavas, pyroclasts and mudflows are interbedded. Hence foundation problems in volcanic sequences arise because weak beds of ash, tuff and mudstone occur within lava piles which give rise to problems of differential settlement and sliding. In addition

Table 11.1 *Some physical properties of igneous and metamorphic rocks*

	Specific gravity	*Unconfined compressive strength (MN/m^2)*	*Point load strength (MN/m^2)*	*Shore scleroscope hardness*	*Schmidt hammer hardness*	*Young's modulus (GN/m^2)*
Mount Sorrel Granite	2.68	176.4	11.3	77	54	60.6
Eskdale Granite	2.65	198.3	12.0	80	50	56.6
Dalbeattie Granite	2.67	147.8	10.3	74	69	41.1
Markfieldite	2.68	185.2	11.3	78	66	56.2
Granophyre (Cumbria)	2.65	204.7	14.0	85	52	84.3
Andesite (Somerset)	2.79	204.3	14.8	82	67	77.0
Basalt (Derbyshire)	2.91	321.0	16.9	86	61	93.6
Slate* (North Wales)	2.67	96.4	7.9	41	42	31.2
Slate† (North Wales)		72.3	4.2			
Schist* (Aberdeenshire)	2.66	82.7	7.2	47	31	35.5
Schist†		71.9	5.7			
Gneiss	2.66	162.0	12.7	68	49	46.0
Hornfells (Cumbria)	2.68	303.1	20.8	79	61	109.3

* Tested normal to cleavage or schistosity
† Tested parallel to cleavage or schistosity

weathering during periods of volcanic inactivity may have produced fossil soils, these being of much lower strength.

The individual lava flows may be thin and transected by a polygonal pattern of cooling joints. They also may be vesicular or contain pipes, cavities or even tunnels.

Pyroclasts usually give rise to extremely variable ground conditions due to wide variations in strength, durability and permeability. Their behaviour very much depends upon their degree of induration, for example, many agglomerates have a high enough strength to support heavy loads such as concrete dams and also have a low permeability. By contrast, ashes are invariably weak and often highly permeable. One particular hazard concerns ashes, not previously wetted, which are metastable and exhibit a significant decrease in their void ratio on saturation. Tuffs and ashes are frequently prone to sliding. Montmorillonite is not an uncommon constituent in the weathered products of basic ashes.

Lumb (1983) carried out a survey of the engineering properties of some igneous rocks of Hong Kong and showed that when their unconfined compressive strengths are plotted against their apparent porosities (this latter parameter increases with increasing decomposition of rock) a trend of decreasing strength with increasing decomposition is discernible. At low apparent porosities (less than 2%) strengths are more or less independent of porosity but for the decomposed rocks strength decreases exponentially. Figure 11.1 shows cumulative distributions of grades I to IV of weathered granites and volcanics, from which it can be seen that there is considerable overlap between grades. The highest unconfined compressive strength obtained was 356 MN/m^2 for fresh volcanic rock.

The modulus ratios of these rocks are about 500 for the fresh volcanics and dyke rocks compared with about 200 for fresh granites. This marked difference in stiffness is attributed to a greater abundance of microfissures in granites than in volcanics.

The average ratio of compressive to tensile strength (Brazilian test) is about 14 for fresh rock but drops to approximately 8 for the more decomposed rocks. The ratio of compressive strength to point load index averages around 22 for fresh rock.

Oddsson (1981) undertook a study of material from two hyaloclastite formations in Iceland, namely, Sigalda and Surtsey. A hyaloclastite consists of a mixture of rock and glass fragments. Glass is generally the main constituent, especially in the finer-grained varieties. The fragments normally possess irregular and angular shapes. Nevertheless rounded particles are common and are probably the result of erosion within a turbulent eruption cloud.

Initially hyaloclastites are loose deposits but as a result of weathering and diagenesis they usually become harder, particularly due to palagonitization, which is a solution–precipitation mechanism whereby the glass is hydrated and ions are leached out of it. The process slows down as the precipitation of authigenic minerals reduces the porosity and permeability of the rock mass. Increase in strength due to palagonitization can occur quickly; for example, loose pyroclastic material can be transformed into quite hard compact rocks within a matter of less than 20 years.

The compressive strengths of the Sigalda and Surtsey hyaloclastites range between 0.9 and 40.1 MN/m^2, and 4.7 and 43.8 MN/m^2 respectively. The results of

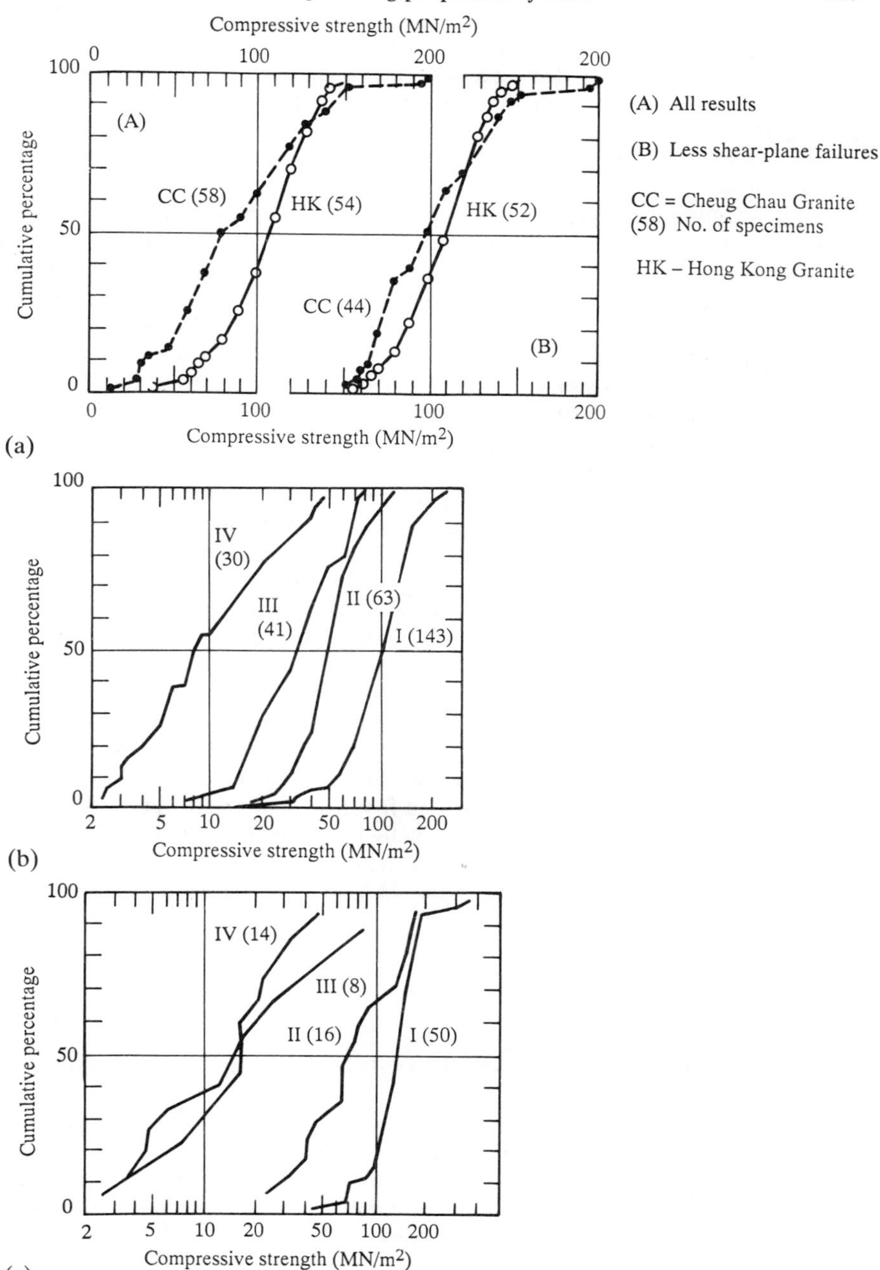

Figure 11.1 *Compressive strength of* (a) *Grade I granite;* (b) *various granites;* (c) *volcanics (after Lumb, 1983)*

triaxial tests showed that the angle of friction increases with grain size and compaction while cohesion decreases with increasing grain size and increases with compaction. Nonetheless even the strongest hyaloclastites tested can only be classified as low-strength rocks. Weaker varieties probably are more frequent than stronger types, often being hidden below the palagonized hyaloclastites forming the uppermost altered beds.

Slates, phyllites and schists are characterized by textures which have a marked preferred orientation. Platey minerals such as chlorite and mica tend to segregate into almost parallel or subparallel bands alternating with granular minerals such as quartz and feldspar. This preferred alignment of platey minerals accounts for the cleavage and schistosity which typify these metamorphic rocks and means that slate, in particular, is notably fissile. Obviously such rocks are appreciably stronger across the lineation than along it (Table 11.1). Not only do cleavage and schistosity adversely affect the strength of metamorphic rocks, they also make them more susceptible to decay. Generally speaking, however, slates, phyllites and schists weather slowly but the areas of regional metamorphism in which they occur have suffered extensive folding so that in places rocks may be fractured and highly deformed. Some schists, slates and phyllites are variable in quality, some being excellent foundations for heavy structures; others, regardless of the degree of their deformation or weathering, are so poor as to be wholly undesirable. For instance. talc, chlorite and sericite schists are weak rocks containing planes of schistosity only a millimetre or so apart. Some schists become slippery upon weathering and therefore fail under a moderately light load.

The engineering performance of gneiss is usually similar to that of granite. However, some gneisses are strongly foliated which means that they possess a texture with a preferred orientation. Generally this will not significantly affect their engineering behaviour. They may, however, be fissured in places and this can mean trouble. For instance, it would appear that fissures opened in the gneiss under the heel of the Malpasset Dam, which eventually led to its failure (Jaeger, 1963).

Fresh, thermally metamorphosed rocks such as quartzite and hornfels are very strong and afford good ground conditions. Marble has the same advantages and disadvantages as other carbonate rocks.

11.2 Engineering properties of arenaceous sedimentary rocks

The process by which a sand is turned into a sandstone is partly mechanical, involving grain fracturing, bending and deformation. However, chemical activity is much more important. The latter includes decomposition and solution of grains, precipitation of material from pore fluids and intergranular reactions. Redistribution of material, such as the solution of quartz at the points of grain contact and its precipitation as grain overgrowths in the void space, cements grains together and reduces the porosity. In sands the porosity may average 30–35%, whereas in a sandstone reduction in pore space may mean that the porosity is halved, or in extreme cases it may be reduced to more or less zero.

The dry density and especially the porosity of a sandstone are influenced by the amount of cement and/or matrix material occupying the pores. Usually the density of a sandstone tends to increase with increasing depth below the surface, the porosity decreasing.

The compressive strength of a sandstone is influenced by its porosity, amount and type of cement and/or matrix material present, as well as the composition of the individual grains. Price (1960) showed that the strength of sandstone with low porosity (less than 3.5%) was controlled by its quartz content and degree of compaction. In those sandstones with a porosity in excess of 6% he found that there was a reasonably linear relationship between dry compressive strength and porosity, that is, for every 1% increase in porosity the strength decreased by approximately 4%. If cement binds the grains together, then a stronger rock is produced than one in which a similar amount of detrital material performs the same function. However, the amount of cementing material is more important than the type of cement, although if two sandstones are equally well cemented, one having a siliceous, the other a calcareous cement, then the former is the stronger. For example, ancient quartz arenites in which the voids are almost completely occupied with siliceous material are extremely strong with crushing strengths exceeding 240 MN/m^2. By contrast poorly cemented sandstones may possess crushing strengths less than 3.5 MN/m^2.

In a consideration of the Fell Sandstones, Bell (1978) found that a highly significant relationship existed between unconfined compressive strength on the one hand and tensile strength and Shore scleroscope and Schmidt hammer hardness measures on the other and a significant relationship existed with Young's modulus. Both Shore scleroscope and Schmidt hammer hardness possessed highly significant relationships with Young's modulus, density and porosity. Bell was unable to demonstrate any statistically significant relationship between mineralogical composition on the one hand and density, hardness, strength and deformability on the other. However, both the packing density and packing proximity had highly significant or significant relationships with most of the index properties investigated.

Singh *et al.* (1978) examined the influence of anisotropy in Chunar Sandstone on its hardness and uniaxial compressive strength. They showed that the average Shore scleroscope hardness normal to the bedding was 46 whereas parallel to the bedding it was 44. The average uniaxial compressive strength normal and parallel to the bedding was respectively 97 and 96 MN/m^2. Incidentally, they found a similar relationship in the Singrauli Coal.

The pore water plays a very significant role as far as the compressive strength and deformation characteristics of a sandstone are concerned. This is illustrated by the Fell Sandstone (Northumberland) and Sherwood (formerly Bunter) Sandstone (Nottinghamshire) which may suffer reductions of uniaxial compressive strength, when a dry specimen is saturated, of nearly 30 and 60% respectively (Table 11.2; Bell and Culshaw, 1992). The difference in the reduction of strength on saturation probably is explained by the difference in cementation, the Sherwood Sandstone being much more poorly cemented and so more porous (Table 11.2).

West (1979) found that Bunter Sandstone obtained from Warrington exhibited a general tendency to decrease in strength with increasing moisture content (the average air-dry strength was 37 MN/m^2 compared with 23 MN/m^2 when saturated). Young's modulus also underwent a reduction from a mean value of 7.2×10^2 MN/m^2 for the sandstone when dry to 5.5×10^2 MN/m^2 for saturated sandstone.

Wissler and Simmons (1985) subjected a number of sandstones to hydrostatic compression up to 200 MN/m^2 in order to observe the elastic and permanent strain response. Specimens were subjected to cyclic loading and unloading, the load being increased to successively higher pressures. They showed that permanent strains are functions of confining pressure and that such strains can account for as much as 30% of the total strain in a sandstone sample. Permanent strains are either compactive or dilational, however, the mechanisms responsible for both probably are active in all samples. Nonetheless compaction is generally dominant.

Dobereiner and De Freitas (1986) conducted a series of undrained triaxial tests with pore pressure measurements on saturated specimens of weak sandstone. They noted that the failure envelopes were non-linear. Weak sandstones such as the Kidderminster (formerly Bunter) Sandstone, under stress behave in an inelastic manner as indicated by their 'S'-shaped stress–strain curves. The nature of the stress–strain curves obtained depends on the confining pressure. Lateral strain begins to increase relative to the rate of axial strain at very low stress levels, as does the onset of dilatancy. As the deviator stress was increased the pore water pressure increased and then decreased (Figure 11.2). Dobereiner and De Freitas assumed that the reduction in pore water pressure was brought about by the increase which occurred in the volume of the specimens due to the onset of dilatancy.

In the case of Kidderminster Sandstone air-dried samples (whose moisture content was probably not zero) yielded an average unconfined compressive strength of 2.3 MN/m^2. On saturation the strength was reduced to 0.45 MN/m^2.

On examination those sandstones with saturated unconfined compressive strengths of less than 20 MN/m^2 were seen to have failed as a result of grain rolling, whereas in stronger sandstones grains were crushed and fractured (cataclasis), with microfracturing occurring prior to macroscopic failure.

Furthermore Dobereiner and de Freitas (1986) recorded that the saturated uniaxial compressive strength and deformability possessed a significant correlation with the amount of grain contact (Figure 11.3). In the Kidderminster Sandstone and Keuper Waterstones sections vertical to the bedding were observed to possess greater areas of grain contact than those parallel to the bedding.

Moore (1974) derived a value of Young's modulus of 1100 MN/m^2 for the Bunter Sandstone from long term (up to 18 months) plate load testing. The modulus was found to increase with depth. At the highest loading, 5.6 MN/m^2, settlement did not exceed 4 mm. Creep accounted for 20–30% of the total settlement at loads varying between 0.3 and 1.5 MN/m^2 but at 3.0 and 5.6 MN/m^2 it was lower. Moore and Jones (1975) concluded that at fairly low stresses the Bunter Sandstone, even though weathered near the surface, provided a sound foundation. Moreover the rapid reduction in settlement with depth presumably means that simple spread foundation structures may be suitable even for sensitive buildings.

Table 11.2 *Some physical properties of arenaceous sedimentary rocks*

	Fell Sandstone (Rothbury)	*Chatsworth Grit (Stanton in the Peak)*	*Sherwood Sandstone (Edwinstowe)*	*Keuper Waterstones (Edwinstowe)*	*Horton Flags* (Helwith Bridge)*	*Bronllwyn Grit* (Llanberis)*
Specific gravity	2.69	2.69	2.68	2.73	2.70	2.71
Dry density (Mg/m^3)	2.25	2.11	1.87	2.26	2.62	2.63
Porosity	9.8	14.6	25.7	10.1	2.9	1.8
Dry unconfined compressive strength (MN/m^2)	74.1	39.2	11.6	42.0	194.8	197.5
Saturated unconfined compressive strength (MN/m^2)	52.8	24.3	4.8	28.6	179.6	190.7
Point load strength (MN/m^2)	4.4	2.2	0.7	2.3	10.1	7.4
Scleroscope hardness	42	34	18	28	67	88
Schmidt hardness	37	28	10	21	62	54
Young's modulus (GN/m^2)	32.7	25.8	6.4	21.3	67.4	51.1
Permeability ($\times 10^{-9}$ m/s)	1740	1960	3500	22.4	–	–

* Greywackes

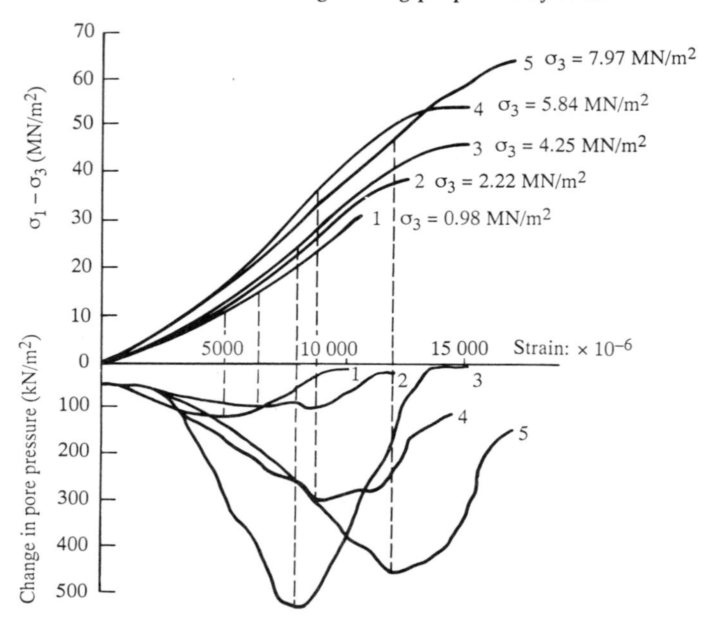

Figure 11.2 *Total stress–strain and pore pressure variation curves of saturated Waterstones tested in triaxial compression with the major principal stress vertical and parallel to the bedding (after Dobereiner and De Freitas, 1986)*

11.3 Engineering properties of argillaceous sedimentary rocks

Shale is the commonest sedimentary rock and is characterized by its lamination. Sedimentary rock of similar size range and composition, but which is not laminated, is usually referred to as mudstone. In fact there is no sharp distinction between shale and mudstone, one grading into the other.

Shale is frequently regarded as an undesirable material to work in. Certainly there have been many failures of structures founded on slopes in shales. Nevertheless many shales have proved satisfactory as foundation rocks. Hence it can be concluded that shales vary widely in their engineering behaviour and that it is therefore necessary to determine the problematic types. This variation in engineering behaviour to a large extent depends upon their degree of compaction and cementation. Indeed Mead (1936) divided shales into compaction and cementation types. The cemented shales are invariably stronger and more durable. A more recent engineering classification of mudrocks based upon composition, grain size, fabric and strength has been provided by Grainger (1984). It is summarized in Figure 11.4.

The degree of packing, and hence the porosity, void ratio and density of shale, depends on its mineral composition and grain size distribution, its mode of

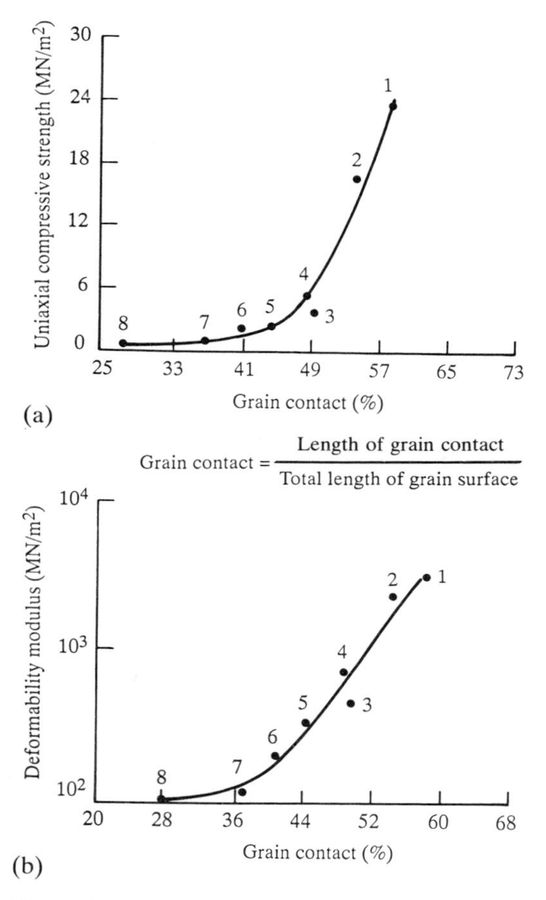

Figure 11.3 *Dependence of strength and deformability on grain contact in weak sandstones: (1) Waterstones (perpendicular to bedding); (2) Waterstones (parallel to bedding); (3) Lahti Sandstone; (4) Bauru Sandstone; (5) Bauru Sandstone; (6) Kidderminster Sandstone; (7) Bauru Sandstone; (8) Kidderminster Sandstone (parallel to bedding) (after Dobereiner and De Freitas, 1986)*

sedimentation, its subsequent depth of burial and tectonic history, and the effects of diagenesis. The porosity of shale may range from slightly under 5% to just over 50%, with natural moisture contents of 3–35%.

The natural moisture content of shales varies from less than 5%, increasing to as high as 35% for some clayey shales. When the natural moisture content of shales exceeds 20% they frequently are suspect as they tend to develop potentially high pore pressures. Usually the moisture content in the weathered zone is higher than in the unweathered shale beneath.

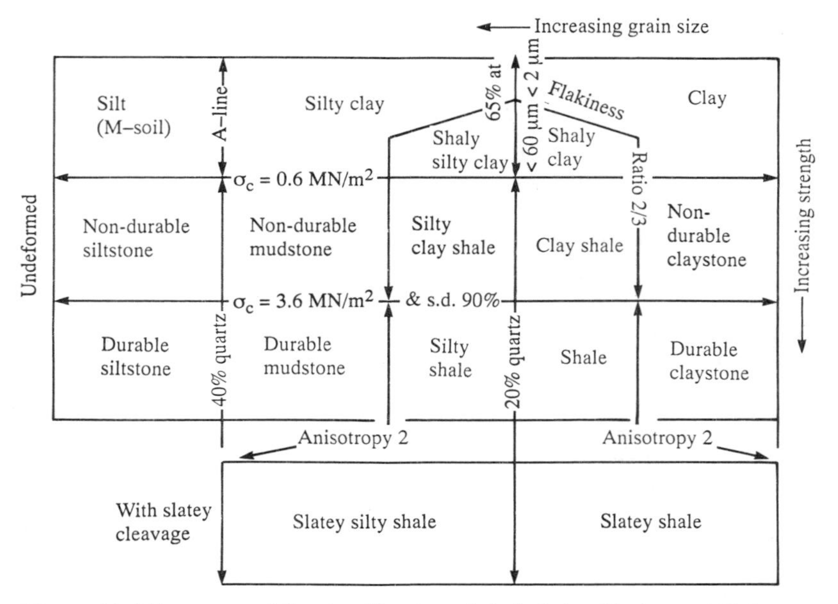

Figure 11.4 *Summary of the classification of shale (after Grainger, 1984)*

Argillaceous materials are capable of undergoing appreciable suction before pore water is removed, drainage commencing when the necessary air-entry suction is achieved (about pF = 2). Under increasing suction pressure the incoming air drives out the water from a shale and some shrinkage takes place in the fabric before air can offer support. Generally as the natural moisture content and liquid limit increase so the effectivenes of soil suction declines.

The swelling properties of certain clay shales have proved extremely detrimental to the integrity of many civil engineering structures. Swelling, especially in such clay shales, is attributable to the absorption of free water by particular clay minerals, notably montmorillonite, in the clay fraction of a shale. Highly fissured overconsolidated shales have greater swelling tendencies than poorly fissured clayey shales, the fissures providing access for water.

Depending upon the relative humidity, many shales slake almost immediately when exposed to air (Kennard *et al.*, 1967). Desiccation of shale, following exposure, leads to the creation of negative pore pressures and consequent tensile failure of the weak intercrystalline bonds. This leads to the production of shale particles of coarse sand, fine gravel size. Alternate wetting and drying causes a rapid breakdown of compaction shales. Low-grade compaction shales, in particular, undergo complete disintegration after several cycles of drying and wetting. Indeed De Graft-Johnson *et al.* (1973) found that the compacted variety of Accra Shale could be distinguished from the cemented variety by wetting and drying tests. The compacted shales generally crumbled to fine material after 2 or 3 cycles, whilst the cemented samples withstood 6 cycles, none of the samples exceeding a loss of 8%. This indicates that well cemented shales are fairly resistant to slaking.

Russell (1981) examined the durability of two major shale formations in Ontario by means of the slake durability test. Although the shales appeared similar, one had a lower durability than the other. This was attributed to the difference in the nature of the microcracks and in the degree of cementation. In other words in the shale with the lower durability, the microcracks were more curved and it was poorly cemented by calcite. Microcracks allow the penetration of water which leads to slaking and the more curved they are, the more they tend to meet on slaking, thereby causing breakdown by the formation of small fragments. Russell noted that durability of compaction shales was influenced by the amount of clay minerals present in the clastic fraction, whereas this was not the case in cemented shales.

According to Olivier (1980) the degradation processes in shales subjected to changes in moisture content are related to partially irreversible anisotropic expansion and shrinkage of the rock as a result of capillary action and drying respectively.

Lamination effects an important control on the breakdown of shales. Other controls on the breakdown of shaley materials include air breakage and dispersal of colloidal material. Badger *et al.* (1956) maintained that the former process occurred only in those shales which were mechanically weak whilst the latter process appeared to be a general cause of disintegration. They also observed that the degree of disintegration of a shale when immersed in different liquids was governed by the manner in which these liquids affected air breakage and ionic dispersion forces. For example, in a liquid with a low dielectric constant little disintegration took place as a result of ionic forces because of the suppression of ionic dissociation from the shale colloids. They also found that the variation in disintegration of different shales in water usually was not connected with the total amount of clay colloid or the variation in the types of clay mineral present. It was rather controlled by the type of exchangeable cations attached to the clay particles and the accessibility of the latter to attack by water which, in turn, depended on the porosity of the shale. Air breakage could assist this process by presenting new surfaces of shale to water.

Mudstones tend to break down along irregular fracture patterns which, when well developed, can mean that these rocks disintegrate within one or two cycles of wetting and drying. Nakano (1967) found that although some mudstones from Japan, when immersed in water, swelled slowly and underwent a consequent decrease in bulk density and strength, they did not disintegrate even after immersion for a lengthy period of time. However, if they were dried and then wetted they disintegrated rapidly into small pieces. It was observed that mudstones started to slake when drying in a relative humidity of 98% and that the drier the mudstones were, compared with their natural state, the greater was the intensity of disintegration in water. This means that such mudstones (the clay fraction in these mudstones consisted of montmorillonite) may deteriorate readily in the zone of fluctuating water table or water vapour pressure.

Venter (1981) investigated the changes which occur in some South African mudrocks when they are subjected to different tempeature and relative humidity conditions. He noted that temperature changes on their own had little effect on volume and moisture content changes. Severe simulated 'night and day' cycles in which both temperature and humidity were changed (from 5°C and 90% RH to 50°C and 21% RH respectively) did not affect the samples to a greater extent than severe humidity changes of the same magnitude without temperature changes. Even small changes in humidity caused fluctuations in volume and moisture content. For example, a maximum shrinkage of 0.34% and a maximum water loss of 0.72% took place with a drop in relative humidity from 88 to 60%, while a maximum shrinkage of 0.49% and a moisture loss of 1.91% occurred with a drop in relative humidity from 90 to 22%. However, changes in temperature and humidity alone did not bring about disintegration of the mudrocks. Indeed the samples returned to their original volume and moisture conditions when the conditions were returned to the original state. Consequently Venter concluded that free water is necessary to start slaking.

After immersion in water, well-stratified samples of mudrock, according to Venter (1981), expanded much more in the direction perpendicular to the bedding than parallel to it. Expansions during subsequent immersions were generally lower although more water was absorbed. The samples which expanded most were those which had been air- or oven-dried. As far as slaking is concerned, those samples which were immersed at their natural moisture contents did not slake to the same extent as those which had been exposed to the atmosphere or which had been dried in an oven. Hence Venter concluded that if mudrock is kept at its natural moisture content without being allowed to lose water, deterioration after exposure will be reduced significantly.

It would appear that the strength of compacted shales decreases exponentially with increasing void ratio and moisture content. However, in cemented shales the amount and strength of the cementing material are the important factors influencing intact strength. Unconfined compressive strength tests on Accra Shales carried out by De Graft-Johnson *et al.* (1973) indicated that the samples usually failed at strains between 1.5 and 3.5%. The compressive strengths varied from 200 kN/m^2 to 20 MN/m^2, with the values of the modulus of elasticity ranging from 6.5 to 1400 MN/m^2. Those samples which exhibited the high strengths were usually cemented types.

Morgenstern and Eigenbrod (1974) carried out a series of compression softening tests on argillaceous materials. They found that the rate of softening of these materials when immersed in water largely depends upon their degree of induration. Poorly indurated materials soften very quickly and they may undergo a loss of up to 90% of their original strength within a few hours. Mudstones, at their natural water content, remain intact when immersed in water. However, they swell slowly, hence decreasing in bulk density and strength. This time-dependent loss in strength is a very significant engineering property of mudstones. A good correlation exists between their initial compressive strength and the amount of strength loss during softening.

The value of shearing resistance of a shale chosen for design purposes lies somewhere between the peak and residual strengths. In weak compaction shales cohesion may be lower than 15 kN/m^2 and the angle of friction as low as 5°. By contrast, Underwood (1967) quoted values of cohesion and angle of friction of 750 kN/m^2 and 56° respectively for dolomitic shales of Ordovician age, and 8–23 MN/m^2 and 45–64° respectively for calcareous and quartzose shales from the Cambrian period. Generally shales with a cohesion less than 20 kN/m^2 and an apparent angle of friction of less than 20° are likely to present problems. The elastic moduli of compaction shales range from less than 140–1400 MN/m^2, whilst those of well cemented shales have elastic moduli in excess of 14 000 MN/m^2 (Table 11.3).

The higher degree of fissility possessed by a shale, the greater is the anisotropy with regard to strength, deformation and permeability. For instance, the influence of lamination on the behaviour of clay shale has been discussed by Wichter (1979). He noted that in triaxial testing the compressive strength parallel to the laminations was some 1.5–2 times less than that obtained at right angles to it, for confining pressures up to 1 MN/m^2. The influence of fissility on Young's modulus can be illustrated by two values quoted by Chappell (1974), 6000 and 7250 MN/m^2, for cemented shale tested parallel and normal to the lamination respectively. Previously Zaruba and Bukovansky (1965) found that the values of Young's modulus were up to five times greater when they tested shale normal as opposed to parallel to the direction of lamination.

According to Burwell (1950) well-cemented shales, under structurally sound conditions, present few problems for large structures such as dams, though their strength limitations and elastic properties may be factors of importance in the design of concrete dams of appreciable height. They, however, have lower moduli of elasticity and generally lower shear values than concrete and therefore in general are unsatisfactory foundation materials for arch dams.

The problem of settlement in shales generally resolves itself into one of reducing the unit bearing load by widening the base of structures or using spread footings. In some cases appreciable differential settlements are provided for by designing articulated structures capable of accommodating differential movements of individual sections without damage to the structure. Severe settlements may take place in low-grade compaction shales. However, compaction shales contain fewer open joints or fissures which can be compressed beneath heavy structures than do cemented shales. Where concrete structures are to be founded on shale and it is suspected that the structural load will lead to closure of defects in the rock, *in situ* tests should be conducted to determine the elastic modulus of the foundation material.

Uplift frequently occurs in excavations in shales and is attributable to swelling and heave. Rebound on unloading of shales during excavation is attributed to heave due to the release of stored strain energy. The conserved strain energy tends to be released more slowly than it is in harder rocks. Shale relaxes towards a newly excavated face and sometimes this occurs as offsets at weaker seams in the shale. The greatest amount of rebound occurs in heavily overconsolidated compaction shales; for example, at Garrison Dam, North Dakota, just over 0.9 m of rebound was measured in the deepest excavation in the Fort Union Clay Shales. What is more high horizontal residual stresses caused saw cuts 75 mm in width to close within 24 h (Smith and Redlinger, 1953). Some 280 mm of rebound was recorded in

Table 11.3 *An engineering evaluation of shales (after Underwood 1967)*

Physical properties			*Probable* in situ *behaviour**						
Laboratory tests and in situ *observations*	*Average range of values*		*High pore pressure*	*Low bearing capacity*	*Tendency to rebound*	*Slope stability problems*	*Rapid sinking*	*Rapid erosion*	*Tunnel support problems*
	Unfavourable	*Favourable*							
Compressive strength (kN/m^2)	350–2070	2070–3500	√	√					√
Modulus of elasticity (MN/m^2)	140–1400	1400–14 000		√					√
Cohesive strength (kN/m^2)	35–700	700–>10 500			√	√			√
Angle of internal friction (deg)	10–20	20–65			√	√			
Dry density (Mg/m^3)	1.12–1.78	1.78–2.56	√					√(?)	
Potential swell (%)	3–15	1–3			√	√		√	√
Natural moisture content (%)	20–35	5–15	√			√			
Coefficient of permeability (m/s)	10^{-7}–10^{-12}	>10^{-7}	←			←	←		
Predominant clay minerals	Montmorillonite or illite	Kaolinite and chlorite	√			√			
Activity ratio	0.75–>2.0	0.35–0.75				√		√	
Wetting and drying cycles	Reduces to grain sizes	Reduces to flakes					√		
Spacing of rock defects	Closely spaced	Widely spaced		√		√		√(?)	√
Orientation of rock defects	Adversely oriented	Favourably oriented		√		√			√
State of stress	> Existing over-burden load	≃ Over-burden load			√	√			√

Note: According to S. Irmay (*Israel Journal of Technology*, 1968 **6**, No. 4, 165–72), the maximum possible $\phi = 47.5°$
* The ticks relate to the unfavourable range of values

the underground workings in the Pierre Shale at Oake Dam, South Dakota (Underwood *et al.*, 1964). Moreover differential rebound occurred in the stilling basin due to the presence of a fault. Differential rebound movements require special decision provision.

The stability of slopes in excavations can be a major problem in shale both during and after construction. This problem becomes particularly acute in dipping formations and in formations containing expansive clay minerals.

Sulphur compounds are frequently present in argillaceous rocks. An expansion in volume large enough to cause structural damage can occur when sulphide minerals such as pyrite and marcasite suffer oxidation and give rise to anhydrous and hydrous sulphates (see Chapter 9). Penner *et al.* (1973) quoted a case of heave in a black shale of Ordovician age in Ottawa which caused displacment of the basement floor of a three-storey building. The maximum movement totalled some 107 mm, the heave rate being almost 2 mm per month. When examined, the shale in the heaved zone was found to have been altered to a depth of between 0.7 and 1 m. Beneath, the unaltered shale contained numerous veins of pyrite, indeed the sulphur content was as high as 1.6%. The heave was attributable to the breakdown of pyrite to produce sulphur compounds which combine with calcium to form gypsum and jarosite. The latter minerals formed in the fissures and between the laminae of the shales in the altered zone. Measurements of the pH gave values ranging from 2.8 to 4.4. It was concluded that the alteration of the shales was the result of biochemical weathering brought about by autotrophic bacteria. The heaving was therefore arrested by creating conditions unfavourable for bacterial growth. This was done by neutralizing the altered zone by introducing a potassium hydroxide solution into the examination pits. The water table in the altered zone was also kept artificially high so that the acids would be diffused and washed away, and to reduce air entry.

Decomposition of sulphur compounds also gives rise to aqueous solutions of sulphate and sulphuric acid which react with tricalcium aluminate in Portland cement to form calcium sulpho-aluminate or ettringite. This reaction is accompanied by expansion. The rate of attack is very much influenced by the permeability of the concrete or mortar and the position of the water table. For example, sulphates can only continue to reach cement by movement of their solutions in water. Thus if a structure is permanently above the water table it is unlikely to be attacked.

By contrast below the water table movement of water may replenish the sulphates removed by reaction with cement, thereby continuing the reaction. Concrete with a low permeability is essential to resist sulphate attack, hence it should be fully compacted. Sulphate resistant cements, that is, those in which the tricalcium aluminate is low can also be used for this purpose (Anon, 1975). Foundations used to be made larger than necessary to counteract sulphate attack but it is now more economical to protect them by impermeable membranes or bituminous coatings.

When a load is applied to an essentially saturated shale foundation the void ratio in the shale decreases and the pore water attempts to migrate to regions of lesser load. Because of its relative impermeability water becomes trapped in the voids in the shale and can only migrate slowly. As the load is increased there comes a point when it is in part transferred to the pore water, resulting in a build-up of pore pressure. Depending on the permeability of the shale and the rate of loading, the pore pressure can more or less increase in value so that it equals the pressure imposed by the load. This greatly reduces the shear strength of the shale and a serious failure can occur, especially in the weaker compaction shales. For instance, high pore pressure in the Pepper Shale was largely responsible for the foundation failure at Waco Dam, Texas (Underwood, 1967). Pore pressure problems are not so important in cemented shales.

Clay shales usually have permeabilities of the order of 1×10^{-8} m/s to 10^{-12} m/s; whereas sandy and silty shales and closely jointed cemented shales may have permeabilities as high as 1×10^{-6} m/s. However, Jumikis (1965) noted that where the Brunswick Shale in New Jersey was highly fissured it could be used as an aquifer. He also noted that the build-up of groundwater pressure along joints could cause shale to lift along bedding planes and lead to slabs of shale breaking from the surface. Hence in this shale formation subsurface water must be drained by an efficient system to keep excavations dry.

Many marls in Britain (e.g. those occurring in the Mercia Mudstone or as it was formerly known, the Keuper Marl Series) consist of between 50 and 90 % clay minerals. The marl may contain thin veins or beds of gypsum. In such cases the groundwaters contain sulphates. According to Dumbleton (1967) illite accounts for 28–56% of the clay minerals and chlorite may total some 39%. Usually more than half the chlorite is of the swelling type. Quartz tends to vary between 5 and 35%. The other minerals include calcite and dolomite (which usually comprise less than 20%), and hematite 1 or 2%. Occasionally other clay minerals such as sepiolite and palygorskite have been found.

The clay particles tend to be aggregated mainly into silt-size units, the aggregated structure being extremely variable (Davis, 1968). Particles composing the aggregate are held together by cement. Sherwood (1967) suggested that silica might be the cementing agent whilst Lees (1965) assumed that the clay particles were bound together by physical forces. The former explanation seems the more likely. Aggregation leads to the lack of corelation between consistency limits and shear strength on the one hand and clay content on the other. Because engineering behaviour is controlled by the aggregates, rather than the individual clay minerals, the plasticity, according to Davis (1967) is lower than would be expected. He consequently proposed the aggregation ratio (A_r) as a means of assessing the degree of aggregation. The aggregation ratio was defined as the percentage weight of clay as determined by mineralogical analysis, expressed as a ratio of the percentage weight of clay particles determined by sedimentation techniques. On the other hand Sherwood (1967) maintained that aggregation did not give anomalous plasticity values for Keuper Marls and that they can generally be classified as materials of low to medium plasticity (Table 11.4). The activity of the Keuper marls increases as does the degree of aggregation.

Certain marls exhibit rapid softening when exposed to wet conditions. Such deterioration can be estimated by assessing their moisture adsorption potential (Kolbuszewski *et al.*, 1965). These marls are fissured, weathering and water

Table 11.4 *Soil classification tests on Keuper Marl (from Sherwood, 1967)*

	Clay content by sedimentation (%)	*Clay content by X-ray analysis (%)*	*Liquid limit (%)*	*Plastic limit (%)*	*Plasticity index (%)*
(1)	26	94	71	40	31
(2)	36	58	33	19	14
(3)	30	87	46	28	25
(4)	12	77	48	29	19

penetrating the fissures and thereby further reducing the strength of the material. The fissures close with increasing depth.

Chandler (1969) recognized five grades of weathering, from unweathered to fully weathered, in the Keuper Marl. He showed that highly and fully weathered marls could be distinguished from material from grades I, II and III by their particle size distribution and plasticity index (Table 11.5).

An extensive review of the Keuper Marl as a foundation material has been provided by Meigh (1976).

Table 11.5 *Some index properties of Keuper Marl (after Chandler, 1969)*

Index property	*Weathering grade*		
	I and II	III	IV
Bulk density (Mg/m^3)	2.5–2.3	2.3–2.1	2.2–1.8
Dry density (Mg/m^3)	2.4–1.9	2.1–1.8	1.8–1.4
Natural moisture content (%)	5–15	12–20	18–35
Liquid limit (%)	25–35	25–40	35–60
Plastic limit (%)	17–25	17–27	17–33
Plasticity index	10–15	10–18	17–35
% clay size (BS 1377)	10–35	10–35	30–50
Aggregation ratio (A_r)	10–2.5	10–2.5	2.5
$c'(kN/m^2)$	$\geqq 27.6$	$\leqq 17.2$	$\leqq 17.2$
ϕ' (deg)	40	42–32	32–25
ϕ'_r (deg)	32–23	29–22	24–18
$^*\dfrac{\tau_{max} - \tau_{res}}{\tau_{max}}(\%)$	55	55–30	35–20

* Percentage reduction from peak residual strength

11.4 Carbonate rocks

Carbonate rocks contain more than 50% carbonate minerals. The term *limestone* is applied to those rocks in which the carbonate fraction exceeds 50%, over half of which is calcite or aragonite. If the carbonate material is made up chiefly of dolomite, then the rock is named dolostone.

The engineering properties of carbonate sediments are influenced by grain size and those post-depositional changes which bring about induration, and thereby increase strength. Consequently these factors have been taken as the bases for engineering classifications of carbonate sediments. A review of classifications of carbonate rocks in relation to engineering has been provided by Burnett and Epps (1979).

Representative values of some physical properties of carbonate rocks are listed in Table 11.6 and a review of the engineering properties of carbonate rocks has been provided by Dearman (1981). It can be seen that generally the density of these rocks increases with age, whilst the porosity is reduced. Diagenetic processes mainly account for the lower porosities of the Carboniferous and Magnesian Limestones quoted in the table.

Al-Jassar and Hawkins (1979) recognized several lithological types in the Carboniferous Limestone formation of the Bristol area and showed that the engineering properties related to lithology. The micritic (microcrystalline calcium carbonate) limestones, for example, possess a higher strength than do the sparitic (coarsely crystalline calcite) types, whilst secondary dolomitization has enhanced the strength of those rocks so affected. In fact the dolostone was found to have the highest strength of the calcareous rocks tested. Al-Jassar and Hawkins noted that all the limestones initially behaved as elastic material, with small plastic deformation occurring in some cases prior to failure (Figure 11.5). On the other hand Attewell (1971), in a study of the Great Limestone of the north of England (again Carboniferous in age), noted that limestone which had been dolomitized had a higher porosity (averaging approximately 7.5%, compared with 4%) and that it possessed a lower unconfined compressive strength (varying from 70 to 165 MN/m^2 compared with 110 to 210 MN/m^2 for limestone which was not dolomitized).

Age often has an influence on the strength and deformation characteristics of carbonate rocks. From Table 11.6 it can be seen that Carboniferous Limestone is generally very strong (Bell, 1981). Conversely the Bath Stone (Great Oolite, Jurassic) is only just moderately strong. Similarly the oldest limestones tend to have the highest values of Young's modulus.

A series of triaxial tests were carried out by Elliott and Brown (1985) to investigate the brittle–ductile behaviour of the Bath Stone. They found that dilatation associated with brittle fracture and cataclastic flow occurred at confining pressures up to 5 MN/m^2. At higher confining pressures, net contractant volumetric strains were measured. This, they assumed, resulted from pore collapse and structural rearrangement.

Thick bedded, horizontally lying limestones relatively free from solution cavities afford excellent foundations. On the other hand thin bedded, highly folded or cavernous limestones are likely to present serious foundation problems. A possibility of sliding may exist in thinly bedded, folded sequences. Similarly if beds are separated by layers of clay or shale, especially when inclined, these may serve as sliding planes and result in failure.

Table 11.6 *Some physical properties of carbonate rocks*

	Carboniferous Limestone (Buxton)	*Magnesian Limestone (Anston)*	*Ancaster Freestone (Ancaster)*	*Bath Stone (Corsham)*	*Middle Chalk (Hillington)*	*Upper Chalk (Northfleet)*
Specific gravity	2.71	2.83	3.70	2.71	2.70	2.69
Dry density (Mg/m^2)	2.58	2.51	2.27	2.30	2.16	1.49
Porosity (%)	2.9	10.4	14.1	15.3	19.8	41.7
Dry unconfined compressive strength (MN/m^2)	106.2	54.6	28.4	15.6	27.2	5.5
Saturated unconfined compressive strength (MN/m^2)	83.9	36.6	16.8	9.3	12.3	1.7
Point load strength (MN/m^2)	3.5	2.7	1.9	0.9	0.4	–
Scleroscope hardness	53	43	38	23	17	6
Schmidt hardness	51	35	30	15	20	9
Youngs modulus (GN/m^2)	66.9	41.3	19.5	16.1	30.0	4.4
Permeability ($\times 10^{-9}$ m/s)	0.3	40.9	125.4	160.5	1.4	13.9

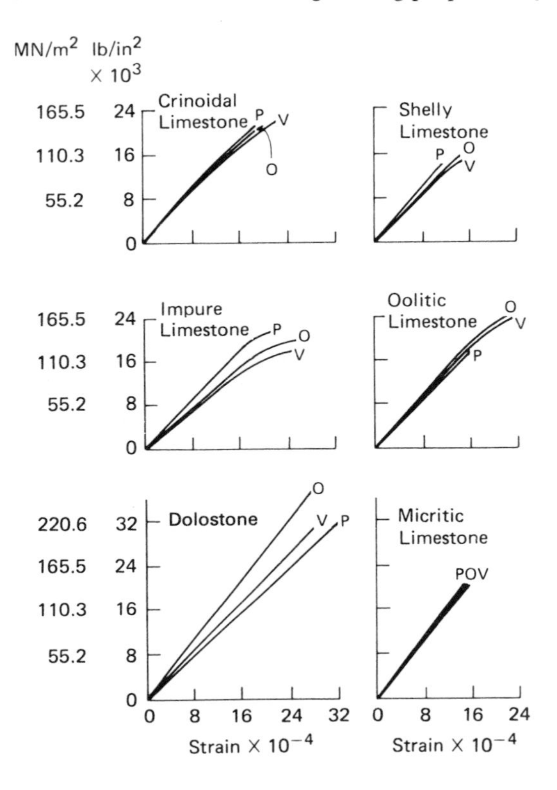

Figure 11.5 *Stress–strain curves for limestones with different lithologies from the Lower Carboniferous of the Bristol area. The tests were carried out parallel (P), vertical (V) and oblique (O) to the bedding (after Al-Jassar and Hawkins, 1979)*

11.4.1 Limestones and solution

Limestones are commonly transected by joints. These have generally been subjected to various degrees of dissolution so that some may gape. Rainwater is generally weakly acidic and further acids may be taken into solution from carbon dioxide or organic or mineral matter in the soil. The degree of aggressiveness of water to limestone can be assessed on the basis of the relationship between the dissolved carbonate content, the pH value and the temperature of the water. At any given pH value, the cooler the water the more aggressive it is. If solution continues its rate slackens and it eventually ceases when saturation is reached. Hence solution is greatest when the bicarbonate saturation is low. This occurs when water is circulating so that fresh supplies with low lime saturation are continually made available. Non-saline water can dissolve up to 400 mg/l of calcium carbonate.

Sinkholes may develop where joints intersect and these may lead to an integrated system of subterranean galleries and caverns. The latter are characeristic of thick massive limestones. The progressive opening of discontinuities by dissolution leads to an increase in mass permeability. Sometimes dissolution produces a highly irregular pinnacled surface on limestone pavements. The size, form, abundance and downward extent of the aforementioned features depends upon the geological structure and the presence of interbedded impervious layers. Solution cavities present numerous problems in the construction of large foundations such as for dams, amongst which bearing strength and watertightness are paramount. Few sites are so bad that it is impossible to construct safe and successful structures upon them, but the costt of the necessary remedial treatment may be prohibitive.

An important effect of solution in limestone is enlargement of the pores, which enhances water circulation thereby encouraging further solution (Sowers, 1975). This brings about an increase in stress within the remaining rock framework which reduces the strength of the rock mass and leads to increasing stress corrosion. On loading the volume of the voids is reduced by fracture of the weakened cement between the particles and by the reorientation of the intact aggregations of rock that become separated by loss of bonding. Most of the resultant settlement takes place rapidly within a few days of the application of load.

Rapid subsidence can take place due to the collapse of holes and cavities within limestone which has been subjected to prolonged solution, this occurring when the roof rocks are no longer thick enough to support themselves. It must be emphasized, however, that the solution of limestone is a very slow process; contemporary solution is therefore very rarely the cause of collapse. For instance, Kennard and Knill (1968) quoted mean rates of surface lowering of limestone areas in the British Isles which ranged from 0.041–0.099 mm annually.

Nevertheless solution may be accelerated by man-made changes in the groundwater conditions or by a change in the character of the surface water that drains into limestone. For instance, James and Kirkpatrick (1980), in a consideration of the location of hydraulic structures on soluble rocks, wrote that if such dry discontinuous rocks are subjected to substantial hydraulic gradients then they will undergo dissolution along these discontinuities, hence leading to rapidly accelerating seepage rates. From experimental work which they carried out on the Portland Limestone they found that the values of solution rate constant (K) increased appreciably at flow velocities which corresponded to a transitional flow regime. They showed that such a flow regime occurred in joints about 2.5 mm in width which experienced a hydraulic gradient of 0.2 (Figure 11.6a). According to these two authors solution takes place along a small joint by retreat of the inlet face due to removal of soluble material (Table 11.7; Figure 11.6b). Dissolution of larger joints gives rise to long tapered enlargements, which enable seepage rates to increase rapidly and runaway situations to develop.

According to Sowers (1975) ravelling failures are the most widespread and probably the most dangerous of all the subsidence phenomena associated with limestones. Ravelling occurs when solution enlarged openings extend upward to a rock surface overlain by soil (Figure 11.7). The openings should be interconnected and lead into channels through which the soil can be eroded by groundwater flow.

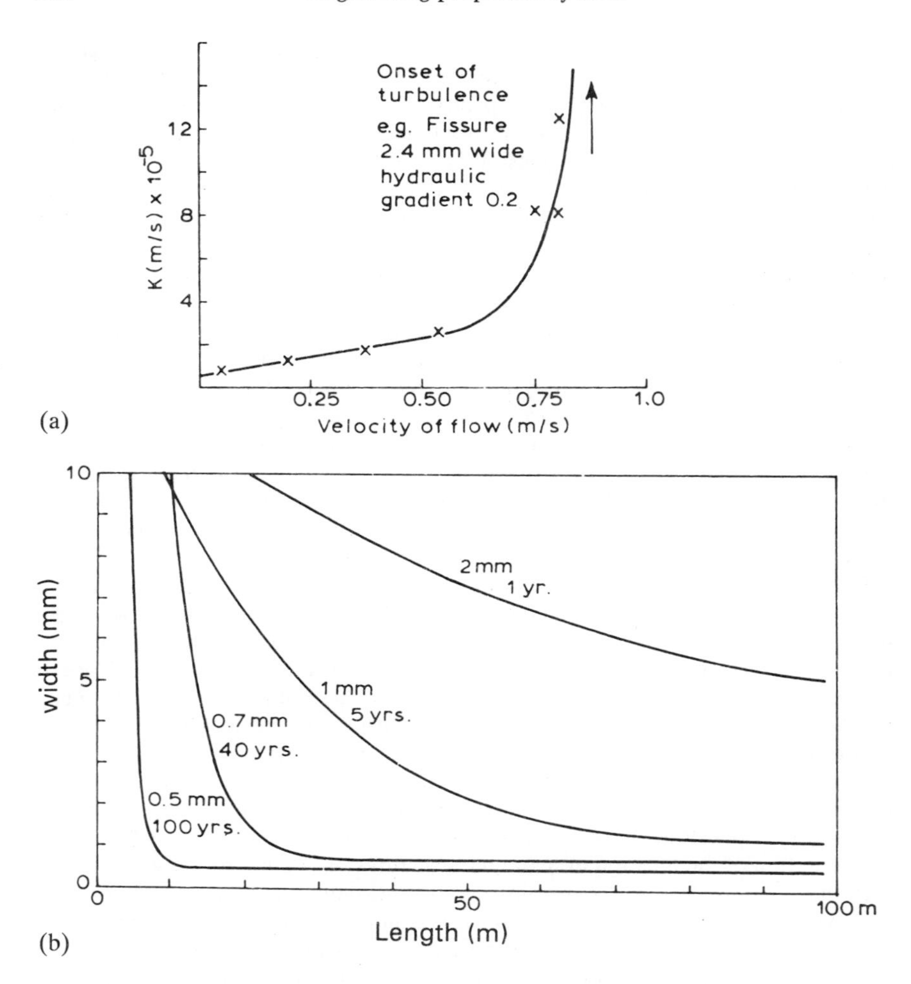

Figure 11.6 (a) *Rate of solution (K) of Portland Stone and its dependent on flow velocity.* (b) *Enlargement of fissures in calcium carbonate rock by pure flowing water (after James and Kirkpatrick, 1980)*

Initially the soil arches over the openings but as they are enlarged a stage is reached when the soil above the roof can no longer support itself and so it collapses. A number of conditions accelerate the development of cavities in the soil and initiate collapse. Rapid changes in moisture content lead to aggravated slabbing or roofing in clays and flow in cohesionless sands. Lowering the water table increases the downward seepage gradient and accelerates downward erosion. It also reduces capillary attraction and increases instability of flow through narrow openings and

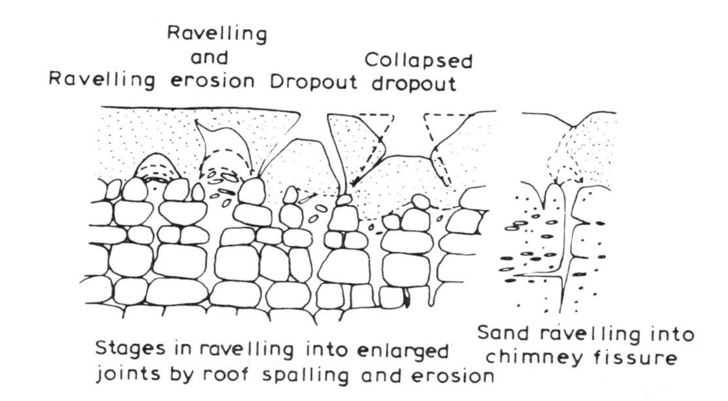

(a)

(b)

Figure 11.7 (a) *Mechanisms of ravelling.* (b) *House being swallowed by ravelling collapse in Central Florida (courtesy Professor George F. Sowers and P. Kehoe)*

Table 11.7 *Solution of soluble rocks (after James and Kirkpatrick, 1980)*

Rock	*(a) Solubility* (c_s) *in pure water* c_s (kg/m^3 at 10°C)*	*(b) Solution rate constant* (*K*) *at 10°C (flow velocity–0.05* m/s) m/s × 10^{-5}*	m^4 (kg/s × 10^{-6})
		θ = 1	θ = 2
Halite	360.0	0.3	
Gypsum	2.5	0.2	
Anhydrite	2.0		0.8
Limestone	0.015	0.4	

* c_s is dependent upon temperatire and the presence of other dissolved salts
* *K* is dependent on temperature, flow velocity and other dissolved salts
θ = order of dissolution reaction

(*c*) Limiting widths of fissures in massive rock*

Rock	(i) *Upper limit for stable inlet face retreat*	(ii) *For a rate of retreat of 0.1 m/year (e.g. cavern formation)*	(iii) *Maximum safe lugeon value†*
Halite	0.05	0.05	
Gypsum	0.2	0.3	50
Anhydrite	0.1	0.2	7
Limestone	0.5	1.5	700

* These values are for pure water, at a fissure spacing in massive rock of one per metre and a hydraulic gradient of 0.2. For water containing 300 mg/l of CO_2, the stable limit width of fissure becomes 0.4 mm in limestone.
† One lugeon unit is equal to a flow of 1 l/m/min at a pressure of 1 MN/m^2 (it is approximately equal to a coefficient of permeability of 10^{-7} m/s).

(*d*) *Solution of particulate deposits**

Rock	(i) *Limiting seepage velocity* (m/s)	(ii) *Width of solution zone* (m)
Halite	6.0×10^{-9}	0.0002
Gypsum	1.4×10^{-6}	0.04
Anhydrite	1.6×10^{-6}	0.09
Limestone	3.0×10^{-4}	2.8

* Rate of movement of solution zone–0.1 m/year. Mineral particles of 50 mm diameter. Pure water.

gives rise to shrinkage cracks in highly plastic clays which weakens the mass in dry weather and produces concentrated seepage during rains. Increased infiltration often initiates failure, particularly when it follows a period when the water table has been lowered.

11.4.2 Chalk

Generally the Chalk is a remarkably pure micritic limestone, containing over 95% calcium carbonate which can be divided into coarse and fine fractions. The coarse fraction, which may constitute 20–30%, falls within the 10–100 μm range. This contains material derived from the mechanical breakdown of large shelled organisms and, to a lesser extent, from foraminifera. The fine fraction, which takes the form of calcite particles about one micron or less in size, is almost entirely composed of coccoliths and may form up to, and sometimes over, 80% of certain horizons.

The Chalk varies in hardness. For example, hardgrounds are present throughout much of the Chalk. They are horizons, like the Chalk Rock and the Melbourne Rock in England, which may be less than one metre or up to 10 m in thickness, which have undergone significant contemporaneous diagenetic hardening and densification. On the other hand the individual particles in the soft chalk, such as that in south-east England, are bound together at their points of contact by thin films of calcite. Such chalk contains only minute amounts of cement. Early cementation prevented gravitational consolidation occurring in soft chalk and helped retain high values of porosity.

In addition to the effects of intrinsic diagenesis, tectonic compaction, pressure solution and late stage solution have altered the engineering behaviour of the Chalk (Clayton, 1983). Stresses due to Alpine earth movements (the most noticeable effects in England are witnessed in the Chalk of the south coast) brought about mechanical disaggregation of the coccolith matrix (followed by reconsolidation), which then was followed by pressure solution and reprecipitation of the calcite under sustained pressures. Such action affected the strength and density of the Chalk. For instance, in areas where the Chalk dips at angles in excess of 30° the density of some of the weakest chalks has been increased to that of the hardest hardground. Nonetheless the density of tectonically modified chalks varies widely over relatively short distances. Pressure solutioning due to overburden pressure has led to strengthening and hardening of the Chalk in Yorkshire, Humberside and Lincolnshire.

There is a wide range of density values displayed by the Chalk. For example, Bell *et al.* (1990) found that dry density varied from 1.35 Mg/m^3 for the Upper Chalk (*M. coranguinum* zone) of Kent to 2.30 Mg/m^3 for the Middle Chalk (*T. lata* zone) of Yorkshire (Table 11.8). In fact distinction often is made between hard and soft chalks on a basis of either dry density (dry density below 1.44 Mg/m^3 indicates soft chalks while values above 1.6 Mg/m^3 indicate harder varieties).

The specific gravity of the Chalk, as can be seen from Table 11.8, varies very little and the values are similar to that of calcite. This testifies to the purity of the Chalk as a formation.

The saturation moisture content also has been used as a means of distinguishing hard from soft chalks, those in which it is less than 15% are hard whilst those in which it exceeds 25% are soft. Table 11.8 also shows a wide range of saturation moisture contents and porosities. The porosity values in soft chalks commonly exceed 35%. In harder chalk the porosity has been reduced by diagenetic processes.

Using mercury injection porosimitry, Price *et al.* (1976) measured median pore diameters of the Middle and Upper Chalk of Yorkshire, obtaining values of 0.39 and 0.41 μm respectively, whilst the corresponding values for southern England and East Anglia were respectively 0.53 and 0.65 μm. Generally larger pores were found in the Upper Chalk and in the southern area. In southern England the median pore diameter in the Lower Chalk was 0.22 μm, a feature attributed, in part, to a high marl content.

The unconfined compressive strength of the Chalk ranges from moderately weak to moderately strong (Table 11.8). However, the unconfined compressive strength of chalk undergoes a marked reduction when it is saturated. For instance, according to Bell *et al.* (1990) some samples of Upper Chalk from Kent suffer a dramatic loss on saturation amounting to almost 70%. Samples from the Lower and Middle Chalk may show a reduction in strength averaging over 50%. In fact the weaker the rock, the greater the loss of strength appears to be on saturation. The pore water also has a critical influence on the triaxial strengths of chalk and obviously its modulus of deformation (Meigh and Early, 1957).

Bell (1977) noted that the mode of failure of the Upper Chalk when tested in triaxial conditions was influenced by the confining pressure. Diagonal shear failure occurred at lower confining pressures but at 5.0 MN/m^2 confining pressure and above, plastic deformation took place, giving rise to barrel-shaped failures in which numerous, small inclined shear planes were developed. Meigh and Early (1957) previously had explained such behaviour, suggesting that at high cell pressures specimens of chalk underwent disaggregation. This they demonstrated by wetting and drying failed specimens which brought about their structural collapse.

Some values of point load strength for chalk are given in Table 11.8. However, Bell *et al.* (1990) pointed out that this test did not prove suitable for soft chalks since the conical platens bit into the material as load was applied.

Carter and Mallard (1974) found that chalk compressed elastically up to a critical pressure, the apparent preconsolidation pressure. Marked breakdown and substantial consolidation occurs at higher pressures. The apparent preconsolidation pressure is influenced by consolidation, cementation and possibly creep. They obtained a coefficient of consolidation (c_v) and volume compressibility (m_v) similar to those found by Meigh and Early (1957) and Wakeling (1965); that is, c_v = 1135 m^2/year, and m_v = 0.019 m^2/MN.

Generally stress–strain curves of chalk show an initial zone of increasing slope with increasing load, a feature which suggests consolidation or closure of bedding planes and pore spaces, before any near-linear deformation occurs. The deformation properties of chalk in the field depend upon its hardness, and the spacing, tightness and orientation of its discontinuities. The values of Young's modulus are also influenced by the amount of weathering chalk has undergone. Using these factors, Ward *et al.* (1968) classified the Middle Chalk at Mundford, Norfolk, into five grades, and showed that the value of Young's modulus varies with grade (Table 11.9). They pointed out that Grades IV and V were largely the result of weathering and were therefore independent of lithology, whereas Grades I and II were completely unweathered so that the difference between them was governed by their lithological character. Grades V, IV and III occur in succession from the surface down whilst Grade I may overlie Grade II or vice versa. Burland and Lord (1970) observed that both the full-scale tank test and plate load tests at Mundford indicated that at low applied pressures even Grade IV chalk behaves elastically. At higher pressures chalk exhibits yielding behaviour. They pointed out that for stresses up to 1 MN/m^2 the plate load tests showed that Grades IV and V exhibit significant creep and in the long term, creep deflections may be considerably larger than immediate deflections. Creep in Grade III is smaller and terminates more rapidly whilst Grades II and I undergo negligible creep.

The dynamically determined value of Young's modulus in the Middle Chalk at Mundford has been shown to increase with increasing depth from 1.8×10^3 MN/m^2 at 5 m to 10.2×10^3 MN/m^2 at 19.3 m (Abbiss, 1979). These values are more than double those of Young's modulus determined by static methods (Table 11.9). The corresponding values of seismic velocities for these five grades of chalk were determined by Grainger *et al.* (1973) and are given in Table 11.9.

Burland *et al.* (1974) found that settlements of a five-storey building founded in soft low-grade chalk at Reading were very small. Their findings agreed favourably with those previously obtained at Mundford.

The permeability of the Chalk is governed by its discontinuity pattern rather than by intergranular flow. As can be seen from the values given in Table 11.8, the Chalk has a high porosity but when the values are compared with intergranular permeability as obtained from laboratory testing low correlation coefficients are obtained. The values of primary permeability obtained by Bell *et al.* (1990) are more or less the same as those found by Ineson (1962). Ineson quoted a range between 0.1×10^{-10} and 25×10^{-9} m/s. The values provided by Bell *et al.* (1990) are as shown in Table 11.10. The reason for the low primary permeability is the small size of the pores and more particularly that of the interconnecting throat areas.

A formation with fluid flow properties as characterized by the values given above could not produce the groundwater yields which the Chalk does. Indeed Price (1987) pointed out that if the Chalk only possessed intergranular and primary fissure component permeabilities, then its transmissivity would be around 20 m^2/day or less. Yet the yields of some larger wells indicate transmissivities in excess of 2000 m^2/day. Hence, secondary fissures, enlarged by solution, are the features which produce the high permeabilities found in many areas of the Chalk.

Chalk is subject to dissolution along discontinuities. However, subterranean solution features tend not to develop in chalk since it is usually softer than limestone and so collapses as solution occurs. Nevertheless, solution pipes and swallow holes are present in the Chalk, being commonly found near the contact of the Chalk with the overlying Tertiary and drift deposits. West and Dumbleton (1972) suggested that high concentrations of water, such as run-off from roads, can lead to the re-activation of swallow holes and the formation of small pipes within a few years. They also recorded that new swallow holes often appear at the surface without warning after periods of heavy rain or following the passage of plant across a site. Moreover voids can gradually migrate upwards through chalk due to

Table 11.8 *Some physical properties of the Chalk of Yorkshire, Norfolk and Kent (from Bell et al. 1990)*

	*Yorkshire**			*Norfolk†*				*Kent‡*
	Lower	*Middle*	*Upper*	*Lower*	*Melbourne Rock*	*Middle*	*Upper*	*Upper*
1. Specific gravity								
Maximum	2.73	2.71	2.72	2.71	2.72	2.74	2.72	2.72
Minimum	2.65	2.69	2.67	2.66	2.68	2.68	2.70	2.65
Mean	2.71	2.70	2.70	2.68	2.70	2.70	2.71	2.69
2. Dry density (Mg/m^3)§								
Maximum	2.13 (L)	2.30 (M)	2.23 (M)	2.17 (L)	2.23 (M)	1.18 (L)	1.70 (VL)	1.61 (VL)
Minimum	1.85 (L)	1.76 (VL)	1.77 (VL)	1.71 (VL)	2.04 (L)	1.62 (VL)	1.54 (VL)	1.35 (VL)
Mean	2.08 (L)	2.14 (L)	2.06 (L)	1.99 (L)	2.17 (L)	1.76 (VL)	1.61 (VL)	1.44 (VL)
3. Saturation moisture content (%)								
Maximum	17.2	19.9	22.9	21.5	11.4	24.8	28.0	37.2
Minimum	10.8	7.2	7.9	9.2	7.5	18.3	21.9	25.2
Mean	11.2	10.2	11.6	13.4	9.5	19.9	25.2	32.6
4. Effective porosity (%)§								
Maximum	30.2 VH	35.0 VH	36.4 VH	34.4 VH	27.0 H	38.2 VH	43.2 VH	45.7 VH
Minimum	17.2 H	16.2 H	17.7 H	19.9 H	16.1 H	30.2 VH	34.3 VH	29.6 H
Mean	20.6 H	21.8 H	23.9 H	26.5 H	19.8 H	34.4 VH	39.9 VH	41.7 VH
5. Dry unconfined compressive strength (MN/m^2)**								
Maximum	32.7 (MS)	36.4 (MS)	34.0 (MS)	30.5 (MS)	38.3 (MS)	25.1 (MS)	12.7 (MS)	6.2 (MS)
Minimum	19.1 (MS)	25.2 (MS)	18.1 (MS)	14.2 (MS)	22.1 (MS)	7.4 (MS)	6.9 (MW)	4.8 (W)
Mean	26.4 (MS)	30.7 (MS)	25.6 (MS)	21.0 (MS)	29.2 (MS)	13.0 (MS)	9.5 (MW)	5.5 (MW)

Property								
6. Saturated unconfined compressive strength (MN/m^2)								
Maximum	16.2	20.4	15.9	13.7	17.5	10.3	5.1	2.2
Minimum	8.6	11.7	7.4	6.2	8.9	3.1	2.8	1.4
Mean	13.7	16.8	11.9	10.7	14.3	5.8	3.6	1.7
7. Point load strength (MN/m^2)††								
Maximum	1.8 (HS)	2.1 (HS)	2.0 (HS)	1.5 (HS)	2.4 (HS)			
Minimum	0.3 (MS)	0.6 (MS)	0.2 (LS)	0.2 (LS)	0.4 (MS)			
Mean	1.4 (MS)	1.7 (HS)	1.2 (HS)	0.8 (MS)	1.7 (HS)			
8. Young's modulus (E_{t50}; GN/m^2)								
Maximum	18.4	21.7	17.1	14.1	18.9	10.4	8.2	4.6
Minimum	7.5	9.1	7.4	6.8	7.3	5.0	4.1	4.2
Mean	12.7	15.2	11.7	8.7	13.5	8.4	6.7	4.4
9. Modulus ratio (E_{t50}/UCS)‡‡								
Maximum	563 (H)	596 (H)	502 (H)	479 (M)	493 (M)	675 (H)	646 (H)	875 (H)
Minimum	393 (M)	361 (M)	409 (M)	462 (M)	330 (M)	414 (M)	594 (H)	741 (H)
Mean	481 (M)	495 (M)	457 (M)	414 (M)	462 (M)	646 (H)	705 (H)	800 (H)

* Yorkshire: Lr. – *H. subglobosus* (? = *S. gracile*) zone near Speeton; Mid – *T. lata* zone, Thornwick Bay; Up. – *M. coranguinum* zone, Selwicks Bay

† Norfolk: Lr. – *S. varians* (? = *M. mantelli*) zone, Hunstanton; M. Rock & Mid. – *T. lata* zone, Hillington; Up. – *M. coranguinum* zone, Burnham Market

‡ Kent: Up. – *M. coranguinum* zone, Northfleet

§ Dry density: VL = very low, less than 1.8 Mg/m^3; L = low 1.8 to 2.2 Mg/m^3; M = moderate, over 2.2 Mg/m^3
Porosity: H = High, 15 to 30%; VH = very high, over 30% (Anon, 1979)

** Unconfined compressive strength: W = weak, 1.25–5 MN/m^2; MW = moderately weak, 5–12.5 MN/m^2; MS = moderately strong, 12.5–50 MN/m^2 (Anon, 1977)

†† Point load strength: LS = low strength, 0.1–0.3 MN/m^2; MS = moderate strength, 0.3–1 MN/m^2; HS = high strength, 1–3 MN/m^2 (Franklin and Broch, 1972)

‡‡ Modulus ratio: H = high, over 500; M = medium, 200–500; L = low, less than 200.

Table 11.9 *Correlation between grades and the mechanical properties of Middle Chalk at Mundford (after Ward et al., 1968)**

Grade	*Description*	*Approx. range of* E(MN/m^2)	*Approx. value of* E_{dyn} MN/m^2 (*after Abbiss, 1979*)	*Range of compression wave velocities* (km/s) (*after Grainger* et al., *1973*)	*Bearing pressure causing yield* (kN/m^2)	*Creep properties*	*SPT N value* (*after Wakeling, 1970*)**	*Rock mass factor* (*after Burland and Lord 1970*)
V	Structureless melange. Unweathered and partly weathered angular chalk blocks and fragments set in a matrix of deeply weathered remoulded chalk. Bedding and jointing are absent	Below 500	Below 500	0.65–0.75	Below 200	Exhibits significant creep	Below 15	0.1
IV	Friable to rubbly chalk. Unweathered or partially weathered chalk with bedding and jointing present. Joints and small fractures closely spaced, ranging from 10–60 mm apart	500–1000	800	1.0–1.2	200–400	Exhibits significant creep	15–20	0.1–0.2

III	Rubbly to blocky chalk. Unweathered medium to hard chalk with joints 60–200 mm apart. Joints open up to 8 mm, sometimes with secondary staining and fragmentary infillings	1000–2000	4000	1.6–1.8	400–600	For pressures not exceeding 400 kN/m^2 creep is small and terminates in a few months	20–25	0.2–0.4
II	Medium hard chalk with widely spaced, closed joints. Joints more than 200 mm apart. Fractures irregularly when excavated, does not break along joints. Unweathered	2000–5000	7000	2.2–2.3	Over 1000	Negligible creep for pressure of at least 400 kN/m^2	25–35	0.6–0.8
I	Hard, brittle chalk with widely spaced, closed joints. Unweathered	Over 5000	Over 2.3	Over 10 000	Over 1000	Negligible creep for presure of at least 400 kN/m^2	Over 35	Over 0.8

* Ward *et al.* emphasized that their classfication was specifically developed for the site at Mundford and hence its application elsewhere should be made with caution

** Wakeling T. R. M. A comparison of the results of standard site investigation methods against the results of a detailed geotechnical investigation in Middle Chalk at Mundford, Norfolk. In *Site Investigations in Soils and Rocks*, British Geotechnical Society, London, 17–22 (1970)

The correlation between SPT N value and grade may be different in the Upper Chalk (see Dennehy, J. P., 'Correlation of the SPT value with chalk grade for some zones of the Upper Chalk'. *Geotechnique,* **26**, 610–14 (1976)

material collapsing. Lowering of the chalk surface beneath overlying deposits due to solution can occur, disturbing the latter deposits and lowering their degree of packing. Hence the chalk surface may be extremely irregular in places.

Chalk during cold weather may suffer frost heave (Lewis and Croney, 1965), ice lenses up to 25 mm in thickness being developed along bedding planes. Higginbottom (1965) suggested that a probable volume increase of some 20–30% of the original thickness of the ground may ultimately result.

11.5 Evaporitic rocks

Evaporitic deposits are quantitatively unimportant as sediments. They are formed by precipitation from saline waters of inland seas or lakes in arid areas.

Representative specific gravities and dry densities of gypsum, anhydrite, rock salt and potash are given in Table 11.11, as are porosity values. Anhydrite according to the classification of unconfined compressive strength (Anon, 1977) is a strong rock, gypsum and potash are moderately strong, whilst rock salt is moderately weak (Table 11.11). Values of Young's modulus are also given in Table 11.11, from which it can be ascertained that gypsum and anhydrite have high values of modulus ratio whilst potash and rock salt have medium values. Evaporitic rocks exhibit varying degrees of plastic deformation before failing. For example, in rock salt the yield strength may be as little as one-tenth the ultimate compressive strength, whereas anhydrite undergoes comparatively little plastic deformation prior to rupture. Creep may account for anything between 20 and 60% of the strain at failure when these evaporitic rocks are subjected to incremental creep tests (Bell, 1981b). Rock salt is most prone to creep.

Langer (1982) has recently proposed a rheological model for the mechanical behaviour of rock salt with time. Initially primary creep occurs as salt is subjected to loading, the creep rate decreasing with time. During the stage of secondary creep, the creep rate is related to the amount of stress and the temperature conditions (Figure 11.7). Under a constant load, failure occurs when the stress, creep rate and temperature are combined in a given manner. The resistance to failure increases with increasing confining conditions.

Justo and Zapico (1975) recorded that the amount of settlement which occurred when gypsum was subjected to plate load testing, the maximum load being 1.2 MN/m^2, was negligible.

Gypsum is more readily soluble than limestone, 2100 mg/l can be dissolved in non-saline waters as compared with 400 mg/l. Sinkholes and caverns can therefore develop in thick beds of gypsum (Eck and Redfield, 1965) more rapidly than they can in limestone. Indeed in the United States such features have been known to form within a few years where beds of gypsum are located beneath dams. Extensive surface cracking and subsidence has occurred in certain areas of Oklahoma and New Mexico due to the collapse of cavernous gypsum (Brune, 1965). The problem is accentuated by the fact that gypsum is weaker than limestone and therefore collapses more readily. Yuzer (1982) described karstic features found in the Sivas area of Turkey where some 180 m of gypsum beds occur in Oligocene-Miocene deposits.

Table 11.10 *Permeability values of chalk ($\times 10^{-9}$ m/s)*

	Yorkshire Lower Chalk	*Norfolk Middle Chalk*	*Kent Upper Chalk*
Maximum	1.2	2.2	37.0
Minimum	0.3	0.5	13.9
Mean	0.9	1.4	27.7

Table 11.11 *Some physical properties of evaporitic rocks*

	Gypsum (Sherburn in Elmet)	*Anhydrite (Sandwith)*	*Rock salt (Winsford)*	*Potash (Loftus)*
Relative density	2.36	2.93	2.2	2.05
Dry density (Mg/m^3)	2.19	2.82	2.09	1.98
Porosity (%)	4.6	2.9	4.8	5.1
Unconfined compressive strength (MN/m^2)	27.5	97.5	11.7	25.8
Point load strength (MN/m^2)	2.1	3.7	0.3	0.6
Scleroscope hardness	27	38	12	9
Schmidt hardness	25	40	8	11
Young's modulus ($\times 10^3$ MN/m^2)	24.8	63.9	3.8	7.9
Permeability ($\times 10^{-10}$ m/s)	6.2	0.3	–	–

Kendal and Wroot (1924) quoted vivid, but highly exaggerated, accounts of subsidences which occurred in the Ripon district, Yorkshire, in the eighteenth and nineteenth centuries due to the solution of gypsum. They wrote that wherever beds of gypsum approach the surface their presence can be traced by broad funnel-shaped craters formed by the collapse of overlying marl into areas from which gypsum has been removed by solution. Apparently these craters only took a matter of minutes to appear at the surface. However, where gypsum is effectively sealed from the ingress of water by overlying impermeable strata such as marl, dissolution does not occur (Redfield, 1968).

The solution rate of gypsum or anhydrite is principally controlled by the area of their surface in contact with water and the flow velocity of water associated with a unit area of the material. Hence the amount of fissuring in a rock mass, and whether it is enclosed by permeable or impermeable beds, is most important. Solution also depends on the sub-saturation concentration of calcium sulphate in solution. According to James and Lupton (1978) the concentration dependence for gypsum is linear whilst that for anhydrite is a square law. The salinity of the water also is influential. For example, the rates of solution of gypsum and anhydrite are increased by the presence of sodium chloride, carbonate and carbon dioxide in solution. It is therefore important to know the chemical composition of the groundwater.

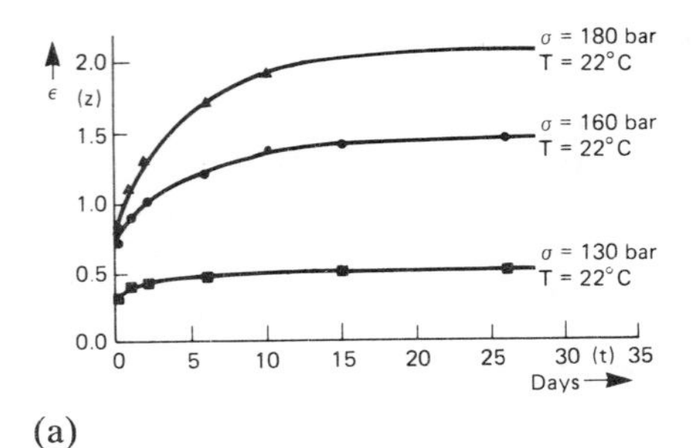

(a)

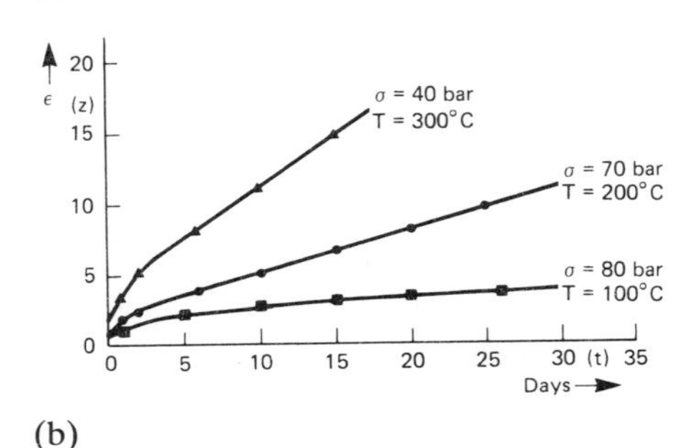

(b)

Figure 11.8 *Creep curves for rock salt* (a) *primary creep,* (b) *secondary creep (after Langer, 1982.*

Massive deposits of gypsum are usually less dangerous than those of anhydrite because gypsum tends to dissolve in a steady manner forming caverns or causing progressive settlements. For instance, if small fissures occur at less than 1 m intervals solution usually takes place by removal of gypsum as a front moving 'downstream' at less than 0.01 m/year. However, James and Lupton (1978) showed that if the following conditions were met:

(1) the rock temperature was 10°C,
(2) the water involved contained no dissolved salts, and
(3) a hydraulic gradient of 0.2 was imposed,

then a fissure, 0.2 mm in width and 100 m in length, in massive gypsum, would, in 100 years, have widened by solution so that a block 1 m^3 in size could be accommodated in the entrance to the fissure. In other words a cavern would be formed. If the initial width of the fissure exceeds 0.6 mm, large caverns would form and a runaway situation would develop in a very short time. In long fissures the hydraulic gradient is low and the rate of flow is reduced so that solutions become saturated and little or no material is removed. Indeed James and Lupton implied that a flow rate of 10^{-3} M/s was rather critical in that if it was exceeded, extensive solution of gypsum could take place. Solution of massive gypsum is not likely to give rise to an accelerating deterioration in a foundation if precautions such as grouting are taken to keep seepage velocities low.

Massive anhydrite can be dissolved to produce uncontrollable runaway situations in which seepage flow rates increase in a rapidly accelerating manner. Even small fissures in massive anhydrite can prove dangerous. If anhydrite is taken in the above example not only is a cavern formed, but the fissure is enlarged as a long tapering section. Within about 13 years the flow rate increases to a runaway situation. However, if the fissure is 0.1 mm in width then the solution becomes supersaturated with calcium sulphate and gypsum is precipitated. This seals the outlet from the fissure and from that moment any anhydrite in contact with the water is hydrated to form gypsum. Accordingly 0.1 mm width seems to be a critical fissure size in anhydrite.

If soluble minerals occur in particulate form in the ground then their removal by solution can give rise to significant settlements. In such situations the width of the solution zone and its rate of progress are obviously important as far as the location of hydraulic structures are concerned (James and Kirkpatrick, 1980; Table 11.7). Anhydrite is less likely to undergo catastrophic solution in a fragmented or particulate form than gypsum. James and Lupton (1978) have provided equations for the determination of these two factors in gypsiferous deposits. Again the solubility of the mineral and its solution rate constant are the most important parameters. Of lesser consequence is information relating to the volumetric proportion of soluble minerals in the ground and details relating to their size, distribution and shape. This does not mean to say that such data should not be obtained, for they obviously aid the determination of the amount of settlement which might take place.

Another point which should be borne in mind, and this particularly applies to conglomerates cemented with soluble material, is that when this is removed by solution the rock is reduced greatly in strength. A classic example of this is associated with the failure of the St Francis Dam in California in 1928. One of its abutments was founded in conglomerate cemented with gypsum which was gradually dissolved, the rock losing strength, and ultimately the abutment failed.

On the other hand anhydrite on contact with water may become hydrated to form gypsum. In so doing there is a volume increase of between 30 and 58% which exerts pressures that have been variously estimated between 2 and 70 MN/m^2 (see Chapter 9). However, Yuzer (1982) claimed that the theoretical maximum swelling pressure is not a realistic value and maintained that such swelling pressures are commonly between 1 and 8 MN/m^2 and rarely exceed 12 MN/m^2. It is believed that no great length of time is required to bring about such hydration.

Salt is even more soluble than gypsum and the evidence of slumping, brecciation and collapse structures in rocks which overlie saliferous strata bear witness to the fact that salt has gone into solution in past geological times. It is generally believed, however, that in areas underlain by saliferous beds, measurable surface subsidence is unlikely to occur except where salt is being extracted (Bell, 1975). Perhaps this is because equilibrium has been attained between the supply of unsaturated groundwater and the salt available for solution. Exceptionally cases have been recorded of rapid subsidence, such as the 'Meade salt sink' in Kansas. Johnson

(1901) explained its formation as due to solution of beds of salt located at depth. This area of water, about 60 m in diameter, formed as a result of rapid subsidence in March 1879.

References

Abbiss, C. P. (1979) 'A comparison of the stiffness of the Chalk at Mundford from a seismic survey and large-scale tank test', *Geotechnique,* **29**, 461–8.

Al-Jassar, S. H. and Hawkins, A. B. (1979). 'Geotechnical properties of the Carboniferous Limestone of the Bristol area', *Proc. 4th Int. Cong. Rock Mech.* (*ISRM*), *Montreux*, A. A. Balkema, Rotterdam, 3–14.

Anon (1970). 'Working Party Report on the Logging of Cores for Engineering Purposes', *Q. J. Engg Geol.*, **3**, 1–24.

Anon (1977). 'The description of rock masses for engineering purposes', Working Party Report, *Quarterly J. Engg Geol.,* **10**, 355–88.

Anon (1979). 'Classification of rocks and soils for engineering geological mapping', *Bull. Int. Assoc. Engg Geol.,* No 19, 364–71.

Attewell, P. B. (1971). 'Geotechnical properties of the Great Limestone in northern England', *Engg Geol.*, **5**, 89–112.

Badger, C. W., Cummings, A. D. and Whitmore, R. L. (1956). 'The disintegration of shale', *J. Inst. Fuel,* **29**, 417–23.

Bell, F. G. (1975). 'Salt and subsidence in Cheshire, England', *Engg Geol.,* **9**, 237–47.

Bell, F. G. (1977). 'A note on the geotechnical properties of chalk', *Engg Geol.,* **11**, 221–2.

Bell, F. G. (1978). 'The physical and mechanical properties of the Fell Sandstone', *Engg Geol.,* **12**, 1–29.

Bell, F. G. (1981a). 'A survey of the physical properties of some carbonate rocks', *Bull. Int. Ass. Engg. Geol.,* No. 24, 105–110.

Bell, F. G. (1981b). 'Geotechnical properties of evaporites', *Bull. Int. Ass. Engg Geol.,* No. 24, 137–44.

Bell, F. G. and Culshaw, M. G. (1992). 'A survey of some relatively weak sandstones of Triassic age', in *Engineering Geology of Weak Rocks*, Cripps, J. C., Culshaw, M. G., Hencher, S. R., Moon, C. F. and Coulthard, J. M. (eds.), Engineering Geology Special Publication No. 8, A. A. Balkema, Rotterdam.

Bell, F. G., Cripps, J. C., Culshaw, M. G. and Edmunds, C. N. (1990). 'Chalk fabric and its relation to certain geotechnical properties', *Proc. Int. Symp. on Chalk*, Brighton, Inst. Civ. Engrs/Inst. Geologists, 187–194.

Brune, G. (1965). 'Anhydrite and gypsum problems in engineering geology', *Bull. Ass. Engg Geologists,* **3**, 26–38.

Burland, J. B. and Lord, J. A. (1970). 'The load deformation behaviour of Middle Chalk at Mundford, Norfolk: a comparison between full-scale performance and *in situ* and laboratory measurements', in *In Situ Investigations in Soils and Rocks*, British Geotechnical Society, London, 3–16.

Burland, J. B., Kee, R. and Burford, D. (1974). 'Short-term settlement of a five storey building on soft chalk', in *Settlement of Structures*, British Geotechnical Society, Pentech Press, London, 259–65.

Burnett, A. D. and Epps, R. J. (1979). 'The engineering geological description of the carbonate suite rocks and soils', *Ground Engg,* **12**, No. 2, 41–8.

Burwell, E. B. (1950). 'Geology in dam construction, Part I', in *Applications of Geology to Engineering Practice*, Berkey Volume, Geol. Soc. Am., 11–31.

Carter, P. G. and Mallard, D. J. (1974). 'A study of the strength, compressibility and density trends within the Chalk of south-east England', *Q. J. Engg Geol.,* **7**, 43–56.

Chandler, R. J. (1969). 'The effect of weathering on the shear strength properties of Keupar Marl', *Geotechnique,* **19**, 321–34.

Chappell, B. A. (1974). 'Deformational response of differently shaped and sized test pieces of shale rock', *Int. J. Rock Mech. Min. Sci.,* **11**, 21–8.

Clayton, C. R. I. (1983). 'The influence of diagenesis on some index properties of chalk in England', *Geotechnique,* **33**, 225–41.

Davis, A. G. (1967). 'On the mineralogy and phase equilibrium of Keuper Marl', *Q. J. Engg Geol.,* **1**, 25–46.

Davis, A. G. (1968). 'The structure of Keuper Marl', *Q. J. Engg Geol.,* **1**, 145–53.

De Graft-Johnson, J. W. S., Bhatia, H. S. and Yeboa, S. L. (1973). 'Geotechnical properties of Accra Shales', *Proc. 8th Int. Conf. Soil Mech. Foundation Engg*, Moscow, **2**, 97–104.

Dobereiner, L. and De Freitas, M. H. (1986). 'Geotechnical properties of weak sandstones', *Geotechnique,* **36**, 79–94.

Dumbleton, M. J. (1967). 'Origin and mineralogy of African red clays and Keuper Marl', *Q. J. Engg Geol.,* **1**, 39–46.

Eck, W. and Redfield, R. C. (1965). 'Engineering geology problems at Sanford Dam, Texas', *Bull. Assoc. Engg Geologists*, **3**, 15–25.

Elliott, G. M. and Brown, E. T. (1985). 'Yield of a soft, high porosity rock', *Geotechnique,* **36**, 79–94.

Franklin, J. A. and Broch, E. (1972). 'The point load strength test', *Int. J. Rock Mech. Min. Sci.,* **9**, 669–97.

Grainger, P. (1984). 'The classification of mudrocks for engineering purposes. *Q. J. Engg Geol.,* **17**, 381–7.

Grainger, P., McCann, D. M. and Gallois, R. W. (1973). 'The application of seismic refraction to the study of fracturing in the Middle Chalk at Mundford. Norfolk', *Geotechnique,* **23**, 219–32.

Higginbottom, I. E. (1965). 'The engineering geology of the Chalk', *Proc. Symp. on Chalk in Earthworks*, Inst. Civil Engrs, London, 1–14.

Ineson, J. (1962). 'A hydrogeological study of the permeability of chalk', *J. Inst. Water Engrs,* **16**, 255–86.

Jaeger, C. (1963). 'The Malpasset Report', *Water Power,* **15**, 55–61.

James, A. N. and Kirkpatrick, I. M. (1980). 'Design of foundations of dams containing soluble rocks and soils', *Q. J. Engg Geol.,* **13**, 189–98.

James, A. N. and Lupton, A. R. R. (1978). 'Gypsum and anhydrite in foundations of hydraulic structures', *Geotechnique,* **28**, 249–72.

Johnson, W. D. (1901). 'The high plains and their utilization', *U.S. Geol. Surv., 21st Annual Rep.*, Part 4, 601–741.

Jumikis, A. R. (1965). 'Some engineering aspects of the Brunswick Shale', *Proc. 6th Int. Conf. Soil Mech. Foundation Engg*, Montreal, **2**, 99–102.

Justo, J. L. and Zapico, L. (1975). 'Compression between measured and estimated settlements at two Spanish aqueducts on gypsum rock' in *Settlement of Structures*, British Geotechnical Society, Pentech Press, London, 266–74.

Kendal, P. F. and Wroot, H. E. (1924). *The Geology of Yorkshire*, Printed privately.

Kennard, M. F. and Knill, J. L. (1968). 'Reservoirs on limestone, with particular reference to the Cow Green Scheme', *J. Inst. Water Engrs,* **23**, 87–113.

Kennard, M. F., Knill, J. L. and Vaughan, P. R. (1967). 'The geotechnical properties and behaviour of Carboniferous shale at Balderhead Dam', *Q. J. Engg Geol.,* **1**, 3–24.

Kolbuszewski, J., Birch, N. and Shojobi, J. O. (1965). 'Keuper Marl research', *Proc. 6th Int. Conf. Soil Mech. Found. Engg*, Montreal, **1**, 59–63.

Langer, M. (1982). 'Geotechnical investigation methods for rock salt', *Bull. Int. Ass. Engg Geol.,* No. 25, 155–64.

Lees, G. (1965). 'Geology of the Keuper Marl', *Proc. Geol. Soc. London*, No. 1621, 46.

Lewis, W. A. and Croney, D. (1965). 'The properties of chalk in relation to road foundations and pavements', *Proc. Symp. on Chalk in Earthworks*, Inst. Civil Engrs, London, 27–42.

Lumb, P. (1983). 'Engineering properties of fresh and decomposed igneous rocks from Hong Kong', *Engg Geol.,* **19**, 81–94.

Mead, W. J. (1936). 'Engineering geology of damsites', *Trans. 2nd Int. Cong. Large Dams*, Washington, D.C., **4**, 183–98.

Meigh, A. C. (1976). 'The Triassic rocks, with particular reference to predicted and observed performance of some major foundations', *Geotechnique,* **26**, 391–452.

Meigh, A. C. and Early, K. R. (1957). 'Some physical and engineering properties of chalk', *Proc. 4th Int. Conf. Soil Mech. Found. Engg*, London, **1**, 68–73.

Moore, J. F. A. (1974). 'A long-term plate test on Bunter Sandstone', *Proc. 3rd Int. Cong. Rock Mech.*, Denver, **2**, 724–32.

Moore, J. F. A. and Jones, C. W. (1975). '*In situ* deformation of Bunter Sandstone' in *Settlement of Structures*, British Geotechnical Society, Pentech Press, London, 311–19.

Morgenstern, N. R. and Eigenbrod, K. D. (1974). 'Classification of argillaceous soils and rocks', *Proc. ASCE. J. Geot. Engg Div.*, GT10, **100**, 1137–56.

Nakano, P. (1967). 'On weathering and change of properties of Tertiary mudstone related to landslide', *Soil and Foundations,* **7**, 1–14.

Oddsson, B. (1981). 'Engineering properties of some weak Icelandic volcanic rocks', *Proc. Int. Symp. on Weak Rock – Soft, Fractured and Weathered Rock*, Tokyo, Akai, K., Hayashi, M. and Nishimatsu, Y. (eds.), A. A. Balkema, Rotterdam, **1**, 197–204.

Penner, E., Eden, W. J. and Gillott, J. E. (1973). 'Floor heave due to biochemical weathering of shale', *Proc. 8th Int. Conf. Soil Mech. Found. Engg*, Moscow, **2**, 151–8.

Price, M. (1987). 'Fluid flow in the Chalk of England', in *Fluid Flow in Sedimentary Basins and Aquifers*, Special Publication No. 34, Goff, J. C. and Williams, B. P. J. (eds), Geological Society, London, 141–56.

Price, M., Bird, M. J. and Forster, S. S. D. (1976). 'Chalk pore size measurements and their significance', *Water Services,* **80**, 596–600.

Price, N. J. (1960). 'The compressive strength of Coal Measures rocks', *Colliery Guardian,* **199**, 283–92.

Redfield, R. C. (1968). 'Brantley reservoir site – an investigation of evaporite and carbonate facies', *Bull. Ass. Engg Geologists,* **6**, 14–30.

Russell, D. J. (1981). 'Controls on shale durability: the response of two Ordovician shales in the slake durability test', *Can. Geot. J.,* **19**, 1–13.

Sherwood, P. T. (1967). 'Classification tests on African red clays and Keuper Marl', *Quarterly J. Engg Geol.,* **1**, 47–56.

Singh, D. P., Nath, R. and Singh, J. B. (Aug. 1978). 'A comparative study of anisotropy of Singrauli Coal and Chunar Sandstone', *J. Mines Metals Fuels,* **26**, No. 8, 283–90.

Smith, C. K. and Redlinger, J. F. (1953). 'Soil properties of the Fort Union clay shale', *Proc. 3rd Int. Conf. Soil Mech. Foundation Engg*, Zurich, **1**, 56–61.

Sowers, G. F. (1975). 'Failures in limestones in the humid subtropics', *Proc. ASCE, J. Geot. Engg Div.,* **101**, GT8, 771–87.

Terzaghi, K. (1946). 'Introduction to tunnel geology', in *Rock Tunnelling with Steel Supports*, Procter, R. and White, T. (eds.), Commercial Shearing and Stamping Co., Youngstown, Ohio, 17–99.

Terzaghi, K. (1962). 'Dam foundations on sheeted rock', *Geotechnique,* **12**, 199–208.

Underwood, L. B. (1967). 'Classification and identification of shales', *Proc. ASCE, Soil Mech. Foundation Engg Div.,* **93**, SM6, 97–116.

Underwood, L. B. Thorfinnson, S. T. and Black, W. T. (1964). 'Rebound in redesign of the Oake Dam hydraulic structures', *Proc. ASCE, J. Soil Mech. Foundation Engg Div.*, **90**, SM2, Paper 3830, 859–68.

Venter, J. P. (1981). 'The behaviour of some South African mudrocks due to temperature and humidity changes, with particular reference to moisture content and volume changes', *Proc. Int. Symp. on Weak Rock – Soft, Fractured and Weathered Rock*, Tokyo, Akai, K., Hayashi, M. and Nishimatsu, Y. (eds.), A. A. Balkema, Rotterdam, **1**, 205–12.

Wakeling, T. R. M. (1965). 'Foundations on chalk', *Proc. Symp. on Chalk in Earthworks*, Inst. Civil Engrs, London, 15–23.

Ward, W. H., Burland, J. B. and Gallois, R. W. (1968). 'Geotechnical assessment of a site at Mundford, Norfolk, for a large proton accelerator', *Geotechnique,* **18**, 399–431.

West, G. (Sept. 1979). 'Strength properties of Bunter Sandstone', *Tunnels and Tunnelling,* **7**, No. 7, 27–9.

West, G. and Dumbleton, M. J. (1972). 'Some observations on swallow holes and mines in the Chalk', *Q. J. Engg Geol.*, **5**, 171–8.

Wichter, L. (1979). 'On the geotechnical properties of a Jurassic clay shale', *Proc. 4th Int. Cong. Rock Mech. (ISRM), Montreux,* **1**, A. A. Balkema, Rotterdam, 319–28.

Wissler, T. M. and Simmons, G. (1985). 'The physical properties of a set of sandstones – Part II. Permanent and elastic strains during hydrostatic compression to 200 MPa', *Int. J. Rock Mech. Min. Sci. Geomech. Abstr.*, **22**, 394–406.

Yuzer, E. (1982). 'Engineering properties and evaporitic formations of Turkey', *Bull. Int. Ass. Engg Geol.*, No. 25, 107–10.

Zaruba, Q. and Bukovansky, M. (1965). 'Mechanical properties of Ordovician shales of Central Bohemia', *Proc. 6th Int. Conf. Soil Mech. Foundation Engg*, Montreal, **3**, 421–4.

Chapter 12

Subsurface water and ground conditions

12.1 The origin and occurrence of groundwater

The principal source of groundwater is meteoric water, that is, precipitation (rain, sleet, snow and hail). However, two other sources are very occasionally of some consequence. These are juvenile water and connate water. The former is derived from magmatic sources whilst the latter represents the water in which sediments were deposited. This was trapped in the pore spaces of sedimentary rocks as they were formed and has not been expelled.

The amount of water that infiltrates into the ground depends upon how precipitation is dispersed, namely, on what proportions are assigned to immediate run-off and to evapotranspiration, the remainder constituting the proportion allotted to infiltration/percolation. Infiltration refers to the seepage of surface water into the ground, percolation being its subsequent movement, under the influence of gravity, to the zone of saturation. In reality one cannot be separated from the other.

The pores within the zone of saturation are filled with water, generally referred to as *phreatic water*. The upper surface of this zone is therefore known as the phreatic surface but is more commonly termed the *water table*. Above the zone of saturation is the zone of aeration in which both air and water occupy the pores. The water in the zone of aeration is commonly referred to as *vadose water*. Meinzer (1942) divided this zone into three belts, those of soil water, the intermediate belt and the capillary fringe. The uppermost or soil water belt discharges water into the atmosphere in perceptible quantities by evapotranspiration. In the capillary fringe, which occurs immediately above the water table, water is held in the pores by capillary action. An intermediate belt occurs when the water table is far enough below the surface for the soil water belt not to extend down to the capillary fringe.

The geological factors which influence percolation not only vary from one rock outcrop to another but may do so within the same one. This, together with the fact that rain does not fall evenly over a given area, means that the contribution to the zone of saturation is variable. This in turn influences the position of the water table, as do the points of discharge. A rise in the water table as a response to percolation is partly controlled by the rate at which water can drain from the area of recharge. Accordingly it tends to be greatest in areas of low transmissivity (see below).

Mounds and ridges form in the water table under the areas of greatest recharge. Superimpose upon this the influence of water draining from lakes, streams and wells, and it can be seen that a water table is continually adjusting towards equilibrium. Because of the low flow rates in most rocks this equilibrium is rarely, if ever, attained before another disturbance occurs.

As pointed out, the water table fluctuates in position, particularly in those climates where there are marked seasonal changes in rainfall. Thus permanent and intermittent water tables can be distinguished, the former marking the level beneath which the water table does not sink whilst the latter is an expression of the fluctuation. Usually water tables fluctuate within the lower and upper limits rather than between them, especially in humid regions, since the periods between successive recharges are small. The position at which the water table intersects the surface is termed the spring line. Intermittent and permanent springs similarly can be distinguished.

A perched water table is one which forms above a discontinuous impermeable layer such as a lens of clay in a formation of sand, the clay impounding a water mound.

An aquifer is the term given to a rock or soil mass which only contains water, but from which water can be readily abstracted in significant quantities. The ability of an aquifer to transmit water is governed by its permeability. Indeed the permeability of an aquifer usually is in excess of 10^{-5} m/s.

By contrast, a formation with a permeability of less than 10^{-9} m/s is one which is regarded as impermeable and is referred to as an aquiclude. For example, clays and shales are aquicludes. Even when such rocks are saturated they tend to impede the flow of water through stratal sequences.

An aquitard is a formation which transmits water at a very slow rate but which, over a large area of contact, may permit the passage of large amounts of water between adjacent aquifers which it separates. Sandy clays provide an example.

An aquifer is described as unconfined when the water table is open to the atmosphere, that is, the aquifer is not overlain by material of lower permeability. Conversely a confined aquifer is one which is overlain by impermeable rocks. Very often the water in a confined aquifer is under piezometric pressure, that is, there is an excess of pressure sufficient to raise the water above the base of the overlying bed when the aquifer is penetrated by a well. Piezometric pressures are developed when the buried upper surface of a confined aquifer is lower than the water table in the aquifer at its recharge area.

12.2 Capillary movement in soil

Capillary movement in a soil refers to the movement of moisture through the minute pores between the soil particles which act as capillaries. It takes place as a consequence of surface tension, therefore moisture can rise from the water table. This movement, however, can occur in any direction, not just vertically upwards. It occurs whenever evaporation takes place from the surface of the soil, thus exerting a 'surface tension pull' on the moisture, the forces of surface tension increasing as evaporation proceeds. Accordingly capillary moisture is in hydraulic continuity with the water table and is raised against the force of gravity, the degree of saturation decreasing from the water table upwards. Equilibrium is attained when the forces of gravity and surface tension are balanced.

The boundary separating capillary moisture from the gravitational water in the zone of saturation is, as would be expected, ill-defined and cannot be determined accurately. That zone immediately above the water table which is saturated with capillary moisture is referred to as the closed capillary fringe, whilst above this, air and capillary moisture exist together in the pores of the open capillary fringe. The depth of the capillary fringe is largely dependent upon the particle size distribution and density of the soil mass, which in turn influence pore size. In other words the smaller the pore size, the greater is the depth. For example, capillary moisture can rise to great heights in clay soils (Table 12.1) but the movement is very slow. In soils which are poorly graded the height of the capillary fringe generally varies whereas in uniformly textured soils it attains roughly the same height. Where the water table is at shallow depth and the maximum capillary rise is large, moisture is continually attracted from the water table, due to evaporation from the surface, so that the uppermost soil is near saturation. For instance, under normal conditions peat deposits may be assumed to be within the zone of capillary saturation. This means that the height to which the water can rise in peat by capillary action is greater than the depth below ground to which the water table can be reduced by drainage. The coarse fibrous type of peat, containing appreciable sphagnum, may be an exception.

Drainage of capillary moisture cannot be effected by the installation of a drainage system within the capillary fringe as only that moisture in excess of that retained by surface tension can be removed, but it can be lowered by lowering the water table. The capillary ascent, however, can be interrupted by the installation of impermeable membranes or layers of coarse aggregate. These two methods can be used in the construction of embankments, or more simply the height of the fill can be raised.

Below the water table the water contained in the pores is under normal hydrostatic load, the pressure increasing with depth. Because these pressures exceed atmospheric pressure they are designated positive pressures. On the other hand the pressures existing in the capillary zone are less than atmospheric and so are termed negative pressures. Thus the water table is usually regarded as a datum of zero pressure between the positive pore pressure below and the negative above.

At each point where moisture menisci are in contact with soil particles the forces of surface tension are responsible for the development of capillary or suction pressure (Table 12.1). The air and water interfaces move into the smaller pores. In so doing the radii of curvature of the interfaces decrease and the soil suction increases. Hence the drier the soil, the higher is the soil suction.

Soil suction is a negative pressure and indicates the height to which a column of water could rise due to such suction. Since this height or pressure may be very large, a logarithmic scale has been adopted to express the relationship between soil suction and moisture content, the latter is referred to as the pF value (Table 12.2).

Table 12.1 *Capillary rises and pressures in soils (after Jumikis, 1968)*

Soil	*Capillary rise* (mm)	*Capillary pressure* (kN/m)
Fine gravel	Up to 100	Up to 1.0
Coarse sand	100–150	1.0–1.5
Medium sand	150–300	1.5–3.0
Fine sand	300–1000	3.0–10.0
Silt	1000–10 000	10.0–100.0
Clay	Over 10 000	Over 100.0

Table 12.2 *Suction pressure and pF values*

pF value	*Equivalent suction*	
	(mm water)	(kN/M^2)
0	10	0.1
1	100	1.0
2	1000	10.0
3	10 000	100.0
4	100 000	1000.0
5	1 000 000	10 000.0

Soil suction tends to force soil particles together and these compressive stresses contribute towards the strength and stability of the soil. There is a particular suction pressure for a particular moisture content in a given soil. In fact as a clay soil dries out the soil suction may increase to the order of several megapascals. However, the strength of a soil attributable to soil suction is only temporary and is destroyed upon saturation. At that point soil suction is zero.

12.3 Porosity and permeability

Porosity and permeability are the two most important factors governing the accumulation, migration and distribution of groundwater. However, both may change within a rock or soil mass in the course of its geological evolution. Furthermore it is not uncommon to find variations in both porosity and permeability per metre of depth beneath the ground surface.

12.3.1 Porosity

The porosity of a rock can be defined as the percentage pore space within a given volume. Total or absolute porosity is a measure of the total void volume and is the excess of bulk volume over grain volume per unit of bulk volume. It is usually determined as the excess of grain density (the same as specific gravity) over dry density, per unit of grain density, and can be obtained from the following expression

$$\text{Absolute density} = \left(1 - \frac{\text{Dry density}}{\text{Grain density}}\right) \times 100 \tag{12.1}$$

The effective, apparent or net porosity is a measure of the effective void volume of a porous medium and is determined as the excess of bulk volume overgrain volume and occluded pore volume. It may be regarded as the pore space from which water can be removed.

The factors affecting the porosity of a rock include particle size distribution, sorting, grain shape, fabric, degree of compaction and cementation, solution effects, and lastly mineralogical composition, particularly the presence of clay particles. In experiments with packing arrangements, Frazer (1935) found that for a given mode of packing of equal-sized spheres porosity was independent of size. Rhombohedral packing (Figure 12.1) was the tightest form and produced a porosity of 25.9% whilst the loosest type of packing gave rise to a porosity of 87.5%.

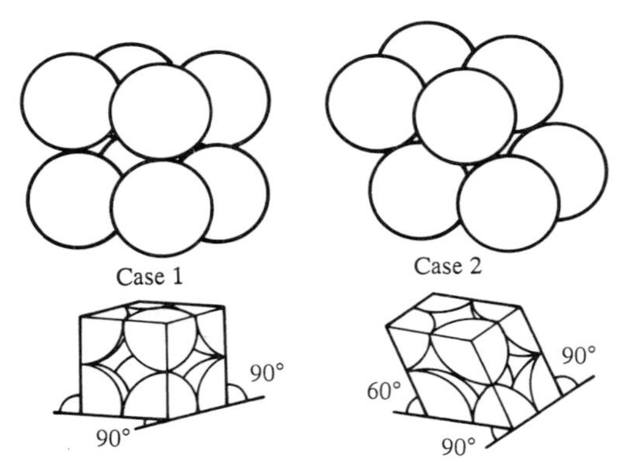

Figure 12.1 *Packing of spherical grains. Unit cells of cubic and rhombohedral packing (after Frazer, 1935)*

However, in natural assemblages as grain sizes decrease so friction, adhesion and bridging become more important because of the higher ratio of surface area to volume. Therefore as the grain size decreases the porosity increases. For example, in coarse sands it ranges from 39 to 41%, medium 41 to 48% and fine 44 to 49%. Whether the grain size is uniform or non-uniform is of fundamental importance with respect to porosity. The highest porosity is commonly attained when all the grains are the same size. The addition of grains of different size to such an assemblage lowers its porosity and this is, within certain limits, directly proportional to the amount added. As would be expected the skewness of the size distribution influences porosity. For example, sands with a negative skewness, that is, an excess of coarse particles in relation to fines, tend to have higher porosities.

Irregularities in grain shape result in a larger possible range of porosity, as irregular forms may theoretically be packed either more tightly or more loosely than spheres. Similarly angular grains may either cause an increase or a decrease in porosity, although the only type of angularity which has been found experimentally to produce a decrease is that in which the grains are mildly and uniformly disc-shaped.

After a sediment has been buried and indurated several additional factors help determine its porosity. The chief amongst these are closer spacing of grains, deformation and granulation of grains, recrystallization, secondary growth of minerals, cementation and, in some cases, dissolution. For instance, when chemical cements are present in sandstones in large amounts their influence on porosity is dominant and masks the control of other factors. Thus, two types of porosity may be distinguished, original and secondary. Orginal porosity is an inherent characteristic of the rock in that it was determined at the time the rock was formed. The process by which a given sediment has accumulated affects its porosity in two ways. First, the nature and variety of the materials deposited affects the entire deposit by controlling the range and uniformity of the sizes present, as well as their degree of rounding. Second, is by the manner in which the material is packed. Hence the original porosity results from the physical impossibility of packing grains in such a way as to exclude interstitial voids of a conjugate nature. On the other hand, secondary porosity results from later changes undergone by the rock which may either increase or decrease its original porosity.

12.3.2 Specific retention and specific yield

Even though a rock or soil may be saturated, only a certain proportion of water can be removed by drainage under gravity or pumping, the remainder being held in place by capillary or molecular forces. The ratio of the volume of water retained, V_{wr}, to that of the total volume of rock or soil, V, expressed as a percentage, is referred to as the specific retention, S_{re}:

$$S_{re} = V_{wr}/V \times 100 \qquad (12.2)$$

The amount of water retained varies directly in accordance with the surface area of the pores and indirectly with regard to the pore space. The specific surface of a particle is governed by its size and shape. For example, particles of clay have far larger specific surfaces than do those of sand. As an illustration, a grain of sand, 1 mm in diameter, has a specific surface of about 0.002 m^2/g, compared with kaolinite, which varies from approximately 10 to 20 m^2/g. Hence clays have a much higher specific retention than sands (Figure 12.2).

The specific yield, S_y, of a rock or soil refers to its water yielding capacity attributable to gravity drainage as occurs when the water table declines. It was defined by Meinzer (1942) as the ratio of the volume of water, after saturation, that can be drained by gravity, V_{wd}, to the total volume of the aquifer, expressed as a percentage; hence

$$S_y = V_{wd}/V \times 100 \tag{12.3}$$

The specific yield plus the specific retention is equal to the porosity. The relationship between the specific yield and particle size distribution is shown in Figure 12.2. In soils the specific yield tends to decrease as the coefficient of uniformity increases. Examples of the specific yield of some common types of soil and rock are given in Table 12.3 (it must be appreciated that individual values of specific yield can vary considerably from those quoted).

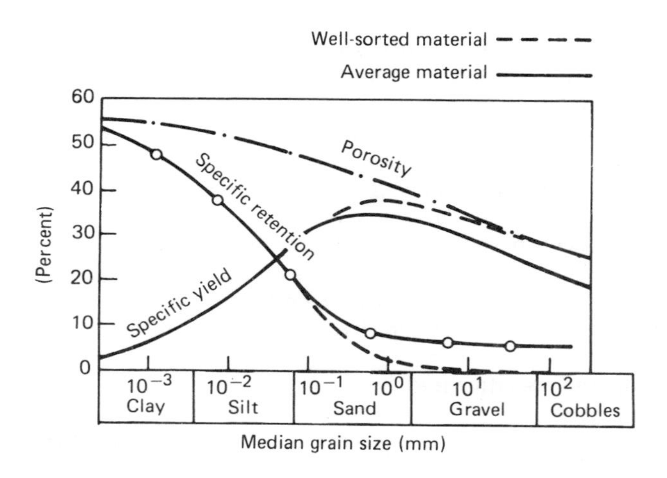

Figure 12.2 *Relationship between grain size, porosity, specific retention and specific yield (from Bear, 1979)*

12.3.3 Permeability

Permeability may be defined as the ability of a rock to allow the passage of fluids into or through it without impairing its structure. In ordinary hydraulic usage a substance is termed permeable when it permits the passage of a measurable quantity of fluid in a finite period of time and impermeable when the rate at which it transmits that fluid is slow enough to be negligible under existing temperature–pressure conditions (Table 12.4). The permeability of a particular material is defined by its coefficient of permeability. The transmissivity or flow in m^3/day through a section of aquifer 1 m wide under a hydraulic gradient of unity is sometimes used as a convenient quantity in the calculation of groundwater flow

Table 12.3 *Some examples of specific yield*

Material	*Specific yield (%)*
Gravel	15–30
Sand	10–30
Dune sand	25–35
Sand and gravel	15–25
Loess	15–20
Silt	5–10
Clay	1–5
Till (silty)	4–7
Till (sandy)	12–18
Sandstone	5–25
Limestone	0.5–10
Shale	0.5–5

instead of the coefficient of permeability. The transmissivity, T, and coefficient of permeability, k, are related to each other as follows:

$$T = kH \tag{12.4}$$

where H is the saturated thickness of the aquifer.

The flow through a unit cross section of material is modified by temperature, hydraulic gradient and the hydraulic conductivity. The latter is affected by the uniformity and range of grain size, shape of the grains, stratification, the amount of

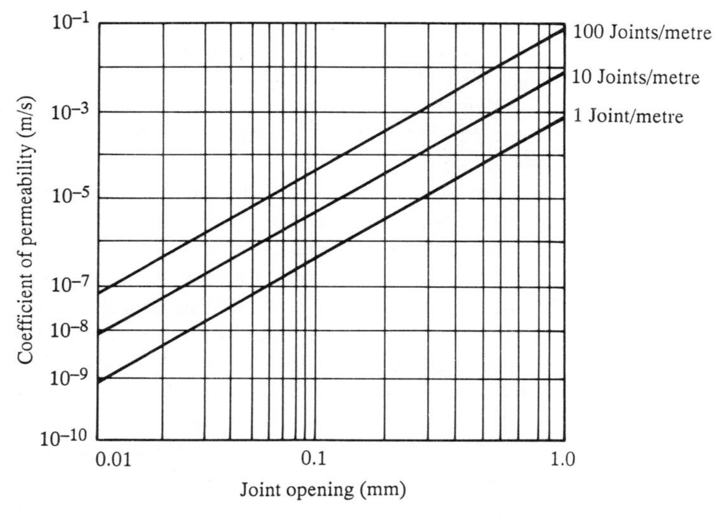

Figure 12.3 *Influence of joint opening and spacing on the coefficient of permeability in the direction of a set of smooth parallel joints in a rock mass (after Hoek and Bray, 1981)*

consolidation and cementation undergone, and the presence and nature of discontinuities. Temperature changes affect the flow rate of a fluid by changing its viscosity. The rate of flow is commonly assumed to be directly proportional to the hydraulic gradient but this is not always so in practice.

Permeability and porosity are not necessarily as closely related as would be expected, for instance, very fine textured sandstones frequently have a higher porosity than coarser ones, though the latter are more permeable. Nevertheless, Griffith (1955) did find that a linear relationship existed between porosity and the logarithm of permeability in sands.

As can be inferred from above, the permeability of a clastic material is also affected by the interconnections between the pore spaces. If these are highly tortuous then the permeability is accordingly reduced. Consequently tortuosity figures importantly in permeability, influencing the extent and rate of free water saturation. It can be defined as the ratio of the total path covered by a current flowing in the pore channels between two given points to the straight line distance between them.

Stratification in a formation varies within limits both vertically and horizontally. It is frequently difficult to predict what effect stratification has on the permeability of the beds. Nevertheless, in the great majority of cases where a directional difference in permeability exists, the greater permeability is parallel to the bedding. For example, the Permo-Triassic sandstones of the Mersey and Weaver Basins are notably anisotropic as far as permeability is concerned, the flow parallel to the bedding being higher than across it. Ratios of 5:1 are not uncommon and occasionally values of 100:1 have been recorded where fine marl partings occur.

The permeability of intact rock (primary permeability) is usually several orders less than the *in situ* permeability (secondary permeability). Although the secondary permeability is affected by the frequency, continuity and openness, and amount of infilling, of discontinuities, a rough estimate of the permeability can be obtained from their frequency (Table 10.7 or Figure 12.3). Admittedly such estimates must be treated with caution and cannot be applied to rocks which are susceptible to solution.

The frictional resistance to flow through joint systems is frequently much lower than that offered by a porous medium, hence appreciable quantities of water may be transmitted. An example is provided by the massive limestones of Lower Carboniferous age of the Pennine area, the permeability of an intact sample being much lower than that obtained by field tests (10^{-16} to 10^{-11} m/s, and 10^{-5} to 10^{-1} m/s respectively). The significantly higher permeability found in the field is attributable to the joint systems and bedding planes which have been opened by dissolution. The mass permeability of sandstones is also very much influenced by the discontinuities. For instance, the average laboratory permeability for the Fell Sandstone Group from Shirlawhope Well near Longframlington, Northumberland, is 17.4×10^{-7} m/s[8] (Bell, 1978). This compares with an estimated value of 2.4×10^{-3} m/s obtained from field tests. From the foregoing examples it can be concluded that as far as the assessment of flow through rock masses is concerned, field tests provide more reliable results than can be obtained from testing intact samples in the laboratory.

Table 12.4 *Relative values of permeabilities*

Rock and soil types	*Porosity*		*Permeability range* (m/s)						*Well yields*			*Type of water-bearing unit*
	Primary (grain)	*Secondary (fracture)**	$>10^{-1}$	$>10^{-3}$	$>10^{-5}$	$>10^{-7}$	$>10^{-9}$	$<10^{-9}$	*High*	*Medium*	*Low*	
			Very high	*High*	*Medium*	*Low*	*Very low*	*Impermeable*				
	(%)											
Sediments, unconsolidated												
Gravel	30 40											Aquifer
Coarse sand	30–40											Aquifer
Medium to fine sand	23–25											Aquifer
Silt	40–50	Occasional										Aquiclude
Clay, till	45–55	Often fissured										Aquiclude
Sediments, consolidated												
Limestone, dolostone	1–50	Solutions joints, bedding planes										Aquifer or aquifuge†
Coarse, medium sandstone	<20	Joints and bedding planes										Aquifer or aquiclude
Fine sandstone	<10	Joints and bedding planes										Aquifer or aquifuge
Shale, siltstone	–	Joints and bedding planes										Aquifuge or aquifer
Volcanic rocks, e.g. basalt	–	Joints and bedding planes										Aquifer or aquifuge
Plutonic and metamorphic rocks		Weathering and joints decreasing as depth increases										Aquifuge or aquifer

* Rarely exceeds 10 per cent
† Aquifuge: rock which neither transmits nor stores water

Dykes often act as barriers to groundwater flow so that the water table on one side may be higher than on the other. Fault planes occupied by clay gouge may have a similar effect. Conversely they may act as conduits where the fault plane is not sealed. The movement of water across a permeable boundary which separates aquifers of different permeabilities leads to deflection of flow, the bigger the difference the larger the deflection. When groundwater meets an impermeable boundary it flows along it and, as noted previously, in some situations, such as the occurrence of a dyke, may be impounded. The nature of a rock mass also influences whether flow is steady or unsteady. Generally it is unsteady since it is usually due to discharge from storage.

12.4 Flow through soils and rocks

Water possesses three forms of energy, namely, potential energy attributable to its height, pressure energy owing to its pressure, and kinetic energy due to its velocity. The latter can usually be discounted in any assessment of flow through soils. Energy in water is usually expressed in terms of head. The head possessed by water in soils or rocks is manifested by the height to which water will rise in a standpipe above a given datum. This height is usually referred to as the piezometric level and provides a measure of the total energy of the water. If at two different points within a continuous area of water there are different amounts of energy, then there will be a flow towards the point of lesser energy and the difference in head is expended in maintaining that flow. Other things being equal, the velocity of flow between two points is directly proportional to the difference in head between them. The hydraulic gradient, i, refers to the loss of head or energy of water flowing through the ground. This loss of energy by the water is due to the friction resistance of the ground material and this is greater in fine- than coarse-grained soils. Thus, there is no guarantee that the rate of flow will be uniform, indeed this is exceptional. However, if it is assumed that the resistance to flow is constant, then for a given difference in head the flow velocity is directly proportional to the flow path.

12.4.1 Darcy's law

Before any mathematical treatment of groundwater flow can be attempted certain simplifying assumptions have to be made, namely, that the material is isotropic and homogeneous, that there is no capillary action and that a steady state of flow exists. Since rocks and soils are anisotropic and heterogeneous, as they may be subject to capillary action and as flow through them is characteristically unsteady, any mathematical assessment of flow must be treated with caution.

The basic law concerned with flow is that enunciated by Darcy (1856) which states that the rate of flow, v, is proportional to the gradient of the potential head, i, measured in the direction of flow

$$v = ki \qquad (12.5)$$

and for a particular rock or soil or part of it, of area, A

$$Q = vA = Aki \tag{12.6}$$

where Q is the quantity in a given time. The ratio of the cross-sectional area of the pore spaces in a soil to that of the whole soil is given by $e/(1 + e)$, where e is the void ratio. Hence the seepage velocity, v_s, is

$$v_s = [(1 + e)/e]ki \tag{12.7}$$

Darcy's law is valid as long as a laminar flow exists. Departures from Darcy's law therefore, occur when the flow is turbulent such as when the velocity of flow is high. Such conditions exists in very permeable media, normally when the Reynolds number* can attain values above four. Accordingly it is usually accepted that this law can be applied to those soils which have finer textures than gravels. Furthermore Darcy's law probably does not accurately represent the flow of water through a porous medium of extremely low permeability, because of the influence of surface and ionic phenomena and the presence of gases.

Apart from an increase in the mean velocity, the other factors which cause deviations from the linear laws of flow include, first, the non-uniformity of pore spaces, since differing porosity gives rise to differences in the seepage rates through pore channels. A second factor is an absence of a running-in section where the velocity profile can establish a steady-state parabolic distribution. Lastly, such deviations may be developed by perturbations due to jet separation from wall irregularities.

Darcy omitted to recognize that permeability also depends upon the density, ρ, and dynamic viscosity of the fluid, μ, involved, and the average size, D_n, and shape of the pores in a porous medium. In fact, permeability is directly proportional to the unit weight of the fluid concerned and is inversely proportional to its viscosity. The latter is very much influenced by temperature. The following expression attempts to take these factors into account:

$$k = CD_{n}^{2}\rho/\mu \tag{12.8}$$

where C is a dimensionless constant or shape factor which takes note of the effects of stratification, packing, particle size distribution and porosity. It is asssumed in this expression that both the porous medium and the water are mechanically and

*Reynolds number, N_R, is commonly used to distinguish between laminar and turbulent flow and is expressed as follows:

$$N_R = \rho\frac{vR}{\mu}$$

where ρ is density, v is mean velocity, R is hydraulic radius and μ is dynamic viscosity. Flow is laminar for small values of Reynolds number.

physically stable, but this may never be true. For example, ionic exchange on clay and colloid surfaces may bring about changes in mineral volume which, in turn, affect the shape and size of the pores. Moderate to high groundwater velocities will tend to move colloids and clay particles. Solution and deposition may result from the pore fluids. Small changes in temperature and/or pressure may cause gas to come out of solution which may block pore spaces.

It has been argued that a more rational concept of permeability would be to express it in terms that are independent of the fluid properties. Thus the intrinsic permeability (k_i) characteristic of the medium alone has been defined as

$$k_i = CD_n^2 \tag{12.9}$$

However, it has proved impossible to relate C to the properties of the medium. Even in uniform spheres it is difficult to account for the variations in packing arrangement. In this context a widely accepted relationship for laminar flow through a permeable medium is that given by Fair and Hatch (1935)

$$k = \frac{1}{m\left[\frac{(1-n)^2}{n^3}\left(\frac{\theta}{100}\sum\frac{P}{D_m}\right)^2\right]} \tag{12.10}$$

where n is the porosity, m is the packing factor found by experiment to have a value of 5, θ is the particle shape factor varying from 6.0 for spherical to 7.7 for angular grains, P is the percentage of particles by weight held between each pair of adjacent sieves, and D_m is the geometric mean opening $(D_1D_2)^{1/2}$ of the pair.

The Kozeny–Carmen equation for deriving the coefficient of permeability also takes the porosity into account as well as the specific surface area of the porous medium, S_a, which is defined per unit volume of solid as

$$k = C_0 \frac{n^3}{(1-n)^2 S_a^2} \tag{12.11}$$

where C_0 is a coefficient, the suggested value of which is 0.2.

12.4.2 General equation of flow

If an element of saturated material is taken, with the dimensions dx, dy and dz (Figure 12.4) and flow is taking place in the x–y plane, then the generalized form of Darcy's Law is

$$v_x = k_x i_x \tag{12.12}$$

$$= k_x \frac{\partial h}{\partial x} \tag{12.13}$$

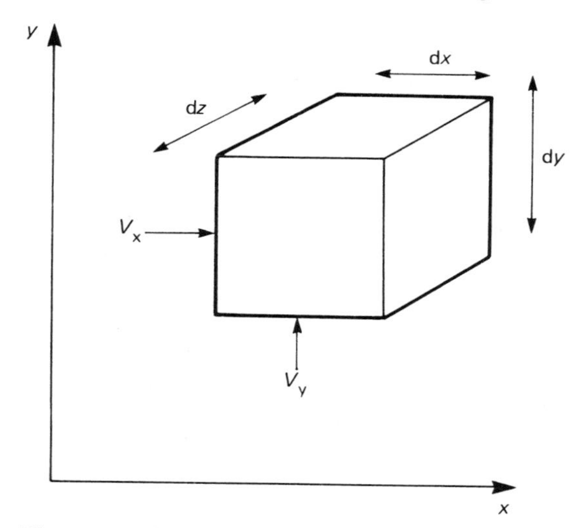

Figure 12.4 *Seepage through a soil element*

and

$$v_y = k_y i_y \quad (12.14)$$

$$_y = k_y \frac{\partial h}{\partial y} \quad (12.15)$$

where h is the total head under steady state conditions and k_x, i_x and k_y, i_y are, respectively, the coefficients of permeability and the hydraulic gradients in the x and y directions. Assuming that the fabric of the medium does not change and that the water is incompressible, then the volume of water entering the element is the same as that leaving in any given time, hence

$$v_x \mathrm{d}y\mathrm{d}z + v_y \mathrm{d}x\mathrm{d}z = \left(v_x + \frac{\partial v_x}{\partial x} dx\right) \mathrm{d}y\mathrm{d}z + \left(v_y + \frac{\partial v_y}{\partial y} \mathrm{d}y\right) \mathrm{d}x\mathrm{d}z \quad (12.16)$$

In such a situation the difference in volume between the water entering and leaving the element is zero, therefore

$$\frac{\partial v_x}{\partial x} + \frac{\partial v_y}{\partial y} = 0 \quad (12.17)$$

Equation (12.17) is referred to as the flow continuity equation. If Equations (12.13) and (12.15) are substituted in the continuity equation, then

$$k_x \frac{\partial^2 h}{\partial x^2} + k_y \frac{\partial^2 h}{\partial y^2} = 0 \tag{12.18}$$

If there is a recharge or discharge to the aquifer ($-w$ and $+w$ respectively) then this term must be added to the right-hand side of the above equation.

If it is assumed that the hydraulic conductivity is isotropic throughout the media so that $k_x = k_y$, then Equation (12.18) becomes

$$\frac{\partial^2 h}{\partial x^2} + \frac{\partial^2 h}{\partial y^2} = 0 \tag{12.19}$$

This is the two-dimensional Laplace equation for steady-state flow in an isotropic porous medium. The partial differential equation governing the two-dimensional unsteady flow of water in an anisotropic aquifer can be written as

$$T_x = \frac{\partial^2 h}{\partial x^2} + T_y \frac{\partial^2 h}{\partial y^2} = S \frac{\partial h}{\partial t} \tag{12.20}$$

where T and S are the coefficients of transmissivity and storage respectively.

12.4.3 Flow through stratified deposits

In a stratified sequence of deposits the individual beds will, no doubt, have different permeabilities, so that vertical permeability will differ from horizontal permeability. Consequently, in such situations, it may be necessary to determine the average values of the coefficient of permeability normal to (k_v) and parallel to (k_h) the bedding. If the total thickness of the sequence is H_T and the thickness of the individual layers are $H_1, H_2, H_3, \ldots, H_n$, with corresponding values of the coefficient of permeability $k_1, k_2, k_3, \ldots, k_n$, then k_v and k_h can be obtained as follows:

$$k_v = \frac{H_T}{H_1/k_1 + H_2/k_2 + H_3/k_3 + \ldots + H_n/k_n} \tag{12.21}$$

and

$$k_h = \frac{H_1k_1 + H_2k_2 + H_3k_3 + \ldots + H_nk_n}{H_T} \tag{12.22}$$

12.4.4 Fissure flow

Generally it is the interconnected systems of discontinuities which determine the permeability of a particular rock mass. Indeed, as mentioned above, the permeability of a jointed rock mass is usually several orders higher than that of intact rock. According to Serafim (1968), the following expression can be used to derive the filtration through a rock mass intersected by a system of parallel sided joints with a given opening, e, separated by a given distance, d:

$$k = \frac{e^3\gamma_w}{12d\mu} \tag{12.23}$$

where γ_w is the unit weight of water and μ its viscosity. The velocity of flow, v, through a single joint of constant gap, e, is expressed by

$$v = \left(\frac{e^2\gamma_w}{12\mu}\right) i \tag{12.24}$$

where i is the hydraulic gradient. Flow through a jointed mass was also considered by Castillo *et al.* (1972).

Wittke (1973) suggested that where the spacing between discontinuities is small in comparison with the dimensions of the rock mass, it is often admissible to replace the fissured rock with regard to its permeability, by a continuous anisotropic medium, the permeability of which can be described by means of Darcy's law. He also provided a resumé of procedures by which three-dimensional problems of flow through rocks under complex boundary conditions could be solved.

Lovelock *et al.* (1975) suggested that it can be shown that the contribution of the fissures (T_f) to the transmissivity of an idealized aquifer can be approximated from the following expression:

$$T_f = \frac{g}{12\mu_k} \sum_{x=1}^{n} b_x^3$$

where b_x is the effective aperture of the xth of n horizontal, parallel-sided smooth-walled openings; g is the acceleration due to gravity; μ_k is the kinematic viscosity of the fluid, and flow is laminar. The third power relationship means that a small variation in effective aperture (b) gives rise to a large variation in the fissure contribution (T_f).

12.5 Pore pressures, total pressures and effective pressures

Subsurface water (contained in pores) is normally under pressure which increases with increasing depth below the water table to very high values. Such pore water pressures have a significant influence on the engineering behaviour of most rock and soil masses and their variations are responsible for changes in the stresses in these masses, which affect their deformation characteristics and failure.

Piezometers are installed in the ground in order to monitor and obtain accurate measurements of pore water pressures. Observations should be made regularly so that changes due to external factors such as excessive precipitation, tides, the seasons, etc. are noted, it being most important to record the maximum pressures which have occurred. Standpipe piezometers allow the determination of the position of the water level (Sherrell, 1976). Hydraulic piezometers, connected to a manometer board, record the changes in pore water pressure. Usually simpler types of piezometer are used in the more permeable soils. The response to piezometers in rock masses can be very much influenced by the incidence and geometry of the discontinuities so that the values of water pressure obtained may be misleading if due regard is not given to these structures.

The efficiency of a soil in supporting a structure is influenced by the effective or intergranular pressure, that is, the pressure between the particles of the soil which develops resistance to applied load. Because the moisture in the pores offers no resistance to shear, it is ineffective or neutral and therefore pore water pressure has also been referred to as neutral pressure. Since the pore water or neutral pressure plus the effective pressure equals the total pressure, reduction in pore pressure increases the effective pressure. Reduction of the pore pressure by drainage consequently affords better conditions for carrying a proposed structure.

The effective pressure at a particular depth is simply obtained by multiplying the unit weight of the soil by the depth in question and subtracting the pore water pressure for that depth. In a layered sequence the individual layers may have different unit weights. The unit weight of each should then be multiplied by its thickness and the pore water pressure involved subtracted. The effective pressure for the total thickness involved is then obtained by summing the effective pressures of the individual layers (Figure 12.5). Water held in the capillary fringe by soil

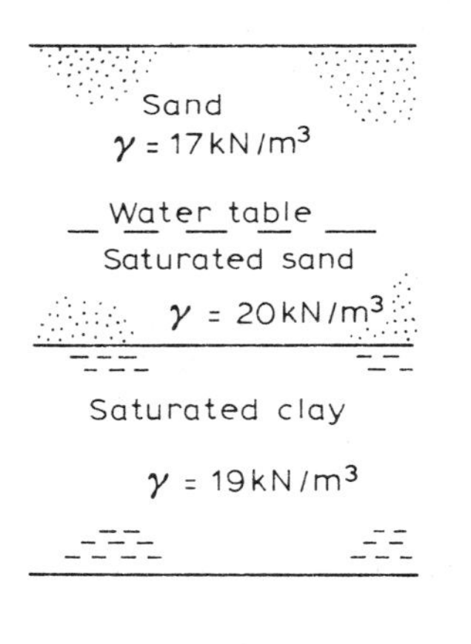

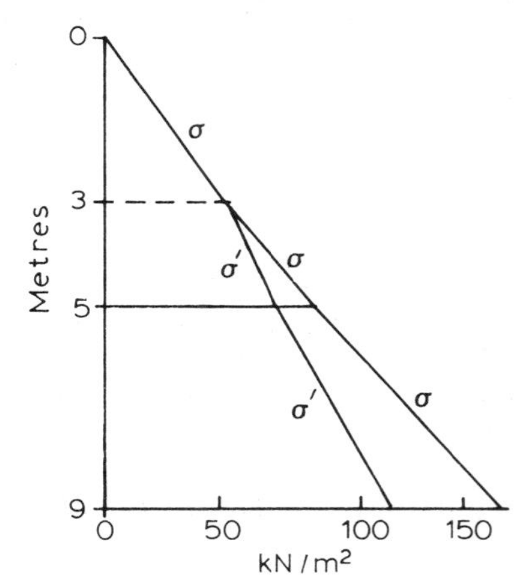

γ = unit weight

Depth (m)	*Total pressure* (σ)		*Pore pressure* (u)		*Effective pressure* (σ') $\sigma' = \sigma - u$
3	3 × 17	= 51	0		51
5	(3 × 17) + (2 × 20)	= 91	2 × 9.8	= 19.6	71.4
9	(3 × 17) + (2 × 20) + (4 × 19)	= 167	6 × 9.8	= 58.8	108.2

Figure 12.5 *Pressure diagram and example of calculation of total and effective overburden pressures*

suction does not affect the values of pore water pressure below the water table. However, the weight of water held in the capillary fringe does increase the weight of overburden and so the effective pressure.

Volume changes brought about by loading compressive soils depend upon the level of effective stress and are not affected by the area of contact. The latter may also be neglected in saturated or near-saturated soils.

In the case of partially saturated soils the void space is occupied by water and air. The pore water pressure (u_w) is less than the pore air pressure (u_a) due to surface tension. Unless the degree of saturation is close to unity the pore air forms continuous channels through the soil and the pore water is concentrated around the interparticle contacts. The boundaries between pore water and pore air will be in the form of menisci whose radii depend on the size of the pore spaces within the soil. The following effective stress equation is used for partially saturated soils:

$$\sigma = \sigma' + u_a - \chi(u_a - u_w) \qquad (12.25)$$

where χ is a parameter related to the degree of saturation of the soil. The term $(u_a - u_w)$ is a measure of the suction in the soil. The limiting values are for fully saturated soil ($S_r = 1$, $\chi = 1$) and for completely dry soil ($S_r = 0$, $\chi = 0$). The value of χ is also influenced, to a lesser extent, by the soil structure and the way the particular degree of saturation was brought about.

There is some evidence which suggests that the law of effective stress as used in soil mechanics, in which the pore pressure is subtracted from all direct stress components, holds true for some rocks. Those with low porosity may, at times, prove the exception. However, Serafim (1968) suggested it appeared that pore pressures have no influence on brittle rocks. This is probably because the strength of such rocks is mainly attributable to the strength of the bonds between the component crystals or grains.

The changes in stresses and the corresponding displacements due, for example, to construction work influence the permeability of a rock mass. For instance, with increasing effective shear stress the permeability increases along discontinuities orientated parallel to the direction of shear stress, whilst it is lowered along those running normal to the shear stress. Consequently the imposition of shear stresses, and the corresponding strains, lead to an anisotropic permeability within joints.

12.6 Critical hydraulic gradient, quick conditions and hydraulic uplift phenomena

As water flows through the soil and loses head, its energy is transferred to the particles past which it is moving, which in turn creates a drag effect on the particles. If the drag effect is in the same direction as the force of gravity, then the effective pressure is increased and the soil is stable. Indeed the soil tends to become more dense. Conversely if water flows towards the surface, then the drag effect is counter to gravity thereby reducing the effective pressure between particles. If the velocity of upward flow is sufficient it can buoy up the particles so that the effective pressure is reduced to zero. This represents a critical condition where the weight of the submerged soil is balanced by the upward acting seepage force. The critical hydraulic gradient (i_c) can be calculated from the following expression

$$i_c = \frac{G_s - 1}{1 + e} \tag{12.25}$$

where G_s is the specific gravity of the particles and e is the void ratio. A critical condition sometimes occurs in silts and sands. If the upward velocity of flow increases beyond the critical hydraulic gradient a quick condition develops.

Quicksands, if subjected to deformation or disturbance, can undergo a spontaneous loss of strength. This loss of strength causes them to flow like viscous liquids. Terzaghi (1925) explained the quicksand phenomenon in the following terms. First, the sand or silt concerned must be saturated and loosely packed. Secondly, on disturbance the constituent grains become more closely packed which leads to an increase in pore water pressure, reducing the forces acting between the grains. This brings about a reduction in strength. If the pore water can escape very rapidly the loss in strength is momentary. Hence the third condition requires that pore water cannot escape readily. This is fulfilled if the sand or silt has a low permeability and/or the seepage path is long. Casagrande (1936) demonstrated that a critical porosity existed above which a quick condition could be developed. He maintained that many coarse-grained sands, even when loosely packed, have porosities approximately equal to the critical conditions whilst medium- and fine-grained sands, especially if uniformly graded, exist well above the critical porosity when loosely packed. Accordingly fine sands tend to be potentially more unstable than coarse-grained varieties. It must also be remembered that the finer sands have lower permeabilities.

Quick conditions brought about by seepage forces are frequently encountered in excavations made in fine sands which are below the water table, as for example, in cofferdam work. As the velocity of the upward seepage force increases further from the critical gradient the soil begins to boil more and more violently. At such a point structures fail by sinking into the quicksand. Liquefaction of potential quicksands may be caused by sudden shocks caused by the action of heavy machinery (notably pile driving), blasting and earthquakes. Such shocks increase the stress carried by the water, the neutral stress, and give rise to a decrease in the effective stress and shear strength of the soil. There is also a possibility of a quick condition developing in a layered soil sequence where the individual beds have different permeabilities. Hydraulic conditions are particularly unfavourable where water initially flows through a very permeable horizon with little loss of head, which means that flow takes place under a great hydraulic gradient.

There are several methods which may be employed to avoid the development of quick conditions. One of the most effective techniques is to prolong the length of the seepage path thereby increasing the frictional losses and so reducing the seepage force. This can be accomplished by placing a clay blanket at the base of an excavation where seepage lines converge. If sheet piling is used in excavation of critical soils, then the depth to which it is sunk determines whether or not quick conditions will develop. Consequently it should be sunk deep enough to avoid a potential critical condition occurring at the base level of the excavation. The hydrostatic head also can be reduced by means of relief wells and seepage can be intercepted by a wellpoint system placed about the excavation. Furthermore a quick condition may be prevented by increasing the downward acting force. This may be brought about by laying a load on the surface of the soil where seepage is discharging. Gravel filter beds may be used for this purpose. Suspect soils also can be densified, treated with stabilizing grouts, or frozen.

When water percolates through heterogeneous soil masses it moves preferentially through the most permeable zones and it issues from the ground as springs. Piping refers to the erosive action of some such springs, where sediments are removed by seepage forces, so forming subsurface cavities and tunnels. In order that erosion tunnels may form, the soil must have some cohesion, the greater the cohesion, the wider the tunnel. In fact fine sands and silts are most susceptible to piping failures. Obviously the danger of piping occurs when the hydraulic gradient is high, that is, when there is a rapid loss of head over a short distance. This may be indicated on a flow net by a close network of squares where the flow is upward. As the pipe develops by backward erosion it nears the source of water supply so that eventually the water breaks into and rushes through the pipe. Ultimately the hole, so produced, collapses from lack of support. Piping has been most frequently noted downstream of dams, the reservoir providing the water source (Penman, 1977). Leaking drains can also give rise to piping.

Subsurface structures should be designed to be stable with regard to the highest groundwater level that is likely to occur. Structures below groundwater level are acted upon by uplift pressures. If the structure is weak this pressure can break it and, for example, cause a blow-out of a basement floor or collapse of a basement wall. If the structure is strong but light it may be lifted, that is, subjected to heave. Uplift can be taken care of by adequate drainage or by resisting the upward seepage force. Continuous drainage blankets are effective but should be designed with filters to function without clogging. The entire weight of structure can be mobilized to resist uplift if a raft foundation is used. Anchors, grouted into bedrock, can provide resistance to uplift.

Moore and Longworth (1979) recorded a failure in a brick pit, 29 m in depth, excavated in Oxford Clay. The failure was brought about by a build-up of hydrostatic pressure in an underlying aquifer (either the Cornbrash or Blisworth Limestone located at depths of 6 and 11 m respectively below the surface of the pit). This initially gave rise to a heave of some 150 mm, and then ruptured the surface clay, thereby allowing the rapid escape of approximately 7000 m^3 of water. The floor of the pit then settled up to 100 mm.

Hydraulic uplift phenomena had previously been reported by Rowe (1968) and occur under a wide range of geological conditions and at differing scales. Horswill and Horton (1976) referred to a case of hydraulic disruption at the base of a shaft, 11 m in diameter, which was sunk in the Upper Lias Clay at Empingham Dam site. The uplift took place when the shaft had reached a point of 15 m above the

Marlstone Rock Bed, the offending limestone aquifer. Within a few hours the clay in the shaft was deformed and fractured, facilitating the inflow of water, thereby flooding the shaft.

12.7 Groundwater abstraction and subsidence

A reduction of the groundwater level can lead to subsidence of the ground surface (Bell, 1988). Subsidence due to the withdrawal of groundwater has developed with most effect in those groundwater basins where there was, or is, intensive abstraction. Such subsidence is attributed to the consolidation of sedimentary deposits in which the groundwater is present, consolidation occurring as a result of increasing effective stress. The total overburden pressure in partially saturated or saturated deposits is borne by their granular structure and the pore water. When groundwater abstraction leads to a reduction in pore water pressure by draining water from the pores, this means that there is a gradual transfer of stress from the pore water to the granular structure. Put another way, the effective weight of the deposits in the dewatered zone increases since the buoyancy effect of the pore water is removed. For instance, if the water table is lowered by 1 m, then this gives rise to a corresponding increase in average effective overburden pressure of $10\,kN/m^2$. As a result of having to carry this increased load the granular structure may deform in order to adjust to the new stress conditions. In particular, the porosity of the deposits concerned undergoes a reduction in volume, the surface manifestation of which is subsidence. Scott (1979) pointed out that subsidence does not necessarily occur simultaneously with the abstraction of water from an underground reservoir. In fact, it occurs usually over a larger period of time than that taken for abstraction.

Consolidation may be elastic or non-elastic depending upon the character of the deposits involved and the range of stresses induced by a decline in the water level. In elastic deformation, stress and strain are proportional, and consolidation is independent of time and reversible. Non-elastic consolidation occurs when the granular structure of a deposit is rearranged to give a decrease in volume, that decrease being permanent. Generally recoverable consolidation represents compression in the pre-consolidation stress range, while irrecoverable consolidation represents compression due to stresses greater than the pre-consolidation pressure.

According to Lofgren (1979) the storage characteristics of compressible formations change significantly during the first cycle of groundwater withdrawal. Such withdrawal is responsible for permanent consolidation of any fine-grained, interbedded formations. The water released by consolidation represents a one-time, and sometimes important, source of water to wells. During a second cycle of prolonged pumping overdraft, much less water is available to the wells and the water table is lowered much more rapidly. As an illustration, drawn down in the San Joaquin Valley takes place during a few months each year, for the rest of the time the water levels are recovering. Because of the cyclic nature of the groundwater abstraction, the elastic component of storage change for a given net decline in water level may be reduced and restored many times while the inelastic component is removed but once. Figure 12.6 illustrates the relationship between fluctuations in water level in semi-confined and confined aquifers, changes in effective stress consolidation, and surface subsidence recorded by Lofgren (1968) at a site near Pixley. It can be seen that each year consolidation commenced during the period of rapid decline in head, continued through the pumping season and ceased when the head began to recover.

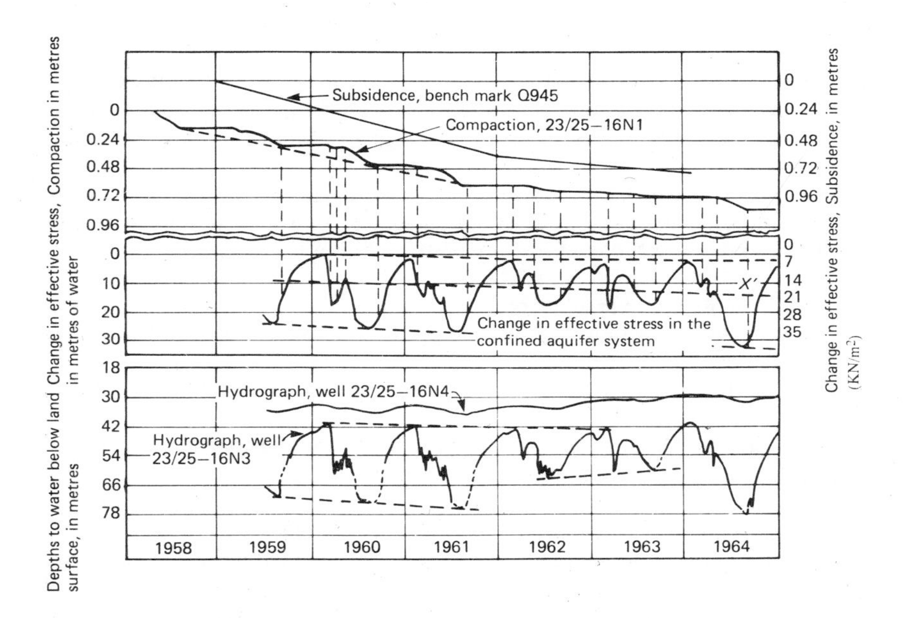

Figure 12.6 *Land subsidence, compaction, water fluctuations and change in effective stress 4.8 km south of Pixley (after Logfren, 1968)*

The amount of subsidence which occurs is governed by the increase in effective pressure, the thickness and compressibility of the deposits concerned, the length of time over which the increased loading is applied, and possibly the rate and type of stress applied. For example, Delflache (1979) reported that the most noticeable subsidence in the Houston–Galveston region of Texas has occurred where the declines in head have been largest and where the thickness of clay in the aquifer system is greatest. Furthermore, the ratio between maximum subsidence and groundwater reservoir consolidation is related to the ratio between depth of burial and the lateral extent of the reservoir. In other words, small reservoirs which are deeply buried do not give rise to noticeable subsidence, even if subjected to considerable consolidation. By contrast, extremely large underground reservoirs may develop appreciable subsidence.

The rate at which consolidation occurs depends on the rate at which the pore water can drain from the system which, in turn, is governed by its permeability. For instance, the low permeability and high specific storage of aquitards and aquicludes under virgin stress conditions means that the escape of water and resultant adjustment of pore water pressures is slow and time-dependent. Consequently in fine-grained beds the increase in stress which accompanies the decline in head becomes effective only as rapidly as the pore water pressures are lowered toward equilibrium with the pressures in adjacent aquifers. The time required to reach this stage varies directly according to the specific storage and the square of the thickness of the zone from which drainage is occurring and inversely according to the vertical permeability of the aquitard. In fact, it may take months or years for fine grained beds to adjust to increases in stress. Moreover, the rate of consolidation of slow-draining aquitards reduces with time and is usually small after a few years of loading.

A number of steps which can be taken to evaluate the potential subsidence likely to occur as a result of withdrawal of groundwater from an aquifer have been outlined by Saxena and Mohan (1979) and are as follows:

(1) Define the *in situ* hydraulic conditions.
(2) Compute the reduction in pore water pressure due to the removal of the required quantity of water.
(3) Convert the reduction in pore water pressure to the equivalent increase in effective pressure.
(4) Determine the deformation in the aquifer zone and the confining layers and translate it into the resultant ground subsidence.

In addition to being the most prominent effect in subsiding groundwater basins, surface fissuring and faulting (Figure 12.7) may develop suddenly and therefore pose a greater potential threat to surface structures. In the USA such fissuring and faulting has occurred especially in the San Joaquin Valley, the Houston–Galveston region and in central Arizona. These fissures, and more particularly the faults, frequently occur along the periphery of the basin. The faults are high-angled, normal faults, with the downthrow on the side towards the centre of the basin. Generally displacements along the faults are not great, less than a metre, but movements may continue over a period of years. Holzer (1976) related the annual variations in the rate of faulting to annual fluctuations in groundwater levels. Geophysical investigations in central Arizona have revealed that upward projections of bedrock lie beneath many fissures.

It is not only falling or low groundwater levels that cause problems, a rising or high water table can be equally troublesome. Since the mid-1960s the rate of abstraction from the Chalk below London has decreased significantly so that water levels are now increasing by as much as 1 m/year in places (Marsh and Davies, 1983). The potential consequences of this includes leaks in tunnels and deep basements and a reduction in pile capacity. Similar problems exist in the Witton area of Birmingham, England where factory basements are being flooded by rising groundwater, and in central Liverpool where British Rail tunnels are affected. In

Figure 12.7 *Earth fissure in south-central Arizona. Fissure results from erosional enlargement of tension fracture caused by differential subsidence. The subsidence is caused by groundwater level decline (courtesy of Dr Thomas L. Holzer, US Geological Survey)*

Louisville, Kentucky, increasing groundwater levels are causing concern over the possibility of structural settlement, damage to basement floors and the disruption of utility conduits (Haggerty and Lippert, 1982).

12.8 Frost action in soil

Frost action in a soil is influenced by the initial temperature of the soil, as well as the air temperature, the intensity and duration of the freeze period, the depth of frost penetration, the depth of the water table, and the type of ground and exposure cover. If frost penetrates down to the capillary fringe in fine grained soils, especially silts, then, under certain conditions, lenses of ice may be developed. The formation of such ice lenses may, in turn, cause frost heave and frost boil which may lead to the break-up of roads, the failure of slopes, etc. Shrinkage, which gives rise to polygonal cracking in the ground, presents another problem when soil is subjected to freezing. The formation of these cracks is attributable to thermal contraction and desiccation. Water which accumulates in the cracks is frozen and consequently helps increase their size. This water also may aid the development of lenses of ice.

12.8.1 Classification of frozen soil

According to Thomson (1980) ice may occur in frozen soil as small disseminated crystals whose total mass exceeds that of the mineral grains. It may also occur as large tabular masses which range up to several metres thick, or as ice wedges. The latter may be several metres wide and may extend to 10 m or so in depth. As a consequence frozen soils need to be described and classified for engineering purposes. A recent method of classifying frozen soils involves the identification of the soil type and the character of the ice (Andersland and Anderson, 1978). First, the character of the actual soil is classified according to the Unified Soil Classification System (see Table 1.8). Second, the soil characteristics consequent upon freezing are added to the description. Frozen soil characteristics are divided into two basic groups based on whether or not segregated ice can be seen with the naked eye (Table 12.6). Third, the ice present in the frozen soil is classified, this refers to inclusions of ice which exceed 25 mm in thickness (Table 12.5).

The amount of segregated ice in a frozen mass of soil depends largely upon the intensity and rate of freezing. When freezing takes place quickly no layers of ice are visible whereas slow freezing produces visible layers of ice of various thicknesses. Ice segregation in soil also takes place under cyclic freezing and thawing conditions.

12.8.2 Mechanical properties of frozen soil

The presence of masses of ice in a soil means that, as far as engineering is concerned, the properties of both have to be taken into account. Ice has no long-term strength, that is, it flows under very small loads. If a constant load is applied to a specimen of ice, instantaneous elastic deformation occurs. This is followed by creep, which eventually develops a steady state. Instantaneous elastic recovery takes place on removal of the load, followed by recovery of the transient creep.

The mechanical properties of frozen soil are very much influenced by the grain size distribution, the mineral content, the density, the frozen and unfrozen water contents, and the presence of ice lenses and layering. The strength of frozen ground develops from cohesion, interparticle friction and particle interlocking, much the same as in unfrozen soils. However, cohesive forces include the adhesion between soil particles and ice in the voids, as well as the surface forces between particles. More particularly, the strength of frozen soils is sensitive to particle size distribution, particle orientation and packing, impurities (air bubbles, salts or organic matter) in the water–ice matrix, temperature, confining pressure and the rate of strain. Obviously the difference in the strength between frozen and unfrozen soils is derived from the ice component.

The relative density influences the behaviour of frozen granular soils, especially their shearing resistance, in a manner similar to that when they are unfrozen (see Chapter 2). The cohesive effects of the ice matrix are superimposed on the latter behaviour and the initial deformation of frozen sand is dominated by the ice matrix. Sand in which all the water is more or less frozen exhibits a brittle type of failure at

Table 12.5 *Description and classification of frozen soils (from Andersland and Anderson, 1978)*

I: Description of soil phase (independent of frozen state)	*Classify soil phase by the unified soil classification system*			
	Major group		*Subgroup*	
	Description	*Designation*	*Description*	*Designation*
II: Description of frozen soil	Segregated ice not visible by eye	N	Poorly bonded or friable	Nf
			Well bonded — No excess ice	Nb n
			Well bonded — Excess ice	Nb e
	Segregated ice visible by eye (ice 25 mm or less thick)	V	Individual ice crystals or inclusions	Vx
			Ice coatings on particles	Ve
			Random or irregularly oriented ice formations	Vr
			Stratified or distinctly oriented ice formations	Vs
III: Description of substantial ice strata	Ice greater than 25 mm thick	ICE	Ice with soil inclusions	ICE + soil type
			Ice without soil inclusions	ICE

low strains, for example, at around 2% strain. However, the presence of unfrozen films of water around particles of soil not only means that the ice content is reduced, but leads to a more plastic behaviour of the soil during deformation. For instance, frozen clay, as well as often containing a lower content of ice than sand, has layers of unfrozen water (of molecular proportions) around the clay particles. These molecular layers of water contribute towards a plastic type of failure.

Lenses of ice are frequently formed in fine grained soils frozen under a directional temperature gradient. The lenses impart a laminated appearance to the soil. In such situations the strength of the bond between soil particles and ice matrix is greater than between particles and adjacent ice lenses. Under very rapid loading the ice behaves as a brittle material, with strengths in excess of those of fine-grained frozen soils. By contrast the ice matrix deforms continuously when subjected to long-term loading, with no limiting long-term strength. The laminated texture of the soil in rapid shear possesses the greatest strength when the shear zone runs along the contact between ice lens and frozen soil.

When loaded, stresses at the point of contact between soil particles and ice bring about pressure melting of the ice. Because of differences in the surface tension of the melt water, it tends to move into regions of lower stress, where it refreezes. The process of ice melting and the movement of unfrozen water are accompanied by a breakdown of the ice and the bonding with the grains of soil. This leads to plastic deformation of the ice in the voids and to a rearrangement of particle fabric. The net result is time-dependent deformation of the frozen soil, namely, creep (Eckardt, 1979; Takegawa *et al.*, 1979). Frozen soil undergoes appreciable deformation under sustained loading, the magnitude and rate of creep being governed by the composition of the soil, especially the amount of ice present, the temperature, the stress and the stress history.

The creep strength of frozen soils is defined as the stress level, after a given time, at which rupture, instability leading to rupture or extremely large deformations without rupture occur (Andersland *et al.*, 1978). Frozen fine-grained soils can suffer extremely large deformations without rupturing at temperatures near to freezing point. Hence the strength of these soils must be defined in terms of the maximum deformation which a particular structure can tolerate. As far as laboratory testing is concerned, axial strains of 20%, under compressive loading, are frequently arbitrarily considered as amounting to failure. The creep strength is then defined as the level of stress producing this strain after a given interval of time.

When strain is plotted against time, three stages of creep are apparent under uniform load (Figure 12.8). At first, strain increases quickly, but then settles at a uniform minimal rate of increase in its second stage. A third, plastic stage is eventually reached during which complete loss of resistance occurs. This feature is well demonstrated in clay at temperatures near the freezing point of water. Sanger and Kaplar (1963) tested an organic silty clay, and showed that failure (in the tertiary stage of creep) occurred at 550 kN/m^2 after 17 h when the sample was maintained at 0°C, but no failure point had been reached after 60 h when tested at −2°C.

In fine grained sediments the intimate bond between the water and the clay particles results in a significant proportion of soil moisture remaining unfrozen at

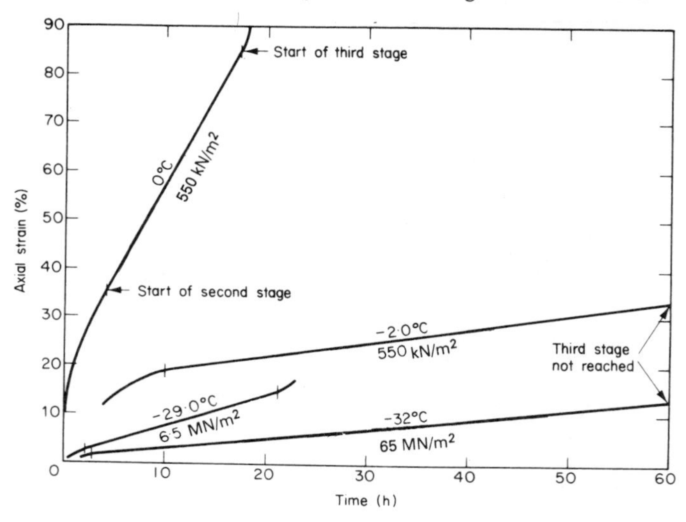

Figure 12.8 *Typical creep curves of samples of silty clays at various applied stresses and temperatures (after Sanger and Kaplar, 1963)*

temperatures as cold as −25°C. The more clay material in the soil, the greater is the quantity of unfrozen moisture. Nonetheless Lovell (1957) measured the unconfined compressive strength of frozen clays and demonstrated that there was a dramatic increase in structural strength with decreasing temperature. In fact, it appears to increase exponentially with the relative proportion of moisture frozen. Using silty clay as an example, the amount of moisture frozen at −18°C is only 1.25 times that frozen at −5°C, but the increase in compressive strength is more than four-fold.

By contrast, the water content of granular soils is almost wholly converted into ice at a very few degrees below freezing point. Hence frozen granular soils exhibit a reasonably high compressive strength only a few degrees below freezing, and there is justification for using this parameter as a design index of their performance in the field, provided that a suitable factor of safety is incorporated. The order of increase in compressive strength with decreasing temperature is shown in Figure 12.9.

Uniaxial compression tests carried out by Parameswaran (1980) on cylindrical specimens of frozen Ottawa Sand containing about 20% by weight, of water, indicated that strength increases with increasing strain rates and decreasing temperatures. The initial tangent modulus (E_i) also increased with increasing strain rates and decreasing temperature. The considerably lower values of modulus and strength at −2°C, as compared with those at lower temperatures, were probably due to larger amounts of water remaining unfrozen at −2°C.

Because frozen ground is more or less impermeable this increases the problems due to thaw by impeding the removal of surface water. What is more, when the thaw occurs the amount of water liberated may greatly exceed that originally present in the melted out layer of the soil (see below). As the soil thaws downwards

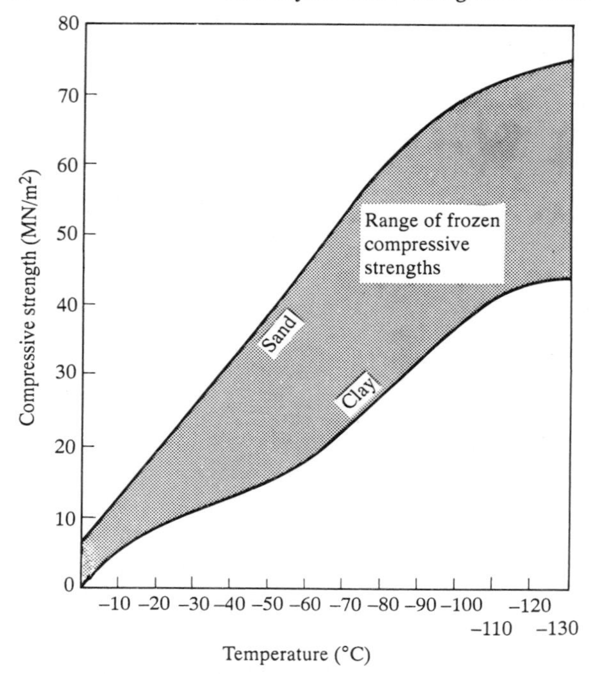

Figure 12.9 *Increase in compressive strength with decreasing temperature*

the upper layers become saturated, and since water cannot drain through the frozen soil beneath, they may suffer a complete loss of strength. Indeed under some circumstances excess water may act as a transporting agent, thereby giving rise to soil flows.

Settlement is associated with thawing of frozen ground. As ice melts, settlement occurs, water being squeezed from the ground by overburden pressure or by any applied loads. Excess pore pressures develop when the rate of ice melt is greater than the discharge capacity of the soil. Since excess pore pressures can lead to the failure of slopes and foundations, both the rate and amount of thaw settlement should be determined. Pore water pressures should also be monitored.

Further consolidation, due to drainage, may occur on thawing. If the soil was previously in a relatively dense state, then the amount of consolidation is small. This situation only occurs in coarse-grained frozen soils containing very little segregated ice. On the other hand, some degree of segregation of ice is always present in fine-grained frozen soils. For example, lenses and veins of ice may be formed when silts have access to capillary water. Under such conditions the moisture content of the frozen silts significantly exceeds the moisture content present in their unfrozen state. As a result when such ice-rich soils thaw under drained conditions they undergo large settlements under their own weight. Methods of predicting the amount and rate of settlement due to thawing of frozen ground have been discussed by Nixon and Ladanyi (1978).

Drainage patterns may be affected when thaw occurs in permafrost. Melt water, because it cannot penetrate the frozen ground, tends to follow any porous paths thawed out in the permafrost. This action is at times responsible for serious soil erosion. Furthermore, in permafrost regions ice may form at the base of shallow or slow-flowing streams, as well as at their surfaces. The development of ice at both the surface and base of a stream impedes or even stops its flow within its channel. Hence water is forced over the banks or into porous formations adjacent to the stream. This overflow water is frozen to form sheets, which gradually produce successive layers across the area surrounding the stream. By the end of the winter such layers of ice (aufeis), may extend up to 4 m in thickness and cover several square kilometres.

Chamberlain and Gow (1979) showed that significant structural changes can be brought about in the fabric of fine-grained soils when they are subjected to alternate freezing and thawing. Their experiments indicated that a reduction took place in the void ratio of the soil and that its vertical permeability was increased. The latter increase was most significant in soils with large plasticity indices, and usually the increase was smaller, the higher the level of applied stress. They attributed the increase in vertical permeability in clayey soils to the formation of polygonal shrinkage cracks, whereas in coarse-grained soils they suggested that it was due to the reduction in the amount of fines in the pores of the coarse fraction.

12.8.3 Frost heave

Jumikis (1968) listed the following factors as necessary for the occurrence of frost heave, namely, capillary saturation at the beginning and during the freezing of the soil, a plentiful supply of subsoil water, and a soil possessing fairly high capillarity together with moderate permeability. According to Kinosita (1979) the ground surface experiences an increasingly larger amount of heave, the higher the initial water table. Indeed it has been suggested that frost heave could be prevented by lowering the water table (Andersland and Anderson, 1978).

Grain size is another important factor influencing frost heave. For example, gravels, sands and clays are not particularly susceptible to heave whilst silts definitely are. The reason for this is that silty soils are associated with high capillary rises but at the same time their voids are large enough to allow moisture to move quickly enough for them to become saturated rapidly. If ice lenses are present in clean gravels or sands, then they simply represent small pockets of moisture which have been frozen. Indeed Taber (1930) gave an upper size limit of 0.007 mm, above which, he maintained, layers of ice do not develop. However, Casagrande (1932) suggested that the particle size critical to heave formation was 0.02 mm. If the quantity of such particles in a soil is less than 1%, no heave is to be expected, but considerable heaving may take place if this amount is over 3% in non-uniform soils and over 10% in very uniform soils. This 0.02 mm criterion has been used by the US Army Corps of Engineers (Anon, 1965), together with data from frost heave tests to develop a frost susceptibility system which is outlined in Figure 12.10. Figure 12.10 shows that soil groups exhibit a range of susceptibilities reflecting variations in particle size distribution, density and mineralogy.

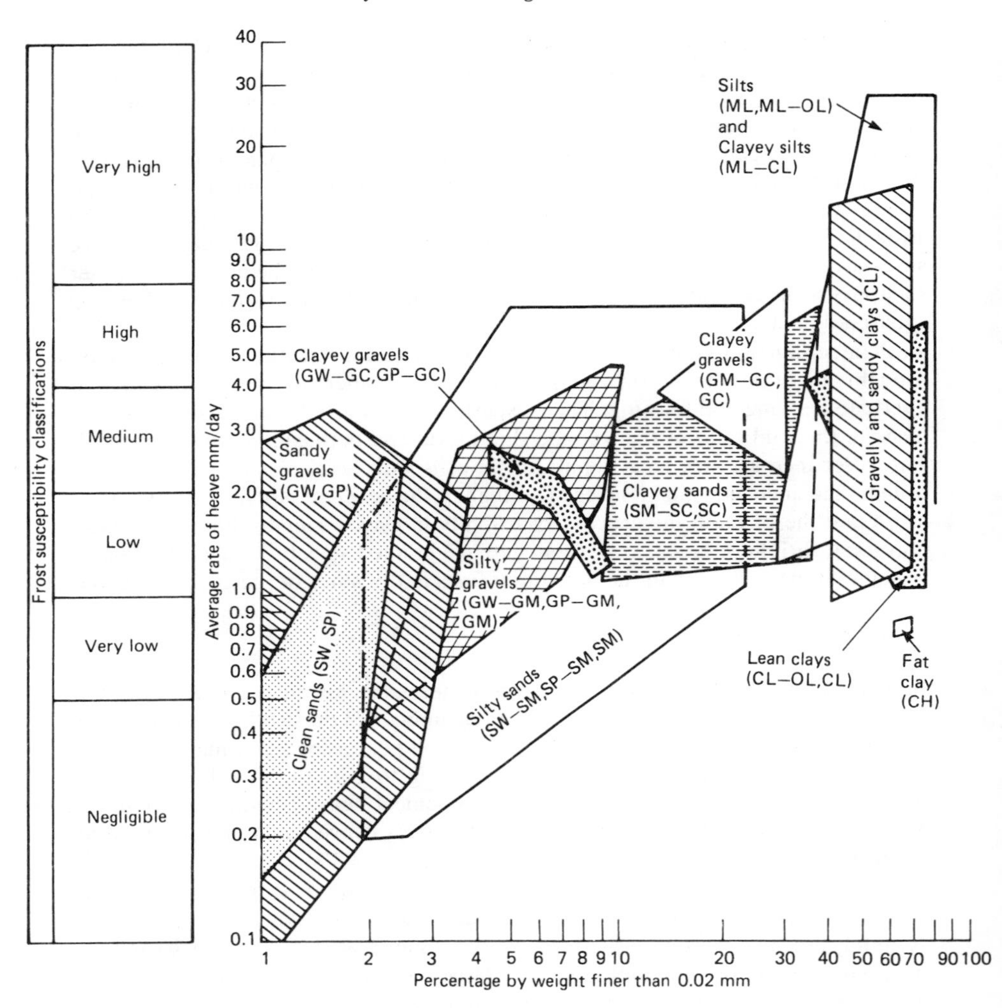

Figure 12.10 *Range in the degree of frost susceptibility of soils according to the US Army Corps of Engineers (Anon, 1965)*

Croney and Jacobs (1967) suggested that under the climatic conditions experienced in Britain well drained cohesive soils with a plasticity index exceeding 15% could be looked upon as non-frost susceptible. They suggested that where the drainage is poor and the water table is within 0.6 m of formation level the limiting value of plasticity index should be increased to 20%. In addition in experiments with sand they noted that as the amount of silt added was increased up to 55% or the clay fraction up to 33%, increase in permeability in the freezing front was the overriding factor and heave tended to increase. Beyond these values the decreasing permeability below the freezing zone became dominant and progressively reduced the heave. This indicates that the permeability below the frozen zone was principally responsible for controlling heave. These two authors also suggested that the permeability of soft chalk is sufficiently high to permit very serious frost heave but in the harder varieties the lower permeabilities minimize or prevent heaving.

Horiguchi (1979) demonstrated, from experimental evidence, that the rate of frost heave increased as the rate of heat removal from the freezing front was increased. However, removal of heat does not increase the rate of heave indefinitely – it reaches a maximum, after which it declines. The maximum rate of heave was shown to be influenced by particle size distribution in that it increased in soils with finer grain size. Horiguchi also found that when the particles were the same size in different test specimens, then the maximum rate of heave depended upon the types of exchangeable cations present in the soil. The rate of heave is also influenced by the thickness of overburden. For instance, Penner and Walton (1979) indicated that the maximum rate of ice accumulation at lower overburden pressures occurred at temperatures nearer to 0°C than at higher overburden pressures. However, it appears that the rate of heave for various overburden pressures tends to converge as the temperature below freezing is lowered. As the overburden pressure increases, the zone over which heaving takes place becomes greater and it extends over an increasingly larger range of temperature.

Maximum heaving, according to Jumikis (1956), does not necessarily occur at the time of maximum depth of penetration of the 0°C line, there being a lag between the minimum air temperature prevailing and the maximum penetration of the freeze front. In fact soil freezes at temperatures slightly lower than 0°C.

As heaves amounting to 30% of the thickness of the frozen layer have frequently been recorded, moisture, other than that initially present in the frozen layer, must be drawn from below, since water increases in volume by only 9% when frozen. In fact when a soil freezes there is an upward transfer of heat from the groundwater towards the area in which freezing is occurring. The thermal energy, in turn, initiates an upward migration of moisture within the soil. The moisture in the soil can be translocated upwards either in the vapour or liquid phase or by a combination of both. Moisture diffusion by the vapour phase occurs more readily in soils with larger void spaces than in fine-grained soils. If a soil is saturated, migration in the vapour phase cannot take place.

In a very dense, closely packed soil where the moisture forms uninterrupted films throughout the soil mass, down to the water table, then, depending upon the texture of the soil, the film transport mechanism becomes more effective than the vapour mechanism. The upward movement of moisture due to the film mechanism, in a freezing soil mass, is slow. Nonetheless a considerable amount of moisture can move upwards as a result of this mechanism during the winter. What is more in the film transport mechanism the water table is linked by the films of moisture to the ice lenses.

Before freezing, soil particles develop films of moisture about them due to capillary action. This moisture is drawn from the water table. As the ice lens grows, the suction pressure it develops exceeds that of the capillary attraction of moisture by the soil particles. Hence moisture moves from the soil to the ice lens. But the capillary force continues to draw moisture from the water table and so the process continues. Jones (1980) suggested that if heaving is unrestrained, the heave (H) can be estimated as follows:

$$H = 1.09kit \qquad (12.26)$$

where k is permeability, i is the suction gradient (this is difficult to derive) and t is time.

An investigation into moisture migration in frozen soils was carried out by Mageau and Morgenstern (1980) who demonstrated that induced moisture migration does occur in frozen soils. They suggested that the rate of migration seems to be governed by the apparent permeability of the soil and the suction force within the frozen fringe. As ice develops in a soil, this reduces its permeability to a critical value at which an ice lens begins to form. It appears that the rate of water uptake by an ice lens is not influenced by the frozen soil above it. A dominant suction pressure develops in the frozen fringe between the warmest ice lens and the frozen–unfrozen interface. Moisture migrates via the interconnected zones of free water within the pores and through the unfrozen films in the frozen soil, under the influence of a temperature gradient. Mageau and Morgenstern concluded that the rate of heave is controlled principally by the frozen fringe of soil between the warmest ice lens and the frozen–unfrozen interface. The amount of unfrozen water in the soil governs the extent of migration. The amount of unfrozen soil is, in turn, controlled by soil type and temperature. For example, Mageau and Morgenstern found that moisture migration in clayey silt was reduced to an insignificant level at a soil temperature of about −2°C.

The frost heave test has been used to predict frost heave (Jacobs, 1965). A critical review of this test has been provided by Jones (1980), who claimed that it gives poor reproducibility of results. Furthermore, this type of test is unfortunately time consuming and so a rapid freeze test has been developed by Kaplar (1971). Approximate predictions of frost heave have also been based on grain size distribution. However, in a discussion of frost heaving, Reed *et al.* (1979) noted that such predictions failed to take account of the fact that soils can exist at different states of density and therefore porosity, yet they have the same grain size distribution. What is more pore size distribution controls the migration of water in the soil and hence, to a large degree, the mechanism of frost heave. They accordingly derived expressions, based upon pore space, for predicting the amount of frost heave (Y) in mm/day:

$$Y = 581.1(X_{3.0}) - 5.46 - 29.46(X_{3.0})/(X_0 - X_{0.4}) \qquad (12.27)$$

where $X_{3.0}$ = cumulative porosity for pores >3.0 μm but <300 μm, X_0 = total cumulative porosity, $X_{0.4}$ = cumulative porosity for pores >0.4 μm but <300 μm.

A simpler expression based on pore diameters rather than cumulative porosity, but which was somewhat less accurate, was as follows:

$$Y = 1.694(D_{40}/D_{80}) - 0.3805 \qquad (12.28)$$

where D_{40} and D_{80} are the pore diameters whereby 40% and 80% of the pores are larger respectively.

Where there is a likelihood of frost heave occurring it is necessary to estimate the depth of frost penetration (Jumikis, 1968). Once this has been done, provision can be made for the installation of adequate insulation or drainage within the soil and to determine the amount by which the water table may need to be lowered so that it is not affected by frost penetration. The base of footings should be placed below the estimated depth of frost penetration as should water supply lines and other services. Frost-susceptible soils may be replaced by gravels. The addition of certain chemicals to soil can reduce its capacity for water absorption and so can influence frost susceptibility. For example, Croney and Jacobs (1967) noted that the addition of calcium lignosulphate and sodium tripolyphosphate to silty soils were both effective in reducing frost heave. The freezing point of the soil may be lowered by mixing in solutions of calcium chloride or sodium chloride, in concentrations of 0.5–3.0% by weight of the soil mixture. The heave of non-cohesive soils containing appreciable quantities of fines can be reduced or prevented by the addition of cement or bituminous binders. Addition of cement both reduces the permeability of a soil mass and gives it sufficient tensile strength to prevent the formation of small ice lenses as the freezing isotherm passes through.

References

Andersland, O. B. and Anderson, D. M. (eds.) (1978). *Geotechnical Engineering for Cold Regions*, McGraw-Hill, New York.

Andersland, O. B., Sayles, F. H. and Ladanyi, B. (1978). 'Mechanical properties of frozen ground', in *Geotechnical Engineering for Cold Regions*, McGraw-Hill, New York, 216–75.

Anon (1965). *Soils and Geology – Pavement Design for Frost Conditions*, Technical Manual TM 5-818-2, US Corps of Engineers, Department of Army, Washington DC.

Bear, J. (1979). *Hydraulics of Groundwater*, McGraw-Hill, New York.

Bell, F. G. (1978). 'Some petrographic factors relating to porosity and permeability in the Fell Sandstone of Northumberland', *Q. J. Engg Geol.,* **11**, 113–26.

Bell, F. G. (1988). 'Subsidence due to the abstraction of fluids', in *Underground Movements and Engineering Geology*, Special Publication in Engineering Geology No. 5, Bell, F. G., Culshaw, M. G., Cripps, J. C. and Lovell, M. A. (eds.), The Geological Society, London, 363–76.

Casagrande, A. (1932). 'Discussion on frost heaving', *Proc. Highway Res. Board*, Bull., No. 12, 169, Washington DC.

Casagrande, A. (1936). 'Characteristics of cohesionless soils affecting the stability of slopes and earth fills', *J. Boston Soc. Civil Engrs,* **23**, 3–32.

Castillo, E., Karadi, G. M. and Krizek, R. J. (1972). 'Unconfined flows through jointed rock', *Water Res. Bull., Am. Water Res. Assoc.,* **8**, No. 2, 266–81.

Chamberlain, E. J. and Gow, A. J. (1979). 'Effect of freezing and thawing on the permeability and structure of soils', *Engg Geol. Spec. Issue on Ground Freezing,* **13**, 73–92.
Croney, D. and Jacobs, J. C. (1967). 'The frost susceptibility of soils and road materials', *Trans. Road Res. Lab.*, Report LR90, Crowthorne.
Darcy, H. (1856). *Les Fontaines Publiques de la Ville de Dijon*, Dalmont, Paris.
Delflache, A. P. (1979). 'Land subsidence versus head decline in Texas', in *Evaluation and Prediction of Subsidence Proc. Speciality Conf.*, ASCE, Gainsville, 320–31.
Eckardt, H. (1979). 'Creep behaviour of frozen soils in uniaxial compression tests', *Engg Geol., Spec. Issue on Ground Freezing,* **13**, 185–96.
Fair, G. M. and Hatch, L. P. (1935). 'Fundamental factors governing the streamline flow of water through sand', *J. Am. Water Works Assoc.,* **25**, 1151–65.
Frazer, H. J. (1935). 'Experimental study of porosity and permeability of clastic sediments', *J. Geol.,* **43**, 910–1010.
Griffith, J. C. (1955). 'Petrography and petrology of the Cow Run Sand, St Mary's. West Virginia', *J. Sedimentary Petrology,* **22**, 15–31.
Haggerty, T. J. and Lippert, K. (1982). 'Rising groundwater – problem or resource', *Ground Water,* **20**, No. 2, 217–23.
Holzer, T. L. (1976). 'Ground failure in areas of subsidence due to groundwater decline in the United States', *Proc. 2nd Int. Symp. Land Subsidence*, Anaheim, Int. Ass. Hydrol. Sci., Unesco Publ. No. 121, 423–32.
Horiguchi, K. (1979). 'Effect of rate of heat removal on rate of frost heaving', *Engg Geol., Spec. Issue on Ground Freezing,* **13**, 63–72.
Horswill, P. and Horton, A. (1976). 'Cambering and valley bulging in the Gwash Valley at Empingham, Rutland', *Phil. Trans. R. Soc. Ser. A.,* **283**, 427–62.
Jacobs, J. C. (1965). 'The Road Research Laboratory frost heave test', *Trans. Road Res. Lab.,* Lab. note LN/766/JCJ, Crowthorne.
Jones, R. H. (1980). 'Frost heave of roads', *Q. J. Engg Geol.,* **13**, 77–86.
Jumikis, A. R. (1956). 'The soil freezing experiment', *Highway Res. Board, Bull. No. 135*, Factors Affecting Ground Freezing, Nat. Acad. Res. Coun. Pub. 425, Washington DC.
Jumikis, A. R. (1968). *Soil Mechanics*, Van Nostrand, Princeton.
Kaplar, C. W. (1971). 'Experiments to simplify frost susceptibility testing of soils', *US Army Corps Engrs*, Cold Regions Res. and Engg Lab., Hanover NH, Tech. Report 223.
Kinosita, S. (1979). 'Effects of initial soil-water conditions on frost heaving characteristics', *Engg Geol., Spec. Issue on Ground Freezing,* **13**, 53–62.
Löfgren, B. E. (1968). *Analysis of Stress Causing Land Subsidence*, US Geol. Surv., Prof. Paper No. 600-B, B219–225.
Löfgren, B. E. (1979). 'Changes in aquifer system properties with groundwater depletion', in *Evaluation and Prediction of Subsidence*, Saxena, S. K. (ed.), *Proc. Speciality Conference*, ASCE, Gainsville, 26–46.
Lovell, C. W. (1957). 'Temperature effects on phase composition and strength of a partially frozen soil', *Highway Res. Board*, Bull. No. 168, Washington DC.
Lovelock, P. E. R., Price, N. and Tate, T. K. (1975). 'Groundwater conditions in the Penrith Sandstone at Cliburn, Westmorland', *J. Inst. Water Engrs,* **29**, 157–74.
Mageau, D. W. and Morgenstern, N. R. (1980). 'Observations on moisture migration in frozen soils', *Can Geot. J.,* **17**, 54–60.
Marsh, T. J. and Davies, P. A. (1983). 'The decline and partial recovery of groundwater levels below London', *Proc. Inst. Civ. Engrs.*, Part I, **74**, 263–76.
Meinzer, O. (1942). 'Occurrence, origin and discharge of groundwater', in *Hydrology*, Meinzer, O. (ed.), Dover, New York, 385–443.
Moore, J. F. A. and Longworth, T. (1979). 'Hydraulic uplift at the base of a deep excavation in Oxford Clay', *Geotechnique,* **29**, 35–46.

Nixon, J. F. and Ladanyi, B. (1978). 'Thaw consolidation', in *Geotechnical Engineering for Cold Regions*, Andersland, D. B. and Anderson, D. M. (eds.), McGraw-Hill, New York, 164–215.

Parameswaran, V. R. (1980). 'Deformation behaviour and strength of frozen sand', *Canadian Geot. J.*, **17**, 74–88.

Penman, A. D. M. (1977). 'The failire of the Teton Dam', *Ground Engg*, **10**, No. 6, 18–27.

Penner, E. and Walton, T. (1979). 'Effects of temperature and pressure on frost heaving', *Engg Geol. Spec. Issue on Ground Freezing*, **13**, 29–40.

Reed, M. A., Lovell, C. W., Altschaeffl, A. G. and Wood, L. E. (1979). 'Frost heaving rate predicted from pore size distribution', *Can. Geot. J.*, **16**, 463–72.

Rowe, P. W. (1968). 'Failure of foundations and slopes on layered deposits in relation to site investigation practice', *Proc. Inst. Civil Engrs Supp. Vol.*, Paper No. 70575, 73–132.

Sanger, F. J. and Kaplar, C. W. (1963). 'Plastic deformation of frozen soils', *Proc. Inst. Conf. Permafrost.*, Lafayette, Ind. NAS-NRC, Publication No. 1281, Washington DC, 305–15.

Saxena, S. K. and Mohan, A. (1979). 'Study of subsidence due to the withdrawal of water below an aquitard', in *Evaluation and Prediction of Subsidence, Proc. Speciality Conf.*, Am. Soc. Civ. Engrs, Gainsville, 332–8.

Scott, R. F. (1979). 'Subsidence – a review', in *Evaluation and Prediction of Subsidence*, Saxena, S. K. (ed.), *Proc. Speciality Conf.*, Am. Soc. Civ. Engrs., Gainsville, 1–25.

Serafim, J. L. (1968). 'Influence of interstitial water on rock masses', in *Rock Mechanics in Engineering Practice*, Stagg, K. G. and Zienkiewicz, O. C. (eds.), Wiley, London, 55–97.

Sherrell, F. W. (1976). 'Engineering geology and ground water', *Ground Engg*, **9**, No. 4, 21–7.

Taber, S. (1930). 'Mechanics of frost heaving', *J. Geol.*, **38**, 303–17.

Takegawa, K., Nakazawa, A., Ryokai, K. and Akagawa, S. (1979). 'Creep characteristics of frozen soils', *Engg Geol., Spec. Issue on Ground Freezing*, **13**, 197–206.

Terzaghi, K. (1925). *Erdbaumechanik auf Bodenphysikalischer Grundlage*, Deuticke, Vienna.

Thomson, S. (1980). 'A brief review of foundation construction in the western Canadian Arctic', *Q. J. Engg Geol.*, **13**, 67–76.

Wittke, W. (1973). 'Percolation through fissured rock', *Bull. Int. Ass. Engg Geol.*, No. 7, 3–28.

Index

A000021654589